This book provides a self-contained introduction to cellular automata and lattice Boltzmann techniques.

Beginning with a chapter introducing the basic concepts of this developing field, a second chapter describes methods used in cellular automata modeling. Following chapters discuss the statistical mechanics of lattice gases, diffusion phenomena, reaction-diffusion processes and nonequilibrium phase transitions. A final chapter looks at other models and applications, such as wave propagation and multiparticle fluids. With a pedagogic approach, the volume focuses on the use of cellular automata in the framework of equilibrium and nonequilibrium statistical physics. It also emphasizes application-oriented problems such as fluid dynamics and pattern formation. The book contains many examples and problems. A glossary and a detailed list of references are also included.

This will be a valuable book for graduate students and researchers working in statistical physics, solid state physics, chemical physics and computer science.

BASTIEN CHOPARD obtained a Masters degree and a PhD in theoretical physics from the University of Geneva. After three years as a research associate at MIT (Cambridge, Mass.) and in the Forschungzentrum in Jülich (Germany), he is now assistant professor in the Department of Computer Sciences (Parallel Computing Group) of the University of Geneva. His research interests concern the modeling and simulation of complex systems and parallel computing.

MICHEL DROZ obtained a Masters degree and a PhD in theoretical physics from the University of Geneva. He spent three years as research associate at Cornell University (Ithaca, NY) and Temple University (Philadelphia, Penn.). He is now at the Department of Theoretical Physics of the University of Geneva where his main fields of research are equilibrium and non-equilibrium statistical mechanics, complex systems and cellular-automata modeling of physical systems.

Collection Aléa-Saclay:
Monographs and Texts in Statistical Physics

General editor: Claude Godrèche

Cellular Automata Modeling of Physical Systems

Bastien Chopard and Michel Droz
University of Geneva

CAMBRIDGE
UNIVERSITY PRESS

CAMBRIDGE UNIVERSITY PRESS
Cambridge, New York, Melbourne, Madrid, Cape Town, Singapore, São Paulo

Cambridge University Press
The Edinburgh Building, Cambridge CB2 2RU, UK

Published in the United States of America by Cambridge University Press, New York

www.cambridge.org
Information on this title: www.cambridge.org/9780521461689

First published 1998
This digitally printed first paperback version 2005

A catalogue record for this publication is available from the British Library

Library of Congress Cataloguing in Publication data

Chopard, Bastien, 1959–
Cellular automata modeling of physical systems / Bastien Chopard and Michel Droz.
p. cm. – (Collection Aléa-Saclay)
Includes bibliographical references and index.
ISBN 0 521 46168 5 (hc: alk. paper)
1. Statistical physics. 2. Cellular automata. 3. Diffusion.
4. Lattice gas. 5. Phase transformations (Statistical physics)
I. Droz, M. (Michel), 1944– . II. Title. III. Series.
QC174.8.C49 1998
530′.13′01′1–dc21 97-28284 CIP

ISBN-13 978-0-521-46168-9 hardback
ISBN-10 0-521-46168-5 hardback

ISBN-13 978-0-521-67345-7 paperback
ISBN-10 0-521-67345-3 paperback

To Viviane
Cléa, Tristan and Daphné

and Dominique
Anne-Virginie and Philippe

Everything should be made as simple as possible but not simpler
A. Einstein

Contents

Preface

The cellular automata approach and the related modeling techniques are powerful methods to describe, understand and simulate the behavior of complex systems. The aim of this book is to provide a pedagogical and self-contained introduction to this field and also to introduce recent developments. Our main goal is to present the fundamental theoretical concepts necessary for a researcher to address advanced applications in physics and other scientific areas.

In particular, this book discusses the use of cellular automata in the framework of equilibrium and nonequilibrium statistical physics and in application-oriented problems. The basic ideas and concepts are illustrated on simple examples so as to highlight the method. A selected bibliography is provided in order to guide the reader through this expanding field.

Several relevant domains of application have been mentioned only through references to the bibliography, or are treated superficially. This is not because we feel these topics are less important but, rather, because a somewhat subjective selection was necessary according to the scope of the book. Nevertheless, we think that the topics we have covered are significant enough to give a fair idea of how the cellular automata technique may be applied to other systems.

This book is written for researchers and students working in statistical physics, solid state physics, chemical physics and computer science, and anyone interested in modeling complex systems. A glossary is included to give a definition of several technical terms that are frequently used throughout the text. At the end of the first six chapters, a selection of problems is given. These problems will help the reader to become familiar with the concepts introduced in the corresponding chapter, or will introduce him to new topics that have not been covered in the text. Some problems are rather easy, although they usually require some programming effort, but other problems are more involved and will demand significant time to complete.

Most of the cellular automata simulations and results presented in this book have been produced on the 8k Connection Machine CM-200 of the University of Geneva. Others have been computed on an IBM SP2 parallel computer, also installed at the University of Geneva. Although a parallel supercomputer is quite useful when considering large scale simulations, common workstations and even modern personal computers are well adapted to perform cellular automata computations, except for on-line display which is alway very desirable. Dedicated hardware is also available but, usually, less flexible than a general purpose machine.

Despite our effort, several errors and misprints are still likely to be present. Please report them to us* (as well as any comment or suggestion).

We would like to thank all the people who have made this book possible and, in particular Claude Godrèche who gave us the opportunity to write it. Special thanks go to Pascal Luthi and Alexandre Masselot who made several original and important simulations which are presented in this book. Other people have played a direct or indirect role in the preparation of the manuscript. Among them, we thank Rodolphe Chatagny, Stephen Cornell, Laurent Frachebourg, Alan McKane, Zoltan Racz and Pierre-Antoine Rey.

Finally we acknowledge the Swiss National Science Foundation who funded some of the research reported here and the Computer Science Department and the Theoretical Physics Department of the University of Geneva for making available the necessary environment and infrastructure for this enterprise.

---

* Bastien.Chopard@cui.unige.ch or Michel.Droz@physics.unige.ch

1
Introduction

1.1 Brief history

Cellular automata (often termed CA) are an idealization of a physical system in which space and time are discrete, and the physical quantities take only a finite set of values.

Although cellular automata have been reinvented several times (often under different names), the concept of a cellular automaton dates back from the late 1940s. During the following fifty years of existence, cellular automata have been developed and used in many different fields. A vast body of literature is related to these topics. Many conference proceedings [1–8]), special journal issues [9,10] and articles are available.

In this section, our purpose is not to present a detailed history of the developments of the cellular automata approach but, rather, to emphasize some of the important steps.

1.1.1 *Self-reproducing systems*

The reasons that have led to the elaboration of cellular automata are very ambitious and still very present. The pioneer is certainly John von Neumann who, at the end of the 1940s, was involved in the design of the first digital computers. Although von Neumann's name is definitely associated with the architecture of today's sequential computers, his concept of cellular automata constitutes also the first applicable model of massively parallel computation.

Von Neumann was thinking of imitating the behavior of a human brain in order to build a machine able to solve very complex problems. However, his motivation was more ambitious than just a performance increase of the computers of that time. He thought that a machine with such a

1

complexity as the brain should also contain self-control and self-repair mechanisms. His idea was to get rid of the difference which exists between processors and the data, by considering them on the same footing. This led him to envisage a machine capable of building itself, out of some available material.

Rapidly, he considered the problem from a more formal viewpoint and tried to define the properties a system should have to be self-replicating. He was mostly interested to find a logical abstraction of the self-reproduction mechanism, without reference to the biological processes involved.

Following the suggestions of S. Ulam [11], von Neumann addressed this question in the framework of a fully discrete universe made up of cells. Each cell is characterized by an internal state, which typically consists of a finite number of information bits. Von Neumann suggested that this system of cells evolves, in discrete time steps, like simple automata which only know of a simple recipe to compute their new internal state. The rule determining the evolution of this system is the same for all cells and is a function of the states of the neighbor cells. Similarly to what happens in any biological system, the activity of the cells takes place simultaneously. However, the same clock drives the evolution of each cell and the updating of the internal state of each cell occurs synchronously. These fully discrete dynamical systems (cellular space) invented by von Neumann are now referred to as *cellular automata*.

The first self-replicating cellular automaton proposed by von Neumann was composed of a two-dimensional square lattice and the self-reproducing structure was made up of several thousand elementary cells. Each of these cells had up to 29 possible states [12]. The evolution rule required the state of each cell plus its four nearest neighbors, located north, south, west and east. Due to its complexity, the von Neumann rule has only been partially implemented on a computer [13].

However, von Neumann had succeeded in finding a discrete structure of cells bearing in themselves the recipe to generate new identical individuals. Although this result is hardly even a very primitive form of life, it is quite interesting because it is usually expected that a machine can only build an object of lesser complexity than itself. With self-replicating cellular automata, one obtains a "machine" able to create new machines of identical complexity and capabilities.

The von Neumann rule has the so-called property of universal computation. This means that there exists an initial configuration of the cellular automaton which leads to the solution of any computer algorithm. This sounds a surprising statement: how will such a discrete dynamics help us to solve any problem? It turns out that this property is of theoretical rather than practical interest. Indeed, the property of universal computing means that any computer circuit (logical gates) can be simulated by the rule of the automaton. All this shows that quite complex and unexpected behavior can emerge from a cellular automaton rule.

After the work of von Neumann, others have followed the same line of research and the problem is still of interest [14]. In particular, E.F. Codd [15] in 1968 and much later C.G. Langton [16] and Byl [17] proposed much simpler cellular automata rules capable of self-replicating and using only eight states. This simplification was made possible by giving up the property of computational universality, while still conserving the idea of having a spatially distributed sequence of instructions (a kind of cellular DNA) which is executed to create a new structure and then entirely copied in this new structure.

More generally, artificial life is currently a domain which is intensively studied. Its purpose is to better understand real life and the behavior of living species through computer models. Cellular automata have been an early attempt in this direction and can certainly be further exploited to progress in this field [18,19].

1.1.2 Simple dynamical systems

In a related framework, it is interesting to remember that it is precisely a simple ecological model that has brought the concept of cellular automata to the attention of wide audience. In 1970, the mathematician John Conway proposed his now famous *game of life* [20]. His motivation was to find a simple rule leading to complex behaviors. He imagined a two-dimensional square lattice, like a checkerboard, in which each cell can be either alive (state one) or dead (state zero). The updating rule of the game of life is as follows: a dead cell surrounded by exactly three living cells comes back to life. On the other hand, a living cell surrounded by less than two or more than three neighbors dies of isolation or overcrowdness. Here, the surrounding cells correspond to the neighborhood composed of the four nearest cells (north, south, east and west) plus the four second nearest neighbors, along the diagonals. Figure 1.1 shows three configurations of the game of life automaton, separated by 10 iterations.

It turned out that the game of life automaton has an unexpectedly rich behavior. Complex structures emerge out of a primitive "soup" and evolve so as to develop some skills. For instance, objects called *gliders* may form (see problems, section 1.4). Gliders correspond to a particular arrangement of adjacent cells that has the property to move across space, along straight trajectories. Many more such structures have been identified in the vast body of literature devoted to the game of life [21,22]. As for the von Neumann rule, the game of life is a cellular automata capable of computational universality.

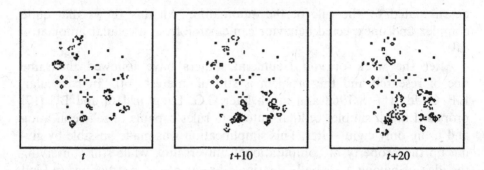

Fig. 1.1. The game of life automaton. Black dots represents living cells whereas dead cells are white. The figure shows the evolution of some random initial configurations.

In addition to these theoretical aspects, cellular automata were used in the 1950s for image processing [23]. It was recognized early on that much tedious picture analysis could be carried out automatically, according to a cellular automata computing model: the pixels of an image can be treated simultaneously, using simple local operations. Special-purpose machines based on cellular automata logic have been developed for noise reduction, counting and size estimation in images obtained from observations with a microscope.

At the beginning of the 1980s, S. Wolfram studied in detail a family of simple one-dimensional cellular automata rules (the now famous Wolfram rules [24,25]). He had noticed that a cellular automaton is a discrete dynamical system and, as such, exhibits many of the behaviors encountered in a continuous system, yet in a much simpler framework. A concept such as complexity could be investigated on mathematical models allowing an exact numerical computer calculation, because of their Boolean nature (no numerical errors nor truncation as in more traditional models). Wolfram's results have contributed to prove that cellular automata are important objects to consider for statistical mechanics studies and, at the present time, Wolfram's rule are still the topic of much research.

1.1.3   A synthetic universe

The property of many cellular automata rules being a universal computer made several authors think that the physical world itself could be a very large cellular automaton. Tommaso Toffoli [26] compares cellular automata to a synthetic model of the universe in which the physical laws are expressed in terms of simple local rules on a discrete space–time structure.

T. Toffoli, N. H. Margolus and E. Fredkin recognized the importance of cellular automata as a modeling environment for physical systems. They were very interested in the analogy that exists between the theory of information as it is used to describe numerical processing in a computer and the laws of physics. Cellular automata provide an excellent framework to develop these ideas. In particular, they showed how to build a fully time-reversible logic from which any numerical operation can be implemented without any loss of information. The so-called *billiard ball* [26] is a cellular automata rule which is an example of such a reversible model of computation.

The possibility of displaying, on a computer screen, the time evolution of large cellular automata systems, at the rate of several updates per second of the complete lattice offers a way of performing experiments live on an artificial universe, whose evolution rules are set up by the observer. By building their first general purpose cellular automata machines CAM-6 in the mid-1980s, Toffoli and Margolus provided a very powerful cellular automata environment with the capability of a supercomputer of that time, at a very affordable price and with unique display facilities. This machine has stimulated many developments of cellular automata techniques and has contributed to the spreading of the main ideas to a wide audience of scientists.

Toffoli and Margolus's book [26]: *Cellular Automata Machines: a New Environment for Modeling,* is a wonderful source of inspiration in the field of cellular automata and provide a complete description of the CAM-6 hardware. More recently, Toffoli, Margolus and coworkers have designed CAM-8, a much more powerful hardware environment: a parallel, uniform, scalable architecture for cellular automata experimentation [27]. This hardware platform offers high performance, a flexible approach, display facilities and is naturally appropriate to work on three-dimensional systems. It has been successfully used for many different applications.

1.1.4 *Modeling physical systems*

It was also in the 1980s that an important step in the theory of cellular automata was accomplished. It was recognized that the so-called HPP [28] lattice gas models developed in the 1970s by Hardy, Pomeau and de Pazzis was in fact a cellular automata. This model consists of a simple and fully discrete dynamics of particles moving and colliding on a two-dimensional square lattice, in a such a way as to conserve momentum and particle number.

The HPP dynamics was initially planned as a theoretical model to study fundamental statistical properties of a gas of interacting particles. The actual implementation of this model as a cellular automata rule and the visualization of the fast moving particle shed a different light on the possibilities of such models: isn't it possible to simulate the behavior of a real system of particles (like a fluid or a gas) as a cellular automata rule? After all, it is well known that the flows of a fluid, a gas or even a granular medium are very similar at a macroscopic scale, in spite of their different microscopic nature. A fully discrete and simplified molecular dynamics could work too, provided the system is considered at an appropriate observation scale.

Of course, the idea of using discrete systems as a model of real phenomena has already been considered for several problems. The Ising model of classical spin is a famous example which will be discussed in more detail in the next chapter. From the fluid side, already at the end of the nineteenth century, Maxwell [29], had proposed a discrete velocity system of interacting particles as a model of a gas. In fact, such *lattice gas*, discrete velocity models have been developed independently from cellular automata theory [30,31].

However, cellular automata provide a new conceptual framework, as well as an effective numerical tool, which retains important aspects of the microscopic laws of physics, such as simultaneity of the motion, locality of the interactions and time reversibility.

Cellular automata rules are viewed as an alternative form of the microscopic reality which bears the expected macroscopic behavior. From a numerical point of view it was expected, at the end of the 1980s, that a wind tunnel could be replaced by a fully discrete computer model. The first cellular automata model to give credit to this possibility is the famous FHP model proposed in 1986 by U. Frisch, B. Hasslacher and Y. Pomeau [32], and almost simultaneously by S. Wolfram [33]. These authors showed that their model, despite its fully discrete dynamics, follows, in some appropriate limits, the behavior prescribed by the Navier–Stokes equation of hydrodynamics.

Note that models like FHP or HPP are often termed lattice gas automata (LGA) to distinguish them from the less specific cellular automata terminology. Clearly, from a mathematical point of view, a lattice gas automata is a cellular automata, but the way one thinks, for instance, of the game of life is quite different from the underlying philosophy of the FHP model. This difference will become clear to the reader as he or she becomes more familiar with the next chapter of this book. Nevertheless, in this book, we will often use cellular automata to designate a LGA.

Since the FHP rule was discovered, lattice gas automata or cellular automata fluids as these kind of particle models are now often referred to, have been developed intensively and several insufficiencies of the initial model corrected. The Ecole Normale Supérieure in Paris has been very active and P. Lallemand and D. d'Humieres, in particular, have played a pioneering role in this field [34–37].

However, contrary to first expectations, lattice gas models of fluids have not been able to surpass the traditional numerical methods of hydrodynamics and compute high Reynolds flows. Their relatively high viscosity, which is only determined by the cellular automata rule (and therefore not adjustable), is a limiting factor to the practical study of many of these flows. The finite spatial resolution of the cellular automata lattice (physical phenomena must occur at a much larger scale than the lattice spacing) is another limitation on the study and modeling of fully developed turbulence, unless the system has such a large scale that the advantage of a cellular automata approach vanishes even on today's fastest computers [38].

However, lattice gas automata have been much more successful in modeling complex situations for which traditional computing techniques are not applicable. Flows in porous media [39–41], immiscible [42–46] flows and instabilities, spreading of a liquid droplet and wetting phenomena [47], microemulsion [48] erosion and transport problems [49] are some examples pertaining to fluid dynamics.

Other physical situations, like pattern formation, reaction-diffusion processes [50], nucleation–aggregation growth phenomena, are very well described by cellular automata dynamics and will be investigated in detail in this book.

1.1.5 *Beyond the cellular automata dynamics: lattice Boltzmann methods and multiparticle models*

Very often, the advantage of the cellular automata (or lattice gas) approach is most apparent when complex boundary conditions are present. Due to the microscopic interpretation of the dynamics, these conditions can be taken into account in a much more natural way than in a continuous description (like a differential equation) in which our basic intuition of the phenomena may be lost.

On the other hand, cellular automata models have several weakness related to their fully discrete nature: statistical noise requiring systematic averaging processes, and little flexibility to adjust parameters of a rule in order to describe a wider range of physical situations. At the end of the 1980s, McNamara and Zanetti [51], and Higueras, Jimenez and Succi [52] showed the advantage of extending the Boolean dynamics of the automaton to directly work on real numbers representing the probability for a cell to have a given state.

This approach, called the *lattice Boltzmann method* (LBM), is numerically much more efficient than the Boolean dynamics and provides a

new computational model much more suited to the simulation of high
Reynolds flows and many other relevant applications (for instance glacier
flow [53]).

Lattice Boltzmann models retain the microscopic level of interpretation of the cellular automata approach but neglect many-body correlation
functions. However, this method now constitutes a very promising approach to modeling physical systems and is discussed on several occasions
throughout this book.

In between the strict cellular automata approach and the more flexible
lattice Boltzmann method, there is room for an intermediate description:
the multiparticle models which are still under development at the present
time. These models preserve the concept of a quantized state but an
infinite set of values is accepted. Consequently, numerical stability is
guaranteed (as opposed to the LBM), and many-body correlations taken
into account. The large number of possible states offers more flexibility
when modeling a physical system and yields less statistical noise. But a
multiparticle dynamics is more difficult to devise and numerically slower
than its lattice Boltzmann counterpart. Examples of this approach will be
presented in this book.

From our point of view, the cellular automata approach is not a
rigid framework. It is rather a philosophy of modeling which should
be considered with some pragmatism. The important issue in cellular
automata modeling is to capture the essential features of given phenomena
and translate them to a suitable form to obtain an effective numerical
model. To this end, it is acceptable (and even beneficial) to relax some
of the constraints of the original definition of a cellular automata. The
introduction of the lattice Boltzmann method is an illustration of this fact.
The point is to conserve the spirit of the approach and its relevant features
rather than its limitations. This remark is particularly in order because
present parallel computers offer an ideal and quite flexible platform to
implement cellular automata models without the restrictions imposed by
dedicated hardware.

1.2   A simple cellular automaton: the parity rule

In this section, we discuss a simple cellular automata rule, in order to
introduce and illustrate the concept. This should slowly familiarize the
reader with a more precise notion of cellular automata. Section 1.3 will
present a more formal definition.

Although it is very basic, the rule we study here exhibits a surprisingly rich behavior. It was proposed initially by Edward Fredkin in the
1970s [54] and is defined on a two-dimensional square lattice.

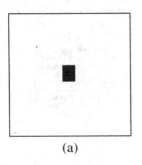
(a)
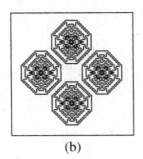
(b)
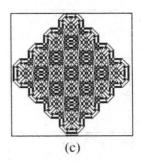
(c)

Fig. 1.2. The $\oplus$ rule on a 256×256 periodic lattice: (a) initial configuration; (b) and (c) configurations after $t_b = 93$ and $t_c = 110$ iterations, respectively.

Each site of the lattice is a cell which is labeled by its position $\vec{r} = (i, j)$ where i and j are the row and column indices. A function $\psi_t(\vec{r})$ is associated to the lattice to describe the state of each cell at iteration t. This quantity can be either 0 or 1.

The cellular automata rule specifies how the states ψ_{t+1} are to be computed from the states at iteration t. We start from an initial condition at time $t = 0$ with a given configuration of the values $\psi_0(\vec{r})$ on the lattice. The state at time $t = 1$ will be obtained as follows

(1) Each site $\vec{r}$ computes the sum of the values $\psi_0(\vec{r}\,')$ on the four nearest neighbor sites $\vec{r}\,'$ at north, west, south and east. The system is supposed to be periodic in both i and j directions (as on a torus) so that this calculation is well defined for all sites.

(2) If this sum is even, the new state $\psi_1(\vec{r})$ is 0 (white), else, it is 1 (black).

The same rule (step 1 and 2) is repeated to find the states at time $t = 2, 3, 4, \dots$.

From a mathematical point of view, this cellular automata parity rule can be expressed by the following relation

$$\psi_{t+1}(i, j) = \psi_t(i + 1, j) \oplus \psi_t(i - 1, j) \oplus \psi_t(i, j + 1) \oplus \psi_t(i, j - 1) \qquad (1.1)$$

where the symbol $\oplus$ stands for the exclusive OR logical operation. It is also the sum modulo 2: $1 \oplus 1 = 0 \oplus 0 = 0$ and $1 \oplus 0 = 0 \oplus 1 = 1$.

When this rule is iterated, very nice geometrical patterns are observed, as shown in figure 1.2. This property of generating complex patterns starting from a simple rule is actually generic of many cellular automata rules. Here, complexity results from some spatial organization which builds up as the rule is iterated. The various contributions of successive iterations combine together in a specific way. The spatial patterns that are observed reflect how the terms are combined algebraically.

(a)   (b)   (c)

Fig. 1.3.   The $\oplus$ rule replicates any initial pattern when the number of iterations is a power of two. Image (a) shows the initial pattern at time $t_a = 0$. Images (b) and (c) show successive iterations at times $t_b = 16$ and $t_c = t_b + 32$.

On closer inspection, we also observe that the initial pattern is replicated at some specific iteration. Figure 1.3 illustrates that point with a more enlightening initial condition. The times at which this happens are a power of two. Another surprising fact occurs when the system size L is a power of two: after $L/2$ iteration the state of each cell vanishes for all possible initial configurations.

These behaviors, as well as the way the pattern builds up can be explained by working out the definition of the rule. Applying the rule 1.1 twice yields ψ_{t+1} as a function of ψ_{t-1}

$$
\begin{aligned}
\psi_{t+1}(i,j) \;=\; & \psi_{t-1}(i+2,j) \oplus \psi_{t-1}(i,j) \oplus \psi_{t-1}(i+1,j+1) \\
& \oplus \psi_{t-1}(i+1,j-1) \oplus \psi_{t-1}(i,j) \oplus \psi_{t-1}(i-2,j) \\
& \oplus \psi_{t-1}(i-1,j+1) \oplus \psi_{t-1}(i-1,j-1) \oplus \psi_{t-1}(i+1,j+1) \\
& \oplus \psi_{t-1}(i-1,j+1) \oplus \psi_{t-1}(i,j+2) \oplus \psi_{t-1}(i,j) \\
& \oplus \psi_{t-1}(i+1,j-1) \oplus \psi_{t-1}(i-1,j-1) \oplus \psi_{t-1}(i,j) \\
& \oplus \psi_{t-1}(i,j-2) \qquad\qquad\qquad\qquad\qquad\qquad\quad (1.2)
\end{aligned}
$$

Since $a \oplus a = 0$ and $a \oplus 0 = a$, for all values of a, one obtains

$$
\psi_{t+2}(i,j) = \psi_t(i+2,j) \oplus \psi_t(i-2,j) \oplus \psi_t(i,j+2) \oplus \psi_t(i,j-2) \quad (1.3)
$$

Thus, after two iterations, the action of the rule is to translate the initial configuration by two lattice sites in the four directions and XOR them.

If now we compute similarly $\psi_{t+3}(i,j)$ as a function of $\psi_t(i,j)$, we no longer get such a simple expression. A relation similar to equation 1.2 is obtained, but without any cancellation. We see that $\psi_{t+3}(i,j)$ is a superposition of 16 translations. As a result, we obtain the rich geometrical structure observed in figure 1.2.

However, it is easy to prove that the behavior of the $\oplus$ rule is simple when the number of iterations performed is a power of two. To show this property, suppose the following relation is true for a given value of T.

$$\psi_t(i, j) = \psi_{t-T}(i+T, j) \oplus \psi_{t-T}(i-T, j) \oplus \psi_{t-T}(i, j+T) \oplus \psi_{t-T}(i, j-T) \quad (1.4)$$

We already know it is true for $T = 1$ and $T = 2$. Then, if we apply this same relation to ψ_{t-T} on the right-hand side, we obtain

$$\begin{aligned}
\psi_t(i, j) =\ & \psi_{t-2T}(i+2T, j) \oplus \psi_{t-2T}(i, j) \oplus \psi_{t-2T}(i+T, j+T) \\
& \oplus \psi_{t-2T}(i+T, j-T) \oplus \psi_{t-2T}(i, j) \oplus \psi_{t-2T}(i-2T, j) \\
& \oplus \psi_{t-2T}(i-T, j+T) \oplus \psi_{t-2T}(i-T, j-T) \\
& \oplus \psi_{t-2T}(i+T, j+T) \oplus \psi_{t-2T}(i-T, j+T) \oplus \psi_{t-2T}(i, j+2T) \\
& \oplus \psi_{t-2T}(i, j) \oplus \psi_{t-2T}(i+T, j-T) \oplus \psi_{t-2T}(i-T, j-T) \\
& \oplus \psi_{t-2T}(i, j) \oplus \psi_{t-2T}(i, j-2T) \\
=\ & \psi_{t-2T}(i+2T, j) \oplus \psi_{t-2T}(i-2T, j) \oplus \psi_{t-2T}(i, j+2T) \\
& \oplus \psi_{t-2T}(i, j-2T) \quad (1.5)
\end{aligned}$$

This result shows that property 1.4 is then also true for $2T$. Therefore, it is true for any lag T which is a power of two. At these particular values of time, the $\oplus$ rule is equivalent to the superposition (in the sense of the addition modulo two) of the initial pattern translated in the four lattice directions by an amount T. When the spatial extension of the initial pattern is small enough, we see it replicated four times. Otherwise destructive interferences show up, which give rise to partial replication.

For a square lattice of size $L = 2^k$, relation 1.4 implies that, after $L/2$ iterations, the result of the $\oplus$ rule is to superpose the initial condition with itself and, therefore, to yield a zero configuration.

For a number of iterations that is not a power of two, some results can also be obtained. In general, the configuration after T steps is the superposition modulo 2 of 4^k different translations of the initial pattern, where k is the number of digits equal to 1 in the binary representation of T.

This property can be proved as follows. First, we notice that the rule is additive, that is any initial pattern can be decomposed into the superposition of one-pixel images. Each of these simple configurations can be evolved independently and the results superposed to obtained the final complete pattern.

We will prove our statement for an initial configuration such that $\psi_0(0, 0) = 1$, only. It is convenient to decompose T as a sum of powers of two. There exists a value of n such that we can write T as

$$T = 2^n + \sum_{\ell=0}^{n-1} a_\ell 2^\ell = 2^n + T' \quad (1.6)$$

where, by construction, $T' \le 2^n - 1$ (equality holds when all the $a_\ell = 1$).

Now, performing T iterations of the rule is equivalent to first do T' steps and then the last 2^n iterations. Clearly, by the definition of the rule, $\psi_{T'}(i,j)$ will contain the terms $\psi_0(i-T',j)$, $\psi_0(i+T',j)$, $\psi_0(i,j-T')$ and $\psi_0(i,j+T')$ and other terms involving only translations by a *smaller* amount. Therefore, the configuration at time T' has a given spatial extension.

When performing the next 2^n iterations, we know from 1.4 that they will result in the superposition of four translations by 2^n of the configuration at time T'. None of these translations will give rise to cancellation because the spatial extension of the configuration at time T' is smaller than the translation length. Indeed, after the left translation by 2^n, the right boundary will move to position $T'-2^n$ with respect to the original pattern. Similarly, the left boundary will move to $-T'+2^n$, due to the right translation. There is no overlap between the patterns generated in this way because, since $T' < 2^n$, one has

$$T' - 2^n \le -T' + 2^n \tag{1.7}$$

Therefore, for each non-zero a_ℓ in expression (1.6), four translations are produced and the final result is composed of 4^k non-over lapping translations, with $k = \sum a_\ell$. When the initial image is not a single pixel, destructive interference is observed.

As a result, we get the rich geometrical structure observed in figure 1.2. As the number of iterations increases more and more terms are generated. Therefore, the algorithmic complexity of the expression becomes larger, thus reflecting the complexity of the cellular automaton configuration. More precisely, k is bounded by the logarithm of T. In order to evaluate the asymptotic complexity of the expression, we write $T \sim 2^k$. The number of translations generated is $4^k = T^2$ and the complexity goes as the square of the number of iterations.

The above discussion has unraveled the mechanisms leading to the complex structures produced by the $\oplus$ rule as being due to the superposition of the initial pattern translated many times by a different amount. From this analysis, we can conclude that this rule is not a self-replicating cellular automaton in the sense imagined by von Neumann.

Finally, it is interesting to note that the $\oplus$ rule is a generalization in two dimensions of the famous rule 90 of Wolfram which will be discusseded in section 2.1.1.

1.3　Definitions

1.3.1　Cellular automata

In this section we shall present a more formal definition of a cellular automaton. In general, a cellular automaton requires

(i) a regular lattice of cells covering a portion of a d-dimensional space;

(ii) a set $\Phi(\vec{r}, t) = \{\Phi_1(\vec{r}, t), \Phi_2(\vec{r}, t), ..., \Phi_m(\vec{r}, t)\}$ of Boolean variables attached to each site $\vec{r}$ of the lattice and giving the local state of each cell at the time $t = 0, 1, 2, ...$;

(iii) a rule $\mathbf{R} = \{R_1, R_2, ..., R_m\}$ which specifies the time evolution of the states $\Phi(\vec{r}, t)$ in the following way

$$\Phi_j(\vec{r}, t+1) = R_j(\Phi(\vec{r}, t), \Phi(\vec{r} + \vec{\delta}_1, t), \Phi(\vec{r} + \vec{\delta}_2, t), ..., \Phi(\vec{r} + \vec{\delta}_q, t)) \quad (1.8)$$

where $\vec{r} + \vec{\delta}_k$ designate the cells belonging to a given neighborhood of cell $\vec{r}$.

The example discussed in the previous section is a particular case in which the state of each cell consists of a single bit $\Phi_1(\vec{r}, t) = \psi_t(\vec{r})$ of information and the rule is the addition modulo 2.

In the above definition, the rule $\mathbf{R}$ is identical for all sites and is applied simultaneously to each of them, leading to a synchronous dynamics. It is important to notice that the rule is *homogeneous*, that is it cannot depend explicitly on the cell position $\vec{r}$. However, spatial (or even temporal) inhomogeneities can be introduced by having some $\Phi_j(\vec{r})$ systematically 1 in some given locations of the lattice to mark particular cells for which a different rule applies. Boundary cells are a typical example of spatial inhomogeneity. Similarly, it is easy to alternate between two rules by having a bit which is 1 at even time steps and 0 at odd time steps.

In our definition, the new state at time $t+1$ is only a function of the previous state at time t. It is sometimes necessary to have a longer memory and introduce a dependence on the states at time $t-1, t-2, ..., t-k$. Such a situation is already included in the definition if one keeps a copy of the previous states in the current state. Extra bits Φ_ℓ can be defined for this purpose. As an example, the one-dimensional second-order rule

$$\Phi_1(r, t+1) = R(\Phi_1(r-1, t), \Phi_1(r+1, t)) \oplus \Phi_1(r, t-1) \quad (1.9)$$

where $\oplus$ is the sum modulo 2, can be expressed as a first-order rule by introducing a new state $\Phi_2(r, t)$ as follows

$$\begin{aligned} \Phi_1(r, t+1) &= R(\Phi_1(r-1, t), \Phi_1(r+1, t)) \oplus \Phi_2(r, t) \quad (1.10) \\ \Phi_2(r, t+1) &= \Phi_1(r, t) \end{aligned}$$

Note that in the first case, the initial condition requires one to specify $\Phi_1(r, t = 0)$ and $\Phi_1(r, t = 1)$. In the second case, there are still two initial conditions to specify, but now they are $\Phi_1(r, t = 0)$ and $\Phi_2(r, t = 0)$.

An expression such as rule 1.9 provides a general way to build a time-reversible dynamics. Indeed, since the operation $\oplus$ is commutative, associative and $a \oplus a = 0$, one can reorganize the terms as

$$\Phi_1(r, t-1) = R(\Phi_1(r-1, t), \Phi_1(r+1, t)) \oplus \Phi_1(r, t+1) \qquad (1.11)$$

which shows that the same rule also allows backward evolution: if one gives $\Phi_1(r, t+1)$ and $\Phi_1(r, t)$ as initial conditions, the system will evolve back to its own past.

1.3.2 Neighborhood

A cellular automata rule is local, by definition. The updating of a given cell requires one to know only the state of the cells in its vicinity. The spatial region in which a cell needs to search is called the *neighborhood*. In principle, there is no restriction on the size of the neighborhood, except that it is the same for all cells. However, in practice, it is often made up of adjacent cells only. If the neighborhood is too large, the complexity of the rule may be unacceptable (complexity usually grows exponentially fast with the number of cells in the neighborhood).

For two-dimensional cellular automata, two neighborhoods are often considered: the von Neumann neighborhood, which consists of a central cell (the one which is to be updated) and its four geographical neighbors north, west, south and east. The Moore neighborhood contains, in addition, second nearest neighbors north-east, north-west, south-east and south-west, that is a total of nine cells. Figure 1.4 illustrates these two standard neighborhoods.

Another useful neighborhood is the so-called Margolus neighborhood which allows a partitioning of space and a reduction of rule complexity. The space is divided into adjacent blocks of two-by-two cells. The rule is sensitive to the location within this so-called Margolus block, namely upper-left, upper-right, lower-left and lower-right. The way the lattice is partitioned changes as the rule is iterated. It alternates between an odd and an even partition, as shown in figure 1.5. As a result, information can propagate outside the boundaries of the blocks, as evolution takes place.

The key idea of the Margolus neighborhood is that when updating occurs, the cells within a block evolve only according to the state of that block and do not immediately depend on what is in the adjacent blocks. To understand this property, consider the difference with the situation of the Moore neighborhood shown in figure 1.4. After an iteration, the cell located on the west of the shaded cell will evolve relatively to its *own* Moore neighborhood. Therefore, the cells within a given Moore neighborhood evolve according to the cells located in a wider region. The purpose of space partitioning is to prevent these "long distance" effects from occurring. The sand rule described in section 2.2.6 give an example of the use of the Margolus neighborhood.

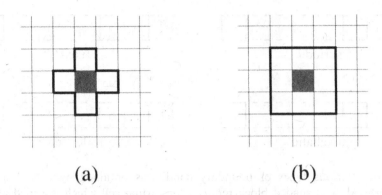

Fig. 1.4. (a) Von Neumann and (b) Moore neighborhoods. The shaded region indicates the central cell which is updated according to the state of the cells located within the domain marked with the bold line.

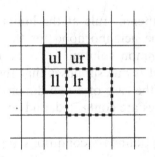

Fig. 1.5. The Margolus neighborhood. The lattice is partitioned in 2×2 blocks. Two partitionings are possible – odd and even partitions, as shown by the blocks delimited by the solid and broken lines, respectively. The neighborhood alternates between these two situations, at even and odd time steps. Within a block, the cells are distinguishable and are labeled *ul*, *ur*, *ll* and *lr* for upper-left, upper-right, lower-left and lower-right. The cell labeled *lr* in the figure will be become *ul*, at the next iteration, with the alternative partitioning.

1.3.3 Boundary conditions

In practice, when simulating a given cellular automata rule, one cannot deal with an infinite lattice. The system must be finite and have boundaries. Clearly, a site belonging to the lattice boundary does not have the same neighborhood as other internal sites. In order to define the behavior of these sites, a different evolution rule can be considered, which sees the appropriate neighborhood. This means that the information of being, or not, at a boundary is coded at the site and, depending on this information, a different rule is selected. Following this approach, it

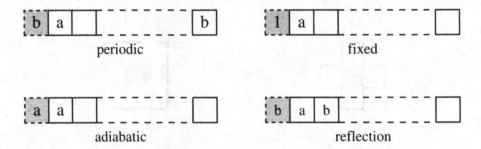

periodic   fixed

adiabatic   reflection

Fig. 1.6. Various types of boundary conditions obtained by extending the neighborhood. The shaded block represents a virtual cell which is added at the extremity of the lattice (left extremity, here) to complete the neighborhood.

is also possible to define several types of boundaries, all with different behavior.

Instead of having a different rule at the limits of the system, another possiblity is to extend the neighborhood for the sites at the boundary . For instance, a very common solution is to assume *periodic* (or cyclic) boundary conditions, that is one supposes that the lattice is embedded in a torus-like topology. In the case of a two-dimensional lattice, this means that the left and right sides are connected, and so are the upper and lower sides.

Other possible types of boundary conditions are illustrated in figure 1.6, for a one-dimensional lattice. We assume that the lattice is augmented by a set of virtual cells beyond its limits. A *fixed* boundary is defined so that the neighborhood is completed with cells having a pre-assigned value. An adiabatic boundary condition (or zero-gradient) is obtained by duplicating the value of the site to the extra virtual cells. A reflecting boundary amounts to copying the value of the other neighbor in the virtual cell.

The nature of the system which is modeled will dictate the type of boundary conditions that should be used in each case.

1.3.4   Some remarks

According to its above definition, a cellular automaton is deterministic. The rule **R** is some well-defined function and a given initial configuration will always evolve the same way. However, as we shall see later on, it may be very convenient for some applications to have a certain degree of randomness in the rule. For instance, it may be desirable that a rule selects one outcome among several possible states, with a probability p. Cellular automata whose updating rule is driven by external probabilities are called *probabilistic* cellular automata. On the other hand, those which strictly comply with the definition given above, are referred to as *deterministic* cellular automata.

In practice, the difference between probabilistic and deterministic cellular automata is not so important. Randomness enters into the rule through an extra bit which, at each time step, is 1 with probability p and 0 with probability $1 - p$, independently at each lattice cell. The question is then how to generate a random bit. We shall see in section 2.1.1 that some very simple *deterministic* cellular automata rules have an unpredictable behavior, that is there is no way to know what state a cell will assume at a future stage, unless the evolution is actually performed. Such a rule can be used to produce a pseudorandom bit which is 1 with probability 1/2. This mechanism can mimic a probabilistic cellular automaton. As a matter of fact, it is interesting to note that rule 30 of Wolfram (see section 2.1.1) has been used to produce very good quality pseudorandom numbers in the Connection Machine parallel computer [55,24].

Probabilistic cellular automata are a very useful generalization because they offer a way to adjust some parameters of a rule in a continuous range of values, despite the discrete nature of the cellular automata world. This is very convenient when modeling physical systems in which, for instance, particles are annihilated or created at some given rate.

As the complexity of a physical system increases, the number of bits necessary to represents the phenomena at each site of the cellular automata lattice may become quite large. Dedicated cellular automata hardware usually imposes a restriction on the maximum number of bits available at each site. Even on a general purpose computer, memory restrictions may apply if the rule is expressed as a *lookup table* in which the updating is computed in advance for all possible neighborhood configurations: the size of a lookup table increases as 2^n, where n is the total number of bits in the entire neighborhood.

However, there is a case where a large number of bits per cell can be easily dealt with using a general purpose computer: when these bits are interpreted as the binary representation of an integer, or even, of a floating point number. Then, the usual arithmetic operations $+, -, *$ and $/$ can be used to build the rule.

Although this is not the original way to consider a cellular automata, the definition we gave allows many interpretations. Strictly speaking, the multiplication, on a computer, of two 32-bit floating points numbers can be seen as a cellular automata rule. This generalization of the original nature of a cellular automata may be quite useful in several applications we have mentioned and which will be discussed in this book: multiparticle automata and lattice Boltzmann models.

Another interesting extension of a cellular automata is to allow an asynchronous updating scheme. Clearly the laws of physics are parallel in the sense that the events do not occur sequentially on a system with several constituents. However, it may be too much to require full synchronism of the processes. It turns out that many physical phenomena are well described by a *master equation* which gives the probability that a process occurs in the time interval $[t, t + dt[$. The master equation is sequential rather than synchronous and it is observed that the same dynamics may converge to different states depending on whether it is implemented sequentially or synchronously. Thus, an asynchronous updating scheme makes sense from the physical point of view as it probably corresponds to reality. Asynchronism can be achieved, for instance by using a random cellular automata: at each site the rule is applied only with a given probability and, if not, the state remains unchanged.

At this stage, the reader may think that any numerical scheme with a discrete space and time can be regardeded as a cellular automaton. This point of view would not be very helpful, however. The richness and interest of the cellular automaton comes from the microscopic contents of its rule: there is in general a clear physical or intuitive interpretation of the dynamics directly at the level of the cell. On the other hand, a numerical scheme like the discretized solution of a differential equation is the result of a mapping from a high level mathematical abstraction to a simpler and more tractable form. The philosophy of cellular automata modeling works in the opposite direction.

1.4 Problems

1.1. Several cellular automata simulators (like *cellsim*) are available. Search the Internet to find some of them*. Packages like Matlab or Mathematica are environments where it is possible to develop you own rules and visualize them [56]. Try to program the rules that are proposed hereafter with Matlab or Mathematica.

1.2. Consider the game of life. The cells can be either 0 (dead) or 1 (alive) and the rule is: (i) a dead cell comes to life if it is sourrounded by exactly three living cells among the eight cells of its Moore neighborhood and (ii) a living cell survives only if it is surrounded by two or three neighbors. Show that an isolated 2×2 block of living cells is a stable configuration when the rule is iterated.

---

* See also http://liinwww.ira.uka.de/~worsch/ca/prog-envs.html

Show that an isolated horizontal segment of three living cells is a period-two oscillating configuration.

1.3. In the game of life, show that the "object" defined by $c_{i,j} = 1$, $c_{i-1,j-1} = 1$, $c_{i-2,j-1} = 1$, $c_{i-1,j-2} = 1$ and $c_{i,j-2} = 1$, where c_{ij} denotes the state of cell (i, j), is a glider; that is it moves at constant speed by continually cycling through different internal configurations.

1.4. Using a programming language you know (C, Fortran, Pascal) express the game of life rule assuming that the state 0 or 1 of each cell is stored in an array of size $n \times n$, where n is the lattice size. Consider periodic boundary conditions.

1.5. Generalize the parity rule by including the central cell in the XOR operation. Does it change qualitatively the behavior? Include other boundary conditions (adiabatic, reflecting). Run this automaton on your computer and see the effect of these boundary conditions.

1.6. Show that the rule

$$C(t+1) = (S(t).and.E(t)).xor.W(t).xor.N(t).xor.C(t)$$

is a random rule which generates a dynamic pattern of 0 and 1, occurring at each site with probability 1/2. C, N, E, W and S denote the central cell of the von Neumann neighborhood and its four neighbors (north, east, west and south).

1.7. Consider the following rule, based on the Margolus neighborhood: the new configuration of a block is obtained by swapping diagonally opposite cells, except when the configuration is $(ul, ll, lr, ur) = (1, 0, 1, 0)$ or $(ul, ll, lr, ur) = (0, 1, 0, 1)$. In these cases, the rule a symmetry with respect to the horizontal line crossing the block, namely $(1, 0, 1, 0) \leftrightarrow (0, 1, 0, 1)$. What is the behavior of this rule (the Margolus blocks alternates between the even and odd partition)?

1.8. Simulate the behavior of the following rule (see [26], time-tunnel rule):

$$Sum(t) = C(t) + N(t) + S(t) + E(t) + W(t)$$

$$C(t+1) = \begin{cases} C(t-1) & \text{if } Sum(t) \in \{0, 5\} \\ 1 - C(t-1) & \text{if } Sum(t) \in \{1, 2, 3, 4\} \end{cases}$$

where C, N, W, E and S have the same meanings as in problem 1.6. Start with an initial configuration where $C(t = 0)$ and $C(t = 1)$ are identical and form a disk in the middle of the lattice.

1.9. Program the parity rule according to the following three methods. There are several techniques to implement a cellular automata rule on a computer. (i) **Direct:** the most obvious way is to define one variable for each automaton state and each lattice site. Then, the rule is computed on the fly, according to some arithmetic or program instructions, over and over for each iteration of the run. (ii) **Lookup table:** another technique is to pre-compute the outputs of the rule for all possible configurations of the neighborhood and store them in a lookup table. This solution is fast since the computation reduces to calculating an index in the lookup table for each lattice site. But it uses quite a lot of memory if the number of bits necessary to specify the state of each cell is large (the table grows exponentially fast). (iii) **Multispin coding:** finally a third method is multispin coding. To save memory and speed, one packs several lattice sites in one computer word (actually, 32 sites in a 32-bit word). Then, bitwise logical operations (like AND, OR,...), can be used to implement the rule and compute at once the new state of 32 sites. Thus, even if one repeats the computation for each iteration, there is a considerable improvement of performance compared with method (i). In general, the bits are packed in the words in the following way (find out why this is advantageous): the first site is stored in the first bit of the first word; the second site goes to the *first* bit of the *second* word and so on. With n words, sites $n+1, n+2,...,n+n$ are stored in the second bit of all the words. Thus, bit $b \in \{0, 1, ..., 31\}$ of word $k \in \{1, .., n\}$ contains one bit of the state of site $l = b * n + k$. For a rule requiring s bits per state, one needs s sets of n words. In a 2D lattice, each line is multispin coded.

1.10. Snowflakes [56]: a simple CA rule can be devised to model the growth of an object whose structure shows some similarities with a snowflake. Crystallization can be viewed as a solidification process in a solid–liquid system. Although crystallization grows from an existing seed, one of the key feature of this process is the *growth inhibition* phenomenon whereby a crystallized site prevents nearby sites from solidifying. This effect can be incorporated in the simple following rule. Each site is in one of two states: solid or liquid; a non-solid site that has exactly one solid neighbor crystallizes. The other sites remain unchanged. Simulate this rule on a hexagonal lattice, starting with one initial solid seed in the middle. Mapping a hexagonal grid on a square lattice can be performed by considering two truncated Moore neighborhoods (see also section 7.2.4): in addition to north, south, east and west, odd line sites of the lattice only see north-west and south-west neighbors, whereas even line sites see north-east and south-east.

2

Cellular automata modeling

2.1 Why cellular automata are useful in physics

The purpose of this section is to show different reasons why cellular automata may be useful in physics. In a first paragraph, we shall consider cellular automata as simple dynamical systems. We shall see that although defined by very simple rules, cellular automata can exhibit, at a larger scale, complex dynamical behaviors. This will lead us to consider different levels of reality to describe the properties of physical systems. Cellular automata provide a fictitious microscopic world reproducing the correct physics at a coarse-grained scale. Finally, in a third section, a sampler of rules modeling simple physical systems is given.

2.1.1 Cellular automata as simple dynamical systems

In physics, the time evolution of physical quantities is often governed by nonlinear partial differential equations. Due to the nonlinearities, solution of these dynamical systems can be very complex. In particular, the solution of these equation can be strongly sensitive to the initial conditions, leading to what is called a chaotic behavior. Similar complications can occur in discrete dynamical systems. Models based on cellular automata provide an alternative approach to study the behavior of dynamical systems. By virtue of their simplicity, they are potentially amenable to easier analysis than continuous dynamical systems. The numerical studies are free of rounding approximations and thus lead to exact results.

Crudely speaking, two classes of problem can be posed. First, given a cellular automaton rule, predicts its properties. Second, find a cellular automaton rule that will have some prescribed properties. These two closely related problems are usually difficult to solve as we shall see on simple examples.

The simplest cellular automata rules are one-dimensional ones for which each site has only two possible states and the rule involves only the nearest-neighbors sites. They are easily programmable on a personal computer and offer a nice "toy model" to start the study of cellular automata.

A systematic study of these rules was undertaken by S. Wolfram in 1983 [24,25]. Each cell (labeled i) has at a given time, two possible states $s_i = 0$ or 1. The state s_i at time $t + 1$ depends only on the triplet (s_{i-1}, s_i, s_{i+1}) at time t:

$$s_i(t + 1) = \Phi(s_{i-1}(t), s_i(t), s_{i+1}(t)) \tag{2.1}$$

Thus to each triplet of sites one associates a value $\alpha_k = 0$ or 1 according to the following list:

$$\underbrace{111}_{\alpha_7} \quad \underbrace{110}_{\alpha_6} \quad \underbrace{101}_{\alpha_5} \quad \underbrace{100}_{\alpha_4} \quad \underbrace{011}_{\alpha_3} \quad \underbrace{010}_{\alpha_2} \quad \underbrace{001}_{\alpha_1} \quad \underbrace{000}_{\alpha_0} \tag{2.2}$$

Each possible cellular automata rule $\mathcal{R}$ is characterized by the values $\alpha_0, ..., \alpha_7$. There are clearly 256 possible choices. Each rule can be identified by an index $\mathcal{N_R}$ computed as follows

$$\mathcal{N_R} = \sum_{i=0}^{7} 2^i \alpha_i \tag{2.3}$$

which corresponds to the binary representation $\alpha_7 \alpha_6 \alpha_5 \alpha_4 \alpha_3 \alpha_2 \alpha_1 \alpha_0$

Giving a rule and an initial state, one can study the time evolution of the system. Some results can be deduced analytically using algebraic techniques, but most of the conclusions follow from numerical iterations of the rules. One can start from a simple initial state (i.e. only one cell in the state 1) or with a typical random initial state. According to their behavior, the different rules have been grouped in four different classes.

(1) Class 1. These cellular automata evolve after a finite number of time steps from almost all initial states to a unique homogeneous state (all the sites have the same value). The set of exceptional initial configurations which behave differently is of measure zero when the number of cells N goes to infinity. A example is given by the rule 40 (see figure 2.1(a)). From the point of view of dynamical systems, these automata evolve towards a simple *limit point* in the phase space.

(2) Class 2. A pattern consisting of separated periodic regions is produced from almost all the initial states. The simple structures generated are either stable or periodic with small periods. An example is given by the rule 56 (see figure 2.1(b)) Here again, some particular initial states (set of measure zero) can lead to unbounded growth. The

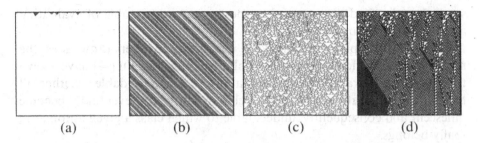

Fig. 2.1. Example of the four Wolfram rules with a random initial configuration. Horizontal lines correspond to consecutive iterations. The initial state is the uppermost line. (a) Rule 40 belonging to class 1 reaches very quickly a fixed point (stable configuration). (b) Rule 56 of class 2 reaches a pattern composed of stripes which move from left to right. (c) Rule 18 is in class 3 and exhibits a self-similar pattern. (d) Rule 110 is an example of a class 4 cellular automaton. Its behavior is not predictable and as a consequence, we observe a rupture in the pattern, on the left part.

evolution of these automata is analogous to the evolution of some continuous dynamical systems to *limit cycles*.

(3) Class 3. These cellular automata evolve from almost all initial states to chaotic, aperiodic patterns. An example is given by the rule 18 (see figure 2.1(c)). Small changes in the initial conditions almost always lead to increasingly large changes in the later stages. The evolution of these automata is analogous to the evolution of some continuous dynamical systems to *strange attractors*.

(4) Class 4. For these cellular automata, persistent complex structures are formed for a large class of initial states. An example is given by the rule 110 (see figure 2.1(d)). The behavior of such cellular automata can generally be determined only by explicit simulation of their time evolution.

The above discussion shows that the "toy rules" considered by Wolfram, although very simple in construction, are capable of very complex behavior. The validity of this classification is not restricted to the simple rules described above but is somehow generic for more complicated rules. For example, one can consider rules for which each cell can have k different states and involve the r neighbors of the sites. In this case, the number of possible rules is $k^{(k^{(2r-1)})}$. Several cases have been studied in the literature and the different rules can be classified in one of the four above classes. Many of the class 4 cellular automata (starting with $k = 2, r = 2$) have the property of computational universality. This means that suitable initial configurations can specify arbitrary algorithmic procedures and thus the system can be used as a general purpose computer capable of evaluating any computable function.

However, this "phenomenological" classification suffers drawbacks, the most serious of which is its non-decidability. Culik and Yu [57] have shown that for the Wolfram rules defined above, it is undecidable whether all the finite configurations of a given cellular automaton eventually become quiescent and consequently, undecidable to which class a given automaton really belongs.

As we shall see later, some of these simple rules are closely related to real physical systems.

2.1.2 *Cellular automata as spatially extended systems*

Spatially extended systems are dynamical systems in which one or several spatial dimensions are added. Typically, they are defined on a lattice with each site characterized by a local state. This state evolves in time according to a given function which usually depends on the values at the site itself and in a local neighborhood. A synchronous updating of all sites is assumed.

According to this definition, a cellular automaton is clearly a spatially extended system. But other systems, termed *coupled map lattices*, also belong to this class. Coupled map lattices can be seen as cellular automata with an *infinite* number of possible states at each site. The only difference between the two types of systems is the continuous character of the local variables. The definition of the rule is replaced by the choice of the local dynamics of the continuous degrees of freedom sitting at each site. Lattice Bolzmann systems, which will be discussed in detail throughout this book, are a particular case of a coupled map lattice.

Extended systems with local interactions and synchronous updating are of great importance to understand the nature of complexity exhibited by several natural processes because they contain the minimal ingredients of many real systems: a dynamics which depends both on space and time. Fully developed turbulence is an example for which basic models are needed to provide some insights on the fundamental mechanisms that are involved. For this reason, coupled map lattices are often use as a toy model for turbulent systems.

An important problem which arises in the study of spatially extended system is the existence of *non-trivial collective behavior*. A non-trivial collective behavior is an unexpected behavior which is observed on macroscopic quantities such as, for instance, the total density (or the sum over all the lattice sites of the local state). A typical example is a periodic or quasi-periodic behavior of this global density, in complete disagreement with equilibrium statistical mechanics or mean-field calculations.

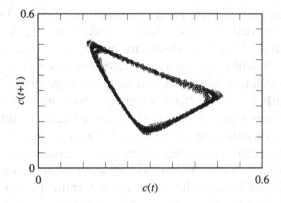

Fig. 2.2. Non-trivial collective behavior for the simple 3D cellular automata rule given by relation (2.4). The plot shows the behavior of the global density $c(t)$ as a function of its previous value.

Usually, non-trivial collective behaviors are likely to appear in high-dimensional lattices. However, there is a simple cellular automata example in three dimensions [58]: one considers a 3D Cartesian lattice with periodic boundary conditions. Each site i can be either in state $s_i = 0$ or in state $s_i = 1$. The evolution rule is the following

$$s_i(t+1) = \begin{cases} 1 & \text{if } \sum_{neighbors} s_j(t) \in \{0, 5\} \\ 0 & \text{otherwise} \end{cases} \qquad (2.4)$$

Here the sum over the neighbors extends to the six nearest neighbors in the cubic lattice, plus the central site i itself. Thus, the sum ranges between 0 and 7. The rule is that the new value $s_i(t+1)$ is 1 if and only if this sum equals 0 or 5.

As time goes on, one can record the density $c(t)$ defined as

$$c(t) = \sum_{lattice} s_i(t)$$

This quantity exhibits a non-trivial behavior in the sense that it does not relax to a constant value at large time. Instead, it follows a low dimensional dynamics which can be represented by the map shown in figure 2.2 where $c(t+1)$ is plotted as a function of $c(t)$.

Such non-trivial collective behaviors in spatially extended systems has been observed on several totalistic cellular automata rules. A totalistic rule means that the new state of the cell i is a function only of the sum of the states s_j for cells j belonging to a given neighborhood of i. Rule (2.4) is an example of a totalistic rule. Totalistic rules can be seen as mean-field-like evolution since in the each site "sees" an average environment.

Non-trivial collective behaviors are not singular phenomena, requiring the fine tuning of some conditions, but appear as general features, robust to small modifications of local rules and initial conditions. The dynamics is usually accompanied by noise which is related to the disorder of the initial configuration. This noise disapears in the asymptotic regime.

Several types of collective behaviors have been observed and characterized for both cellular automata and coupled map models. A uniform treatment of both cases has been given by Chaté and Manneville [59] and their discussion leads to the classification of cellular automata rules according to the various types of behaviors that are observed.

An important question concerns the minimum requirement for producing such non-trivial collective behaviors and several conclusions have been drawn by Chaté and Manneville. First, the deterministic character of the evolution is not essential. Collective behaviors appear both for deterministic and probabilistic evolutions. Second, the synchronicity is essential. To exhibit non-trivial behavior, the system requires a clock providing an "instantaneous" global communication channel. Thus, the analogy with more traditional systems studied in statistical mechanics (sequential Monte-Carlo updating for example) should be made with great care.

2.1.3   Several levels of reality

The previous paragraph illustrated the point that cellular automata have a strong similarity with complex systems, despite the simplicity of their elementary dynamics. It turns out that the reason for such a similar behavior is quite general and will be used all the time throughout this book: often, the macroscopic behavior of a system composed of many interacting constituents depends very little on the microscopic details of the interactions. For instance, it is well known that the Navier–Stokes equation governing the motion of a fluid is first of all an expression of momentum conservation during the elementary interactions between the molecules. The details of these interactions shows up in the coefficients of the equation (for instance, the viscosity), but not in its algebraic structure. We shall come back to this point many times and make it more explicit in various examples.

This is actually one of the most important points behind the concept of cellular automata modeling. Several levels of reality exist when studying a physical system. Roughly speaking, there are two main levels, corresponding to a microscopic and macroscopic scale of observation. The same system may look quite different when described at these various levels. At a microscopic level, interactions may be governed by complicated potentials and, sometimes, require the use of quantum mechanics to be described properly. At a macroscopic level, the properties of the system are dominated by the aggregate effect of all the microscopic interactions. The resulting behavior, when seen at a large scale of observation, is mostly related to the generic features of the microscopic interactions. The complexity of the macroscopic world is apparently disconnected from that of the microscopic world, although the former is driven by the latter. The microscopic details are lost when the system is looked at through a macroscopic filter and new collective properties emerge.

A well-known example of these two levels of reality is given by the study of the magnetic properties of ferromagnetic materials. At a microscopic level, the phenomena is driven by the distribution of electrons in atoms, which, in principle, is described by the laws of quantum mechanics.

On the other hand, at a macroscopic level, one observes a collective behavior: below some critical temperature, the magnetic dipoles associated to each atom align to form macroscopic magnetic domains. This property, which makes sense only at a macroscopic scale of observation, is not related to the detailed form of the ferromagnetic interaction but rather comes from the very general fact that the local energy of interaction is lowered when two magnetic dipoles are oriented along the same direction. As a matter of fact, the macroscopic behavior described here can be reproduced by a very simple dynamical model, namely the famous Ising model. In this model, all the microscopic complexity is reduced to simple interactions between spins located at each site of a regular Cartesian lattice. Spins are either 1 or -1, and each of them accounts for the resulting magnetic dipole of the atom. It flips from up to down (or the converse) according the local energy gain this movement results in. The Ising model offers a very poor description of real microscopic interactions but retains the significant aspects that show up when the system is considered at a larger scale.

Ising-like systems are very close in spirit to a cellular automata model. They are based on a fictitious microscopic universe, which aims at capturing the essential features of real microscopic interactions, but no more. Kinetic Ising models are defined on a discrete space, with a discrete time, as is the case of cellular automata. However, here the spin updating is done sequentially, whereas, in a cellular automata, it is done synchronously. This difference is of importance because a system of spins may not converge to the same final configuration when the updating is done sequentially or in parallel. We shall come back to this point in section 2.2.3.

2.1.4 A fictitious microscopic world

The cellular automata approach works when the simplification of the microscopic laws pertaining to a given phenomena is not relevant at a macroscopic scale of observation. We learn from statistical mechanics that this fact is often true for systems whose complexity comes from a collective behavior rather than from some distinctive aspects of the microscopic interactions.

From the observation that the macroscopic behavior of many systems (which is precisely the level of reality we are interested in) has little to do with the true microscopic nature, it is a clear advantage to invent a much simpler microscopic reality, which is more appropriate to our numerical means of investigation. The following Einstein quotation:

Everything should be made as simple as possible but not simpler

reflects very well the spirit of cellular automata modeling. The important point is to capture the essential and relevant ingredients of a real phenomena and consider them as the fundamental physical laws of a new imaginary system. In that sense, cellular automata are a caricature of the real world rather than its portrait.

Modeling a system at a microscopic level of description has significant advantages. The interpretation of cellular automata dynamics in terms of simple microscopic rules offers a very intuitive and powerful approach to modeling phenomena that are very difficult to include in more traditional approaches (such as differential equations). For instance, boundary conditions are often naturally implemented in a cellular automata model because it has a natural interpretation at this level of description (e.g. particles bouncing back from an obstacle). The phenomena of wetting of a solid substrate by a spreading liquid illustrates the difficulty in defining appropriate boundary conditions (see section 7.2).

The design of correct cellular automata models implies that the essential aspects of a complex phenomena have been recognized and reduced to a simple tractable form. This process of reducing a complex behavior to the sum of simple mechanisms is an essential step in any scientific investigation. Therefore, cellular automata models provide a very powerful tool in fundamental research, real problems and pedagogical applications.

The numerical simplicity of many cellular automata rules makes them well suited to massive computer simulations. The Boolean nature of the operations that are to be performed leads to an exact evolution where no truncation is necessary and no numerical instabilities are present. As a result, the computer simulation is a totally faithful implementation of the mathematical model expressed by the cellular automata rule. This is very interesting when modeling N-body problems, because the many-particle correlation functions are not corrupted by the numerical scheme.

2.2   Modeling of simple systems: a sampler of rules

The purpose of this section is to illustrate the general concept we have presented so far. We will introduce a few simple cellular automata rules aimed at describing the behavior of several physical process. From these examples, we shall develop our intuition and also identify some problems related to the cellular automata approach. In particular, we shall see the importance of developing techniques to assess the range of validity of a model and its limitations.

2.2.1 *The rule 184 as a model for surface growth*

Wolfram's rule 184, can be used to model the formation of spatio-temporal structures for deterministic surface growth. An important surface growth mechanism is through ballistic deposition. Krug and Spohn [60] studied the following simple lattice model. Consider a one-dimensional surface configuration parallel to the x axis of a square lattice. Particles move on straight lines along the y axis and become part of the deposit when they reach the empty nearest neighbor of an occupied site. Thus an incoming particle may stick either to the side edge (with a rate Γ_s) or to the top edge of an existing column (with a rate Γ_t). The limit $\rho \equiv \Gamma_t/\Gamma_s \to \infty$ corresponds to the so-called random deposition model. One then considers the opposite limit $\rho \to 0$. With simultaneous updating, the interface dynamics can be written is the following simple form:

$$h_i(t+1) = \max\left(h_{i-1}, h_i, h_{i+1}\right)(t) \qquad (2.5)$$

for all sites i, where $h_i(t)$ denotes the height of the surface at site i and time t.

One can then consider a slightly modified version of the model, namely deposition onto a one-dimensional surface parallel to the $(1,1)$ direction of the square lattice. The particles are allowed to stick only at the minima of the surface as shown on figure 2.3. Hence, the surface configurations generated have a local gradient ± 1. The particles have a simple dynamics. With rate 1, they jump to the right except when the final site is occupied, in which case they stay (hard-core repulsion). Thus the configuration at time $t+1$ is obtained from the configuration at time t by a local rule which is simply Wolfram's rule 184. For long times, the dynamics leads to a trivial limit cycle. However, the approach to the final state has a rich space–time structure when interpreted as a surface.

2.2.2 *Probabilistic cellular automata rules*

The Wolfram rules can be considered as the deterministic limit of probabilistic cellular automaton rules which can exhibit even more complex behavior as a function of the free parameters entering into the rule. A simple, physically important example of a probabilistic rule is provided by the problem of directed percolation. Consider the square lattice of

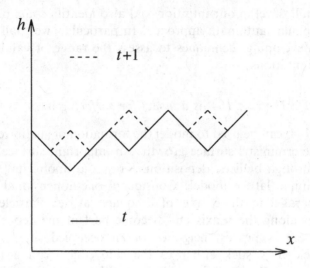

Fig. 2.3. Deterministic growth from a rough surface parallel to the (1, 1) direction. At each time step, all local surface minima are filled with new particles. The corresponding lattice gas evolves according to the automata rule 184.

figure 2.4. Any lattice site may be present with a probability p or absent with probability $(1 - p)$. Moreover, any bond may be present with probability q and absent with probability $(1 - q)$. Assume that a single site is "wet" at time $t = 0$. Present bonds conduct "water" only in the downward (increasing time) direction, and when both sites connected by the bond are present.

For $q = 1$, one speaks of *site-directed percolation*, and for $p = 1$, of *bond-directed percolation*. Given the two probabilities p and q, one can ask several questions: what is the probability of finding wet sites at (time) level t and what is the typical size of the wet cluster? In each case, there are critical values (p_c, q_c), called the percolation thresholds, above which the wet cluster spans the whole system. Directed percolation is a well-known problem of "geometrical phase transition" for which scaling theory and the renormalization group approach have been developed.

This problem can easily be formulated in term of probabilistic cellular automaton. Consider a linear chain or ring of sites; with each site one associates a variable $\psi_i(t) = 0$ or 1. At odd (even) times, odd (even) indexed sites change their state according to a probabilistic rule, and even (odd) indexed sites stay in the same state. The full space–time history can be presented on a two-dimensional lattice. The rules are defined by the conditional probabilities $P(\psi_i(t + 1)|\psi_{i-1}(t), \psi_{i+1}(t))$. For directed

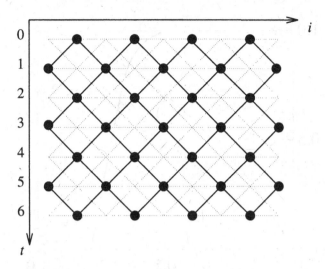

Fig. 2.4. Two-dimensional square lattice for directed percolation. The dots represent the sites which can be present or absent. The heavy lines between the sites represent the bonds. The vertical axis corresponds to time.

percolation, we have:

$$P(1|00) = 0, \quad y \equiv P(1|11) = pq, \quad z \equiv P(1|01) = P(1|10) = pq(2 - q)$$
(2.6)

The situation in the parameter space (p, q) is represented in figure 2.5.

At the transition point, the percolation cluster is a fractal object. Moreover, near the transition point, many physical quantities have a power law behavior in terms of $p - p_c$ or $q - q_c$, with universal critical exponents. Here again, one sees that a very simple cellular automaton rule leads to a complex structure.

Another interesting and simple example of a probabilistic cellular automata rule is related to forest fire models [61,62]. This probabilistic cellular automata model is defined on a d-dimensional hypercubic lattice. Initially, each site is occupied by either a tree, a burning tree or is empty. The state of the system is parallel updated according to the following rule:

1. A burning tree becomes an empty site.

2. A green tree becomes a burning tree if at least one of its nearest neighbors is burning.

3. At an empty site, a tree grows with probability p.

4. A tree without a burning nearest neighbor becomes a burning tree during one time step with probability f (lightning).

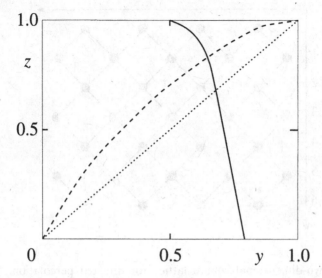

Fig. 2.5. Parameter space $y = pq; z = pq(2 - q)$ defining the cellular automata rules for generalized directed percolation. The dashed and dotted lines correspond, respectively, to the bond and site percolation. The system percolates to the right of the (solid) transition line.

Figure 2.6 illustrates the behavior of this rule, in a two-dimensional situation. Provided that the time scales of tree growth and burning down of forest clusters are well separated (i.e. in the limit $f/p \to 0$), this models has self-organized critical states [62]. This means that in the steady state, several physical quantities characterizing the system have a power law behavior. For example, the cluster size distribution $\mathcal{N}(s)$ and radius of a forest cluster $\mathcal{R}(s)$ vary with the number of trees s in the forest cluster as:

$$\mathcal{N}(s) \sim s^{-\tau} \mathcal{C}(s/s_{max}) \tag{2.7}$$

and

$$\mathcal{R}(s) \sim s^{1/\mu} \mathcal{S}(s/s_{max}) \tag{2.8}$$

Scaling relations can be established between the critical exponents τ and μ, and the scaling functions $\mathcal{C}$ and $\mathcal{S}$ can be computed.

Note that there is a more general correspondence between d-dimensional probabilistic cellular automata and $d + 1$ equilibrium statistical physics problems. To each d-dimensional probabilistic cellular automata one can associate a classical $(d + 1)$-dimensional statistical model [63].

Fig. 2.6. The forest fire cellular automata rule: grey sites correspond to a grown tree, black pixels represent burned sites and the white color indicates a burning tree. The snapshot given here represents the situation after a few hundred iterations. The parameters of the rule are $p = 0.3$ and $f = 6 \times 10^{-5}$.

2.2.3 The Q2R rule

The Q2R rule was proposed by G. Vichniac [64] in the 1980s as a model for Ising spin dynamics. The problem is to find a cellular automata rule modeling the behavior described in section 2.1.3.

We consider a two-dimensional square lattice such that each site holds a spin s_i which is either up ($s_i = 1$) or down ($s_i = 0$). During their discrete time evolution, the spins will flip (or not flip) according to a cellular automata rule. Following our modeling philosophy, this rule will be built up from very simple and general principles. In the present case, it will be local energy conservation. This means we are considering a system which cannot exchange energy with its surroundings. The model will be a microcanonical cellular automata simulation of Ising spin dynamics, without a temperature but with a critical energy.

In an Ising model, the energy of a spin s_i comes from couplings with the spins s_j located in its von Neumann neighborhood (north, west, south and east). The contribution of the pair (s_i, s_j) to this energy is some value $-J$ when two spins are aligned, and J when they are not aligned.

A spin s_i can flip at time t to become $1 - s_i$ at time $t + 1$ if and only if this move does not cause any energy change. Accordingly, spin s_i will flip if the number of its neighbors with spin up is the same as the number of its neighbors with spin down. However, one has to remember that the motion of all spins is simultaneous in a cellular automata. The decision to flip is based on the assumption that the neighbors are not changing. If they flip too (because they obey the same rule), then energy may not be conserved. This is simply illustrated by a one-dimensional chain of spins. Suppose one has the following spin configuration at time t

$$\text{time } t \quad : \qquad \ldots 0 \underbrace{\quad}_{+J} 1 \underbrace{\quad}_{+J} 0 \underbrace{\quad}_{-J} 0 \underbrace{\quad}_{+J} 1 \underbrace{\quad}_{+J} 0 \ldots \qquad (2.9)$$

where the coupling energy is indicated for each pair of spin. Now, if all spins flip according to our prescription, we obtain the situation

$$\text{time } t+1 : \qquad \ldots 1 \underbrace{\quad}_{-J} 1 \underbrace{\quad}_{-J} 1 \underbrace{\quad}_{-J} 1 \ldots \qquad (2.10)$$

As we can see by counting the coupling energy for the middle spins, the energy balance is incorrect. This is due to the simultaneity of the update. A way to cure this problem is to split the updating into two phases: first, one flips the spins located at an odd position along the chain, according to the configuration of the even spins. In the second phase, the even sublattice is updated according to the odd one. In a two-dimensional lattice, the same procedure has to be carried out: the original lattice is decomposed into two sublattices, as given by the white and black squares of a chess board.

This strategy prevents the conflict of motion between adjacent spins. However, at first sight, this method contradicts our definition of a cellular automaton, which stipulates that all sites be updated simultaneously. To conciliate the requirements of the problem with this definition, one can simply extend the basic rule to include some spatial sensitivity: initially, lattice sites hold an extra bit b of information whose value is 0 for the odd sublattice and 1 for the even sublattice. For a two-dimensional system the spatial "texture" given by b at time $t = 0$ would be

$$
\begin{array}{ccccccc}
\ldots & 1 & 0 & 1 & 0 & 1 & 0 & \ldots \\
\ldots & 0 & 1 & 0 & 1 & 0 & 1 & \ldots \\
\ldots & 1 & 0 & 1 & 0 & 1 & 0 & \ldots \\
\ldots & 0 & 1 & 0 & 1 & 0 & 1 & \ldots \\
\ldots & 1 & 0 & 1 & 0 & 1 & 0 & \ldots \\
\ldots & 0 & 1 & 0 & 1 & 0 & 1 & \ldots
\end{array}
\qquad (2.11)
$$

The flipping rule described earlier is then regulated by the value of b. It takes place only for those sites for which $b = 1$. Of course, the value of b is also updated at each iteration according to $b(t+1) = 1 - b(t)$, so that at the next iteration, the other sublattice is considered. Thus, at time $t = 1$, the space parity b of the system changes to

$$
\begin{array}{ccccccc}
\dots & 0 & 1 & 0 & 1 & 0 & 1 & \dots \\
\dots & 1 & 0 & 1 & 0 & 1 & 0 & \dots \\
\dots & 0 & 1 & 0 & 1 & 0 & 1 & \dots \\
\dots & 1 & 0 & 1 & 0 & 1 & 0 & \dots \\
\dots & 0 & 1 & 0 & 1 & 0 & 1 & \dots \\
\dots & 1 & 0 & 1 & 0 & 1 & 0 & \dots
\end{array}
\tag{2.12}
$$

In two dimensions, the Q2R rule can be the expressed by the following relations

$$
s_{ij}(t+1) = \begin{cases} 1 - s_{ij}(t) & \text{if } b_{ij} = 1 \text{ and } s_{i-1,j} + s_{i+1,j} + s_{i,j-1} + s_{i,j+1} = 2 \\ s_{ij}(t) & \text{otherwise} \end{cases}
\tag{2.13}
$$

and

$$
b_{ij}(t+1) = 1 - b_{ij}(t) \tag{2.14}
$$

where the indices (i, j) label the Cartesian coordinates and $s_{ij}(t = 0)$ is either one or zero.

The question is now how well does this cellular automata rule perform to describe an Ising model. Figure 2.7 shows a computer simulation of the Q2R rule, starting from an initial configuration with approximately 11% of spins $s_{ij} = 1$ (figure 2.7 (a)). After a transient phase (figures (b) and (c)), the system reaches a stationary state where domains with "up" magnetization (white regions) are surrounded by domains of "down" magnetization (black regions).

In this dynamics, energy is exactly conserved because that is the way the rule is built. However, the number of spins down and up may vary. Actually, in the present experiment, the fraction of spins up increases from 11% in the initial state to about 40% in the stationary state. Since there is an excess of spins down in this system, there is a resulting macroscopic magnetization.

It is interesting to study this model with various initial fractions ρ_s of spins up. When starting with a random initial condition, similar to that of figure 2.7 (a), it is observed that, for many values of ρ_s, the system evolves to a state where there is, on average, the same amount of spin down and up, that is no macroscopic magnetization. However, if the initial configuration presents a sufficiently large excess of one kind of spin, then a macroscopic magnetization builds up as time goes on. This means there is a phase transition between a situation of zero magnetization and a situation of positive or negative magnetization.

It turns out that this transition occurs when the total energy E of the system is low enough (low energy means that most of the spins are aligned

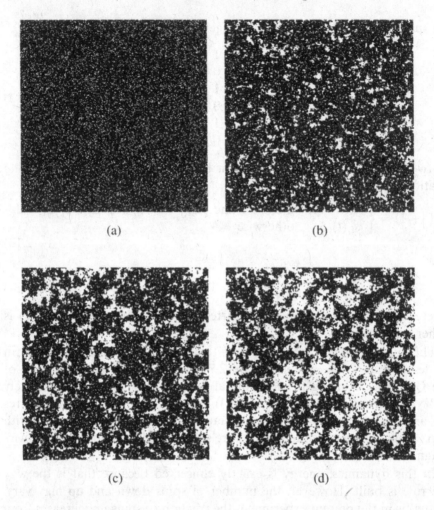

Fig. 2.7. Evolution of a system of spins with the Q2R rule. Black represents the spins down $s_{ij} = 0$ and white the spins up $s_{ij} = 1$. The four images (a), (b), (c) and (d) show the system at four different times $t_a = 0 < t_b < t_c \ll t_d$.

and that there is an excess of one species over the other), or more precisely when E is smaller than a critical energy E_c.

Therefore, the Q2R rule captures an important aspect of a real magnetic system, namely a non-zero magnetization at low energy (which can be related to a low temperature situation) and a transition to a non-magnetic phase at high energy. However, Q2R also exhibits an unexpected behavior that is difficult to detect from a simple observation. There is a loss of ergodicity: a given initial configuration of energy E_0 evolves without visiting completely the region of the phase space characterized by $E = E_0$.

This is illustrated by the following simple example, where a ring of four spins is considered. For the sake of convenience, we shall represent it as a linear chain with periodic boundary conditions (the first and last spins are nearest neighbors).

$$
\begin{array}{ll}
t: & 1001 \\
t+1: & 1100 \\
t+2: & 0110 \\
t+3: & 0011 \\
t+4: & 1001
\end{array}
\tag{2.15}
$$

After four iterations, the system cycles back to its original state. The configuration of this example has $E_0 = 0$. As we observe, it never evolves to 0111, which is also a configuration of zero energy.

This non-ergodicity means that not only energy is conserved during evolution of the automaton, but also other quantities which partition the energy surface in independent regions. As we shall see, the study of the dynamical invariants is very important when studying cellular automata models.

2.2.4 The annealing rule

A natural class of cellular automata rules consists of the so-called *majority rules*. Updating selects the new state of each cell so as to conform to the value currently hold by the majority of the neighbors. Typically, in these majority rules, the state is either 0 or 1.

A very interesting behavior is observed with the twisted majority rule proposed by G. Vichniac [64]: in two dimensions, each cell considers its Moore neighborhood (i.e itself plus its eight nearest neighbors) and computes the sum of the cells having a value 1 in the Moore neighborhood. This sum can be any value between 0 and 9. The new state $s_{ij}(t+1)$ of each cell is then determined from this local sum, according to the following table

$$
\begin{array}{ll}
\text{sum}_{ij}(t) & 0\ 1\ 2\ 3\ 4\ 5\ 6\ 7\ 8\ 9 \\
s_{ij}(t+1) & 0\ 0\ 0\ 0\ 1\ 0\ 1\ 1\ 1\ 1
\end{array}
\tag{2.16}
$$

As opposed to the plain majority rule, here, the two middle entries of the table have been swapped. Therefore, when there is a slight majority of 1s around a cell, it turns to 0. Conversely, if there is a slight majority of 0s, the cell becomes 1.

Surprisingly enough this rule describes the interface motion between two phases, as illustrated in figure 2.8. Vichniac has observed that the normal velocity of the interface is proportional to its local curvature, as required by the Allen–Cahn [65] equation. Of course, due to its local nature, the rule cannot detect the curvature of the interface directly. However, as the

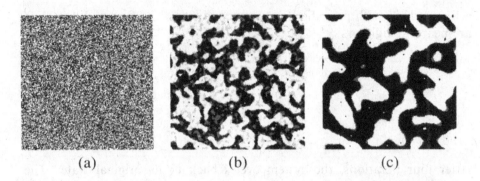

Fig. 2.8. Evolution of the annealing rule. The inherent "surface tension" present in the rule tends to separate the black phases $s = 1$ from the white phase $s = 0$. The snapshots (a), (b) and (c) correspond to $t = 0$, $t = 72$ and $t = 270$ iterations, respectively. The extra gray levels indicate how "capes" have been eroded and "bays" filled: dark gray shows the black regions that have been eroded during the last few iterations and light gray marks the white regions that have been filled.

rule is iterated, local information is propagated to the nearest neighbors and the radius of curvature emerges as a collective effect.

This rule is particularly interesting when the initial configuration is a random mixture of the two phases, with equal concentration. Otherwise, some pathological behavior may occur. For instance, an initial black square surrounded by the white phase will not evolve: right angles are not eroded and are stable structures.

2.2.5 The HPP rule

The HPP rule is the first of an important class of cellular automata models. The basic working objects of such models are point particles that move on a lattice, according to appropriate rules. The state of each cell represents the presence or the absence of such particles.

In order to make this kind of fully discrete "molecular dynamics" compatible with cellular automata dynamics, one should prevent an arbitrary number of particles being present simultaneously on a given site. This is due to the definition of a cellular automaton, in which the states are described by a finite number of bits. The rule should be defined so as to guarantee an exclusion principle.

The purpose of the HPP rule is to model a gas of colliding particles. The essential features that are borrowed from the real microscopic interactions are the conservation laws, namely local conservation of momentum and local conservation of particle number.

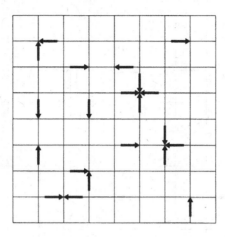

Fig. 2.9. Example of a configuration of HPP particles.

The HPP lattice gas automata is traditionally defined on a two-dimensional square lattice. Particles can move along the main directions of the lattice, as shown in figure 2.9. The model limits to 1 the number of particles entering a given site with a given direction of motion. This is the exclusion principle we were mentioning. Consequently, four bits of information in each site are enough to describe the system during its evolution. For instance, if at iteration t site $\vec{r}$ has the following state $s(\vec{r}, t) = (1011)$, it means that three particles are entering the site along direction 1,3 and 4, respectively.

The cellular automata rule describing the evolution of $s(\vec{r}, t)$ is usually split into two steps: collision and motion. The collision phase specifies how particles entering the same site will interact and change their trajectories. During the motion phase, or propagation, the particles are actually moved to the nearest neighbor site they were traveling to. Figure 2.10 illustrates the HPP rules. This decomposition in two phases is another way to partition space, as in the Margolus neighborhood. According to our Boolean representation of the particles at each site, the collision part for a two-particle head-on collision is expressed as

$$(1010) \rightarrow (0101) \qquad (0101) \rightarrow (1010) \qquad (2.17)$$

all the other configuration being unchanged. During the propagation phase, the first bit of the state variable is shifted to the east neighbor cell, the second bit to the north and so on.

Since the collision phase amounts to rearranging the particles in a different direction, it ensures that the exclusion principle will be satisfied, provided that it was at time $t = 0$.

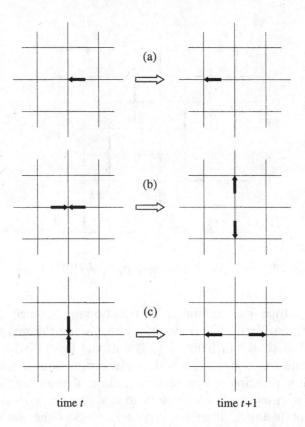

Fig. 2.10. The HPP rule: (a) a single particle has a ballistic motion until it experiences a collision. (b) and (c) The two non-trivial collisions of the HPP model: two particles experiencing a head-on collision are deflected in the perpendicular direction. In the other situations, the motion is ballistic, that is the particles are transparent to each other when they cross the same site.

The aim of this rule is to reproduce some aspects of the real interaction of particles, namely that momentum and particle number are conserved during a collision. From figure 2.10, it is easy checked that these properties are obeyed: for the two situations (b) and (c) where the particles do actually experience a collision, a pair of zero momentum particles along a given direction is transformed into another pair of zero momentum along the perpendicular axis.

The HPP rule captures another important ingredient of the microscopic nature of a real interaction: invariance under time reversal. Figures 2.10 (b) and (c) show that, if at some given time, the direction of motion of all particles are reversed, the system will just trace back its own history. Since the dynamics of a deterministic cellular automaton is exact, this allows

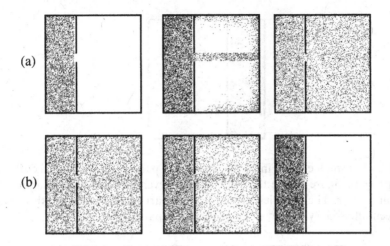

(a)

(b)

Fig. 2.11. Time evolution of a HPP gas. (a) From the initial state to equilibrium. (b) Illustration of time reversal invariance: in the rightmost image of (a), the velocity of each particle is reversed and the particles naturally return to their initial position.

us to demonstrate that physical systems return to their original situation when all particles reverse their velocity.

Figure 2.11 illustrates the time evolution of a HPP gas initially confined in the left compartment of a container. There is an aperture in the wall of the compartment and the gas particles will flow so as to fill the entire space available to them. In order to include a solid boundary in the system, the HPP rule is modified as follows: when a site is a wall (indicated by an extra bit), the particles no longer experience the HPP collision but bounce back to where they come from. Therefore, particles cannot escape a region delimited by such a reflecting boundary.

If the system of figure 2.11 is evolved, it reaches an equilibrium after a long enough time and no trace of its initial state is any longer visible. However, no information has been lost during the process (no numerical dissipation) and the system has the memory from where it comes from. Reversing all the velocities and iterating the HPP rule makes all particle go back to the compartment in which they were initially located.

This behavior is only possible because the dynamics is perfectly exact and no numerical errors are present in the numerical scheme. If one introduces externally some errors (for instance, one can add an extra particle in the system) before the direction of motion of each particle is reversed, then the reversibility is lost. The result of such a change is shown in figure 2.12 which is a snapshot of the HPP gas at a time all particles would have returned to the left compartment if no mistake had been introduced.

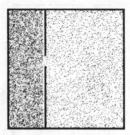

Fig. 2.12.   If a small error is introduced (e.g. one particle is added) in the situation of the previous figure before the velocities are reversed, the system cannot return to its initial state. The particles left in the right part of the system are those which have been affected by the presence of the extra particle.

Fig. 2.13.   Propagation of a compression wave in the HPP model. Initially, a high concentration of particle is created inside a circle, at the middle of the system. The two images represent snapshots at two time steps $t_1 < t_2$. We can observe how anisotropy builds up during time evolution.

The HPP rule is important because it contains the basic ingredients of many models we are going to discuss in this book. We shall see in chapter 3 that the capability of this rule to model a real gas of particle is poor, due to a lack of isotropy. This problem is illustrated in figure 2.13, where a compression wave propagates in the gas. Starting from an isotropic initial condition of high particle density in the middle, we observe that the wave does not propagate identically in all directions (the lattice axis show up in the figure). A remedy to this problem is to use a different lattice (see chapter 3).

2.2.6   *The sand pile rule*

The physics of granular materials has attracted much interest, recently [66–69]. It is possible to devise a simple cellular automata rule to model basic piling and toppling of particles like sand grains. The idea is that grains

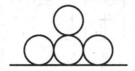

Fig. 2.14.   A stable pile of sand grains.

Fig. 2.15.   A conflict of motion results from the use of the Moore neighborhood. The two gray particles are unstable and both want to move to the same new location.

can stack on top of each other if this arrangement is stable. Of course, real sand grains do not stand on a regular lattice and the stability criteria are expected to depend on the shape of each grain. Despite this microscopic complexity, the result is that sand piles that are too high topple.

In the framework of cellular automata modeling, the toppling mecha-nisms can be captured by the following rule: a grain is stable if there is a grain underneath and two other grains preventing it falling to the left or right, as depicted in figure 2.14. In the Moore neighborhood, it means that a central grain will be at rest if the south-west, south and south-east neighbors are occupied. Otherwise the gain topples downwards to the nearest empty cell.

For instance, suppose one has the situation shown in figure 2.15. Since the updating of all cells is simultaneous, the two shaded grains will move to the same empty cell: there is a conflict. To prevent conflict, each particle should know about the presence of the other one before they move, and take a common decision as to which one is going to fall. This requires a larger neighborhood than the Moore neighborhood and a more complicated rule.

The Margolus neighborhood gives a simple way to deal with the syn-chronous motion of all particles. Piling and toppling is determined by the situation within 2×2 adjacent blocks. Of course (see section 1.3.2), for all other steps, the space partitioning is shifted one cell down and one cell left, so that, after two iterations, each cell sees a larger neighborhood.

Figure 2.16 gives a possible implementation of a sand rule. It shows how the configurations evolve due to toppling.

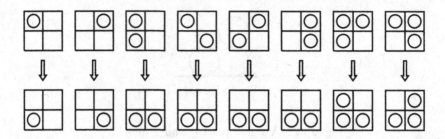

Fig. 2.16. The sand rule expressed for the Margolus neighborhood. Only the configuration that are modified are shown.

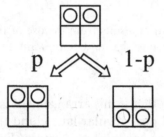

Fig. 2.17. The sand rule allows for blocking (friction) with probability p when two grains occupy the upper part of a Margolus block.

The configuration in which the upper part of a block is occupied by two particles while the lower part is empty, is not listed in this figure, although it certainly yields some toppling. Treating this configuration with a probabilistic rule turns out to produce a more realistic behavior. When this configuration occurs, the evolution rule is given in figure 2.17. The motivation for introducing a probabilistic toppling is that some friction may be present between grains and some "arches" may appear to delay collapse. Of course, the toppling of other configurations could also be controlled by a random choice.

When no blocking occurs, the sand rule can be expressed by the following algebraic relations. We call s_{ul}, s_{ur}, s_{ll} and s_{lr} the state before updating, where the subscripts stand for "upper-left," "upper-right," "lower-left" and "lower-right."

$$s_{ul}(t+1) = s_{ul}s_{ll}(s_{lr} + (1 - s_{lr})s_{ur}) \qquad (2.18)$$

$$s_{ur}(t+1) = s_{ur}s_{lr}(s_{ll} + (1 - s_{ll})s_{ul}) \qquad (2.19)$$

$$s_{ll}(t+1) = s_{ll} + (1 - s_{ll})[s_{ul} + (1 - s_{ul})s_{ur}s_{lr}] \qquad (2.20)$$

$$s_{lr}(t+1) = s_{lr} + (1 - s_{lr})[s_{ur} + (1 - s_{ur})s_{ul}s_{ll}] \qquad (2.21)$$

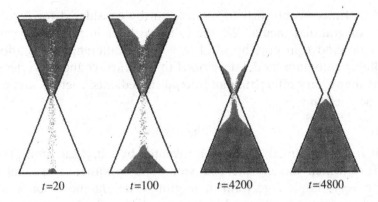

$$t=20 \qquad t=100 \qquad t=4200 \qquad t=4800$$

Fig. 2.18. Evolution of the sand rule for an hour-glass simulation.

It is easy to check that the number of particles is conserved during the evolution, namely that

$$s_{ul}(t+1) + s_{ur}(t+1) + s_{ll}(t+1) + s_{lr}(t+1) = s_{ul} + s_{ur} + s_{ll} + s_{lr} \quad (2.22)$$

Equation 2.18 means that an upper-left particle will fall down unless both the lower-left and the lower-right cells are occupied, or both an upper-right and a lower-left particle are present. The other equations follow from the same kind of arguments.

In order to make this rule more interesting, we must introduce the ground, on which the grain will stop falling. This is necessary to observe piling and then toppling. The ground will be represented as an extra bit g in the state of each cell which, of course, does not evolve during updating. A sand grain located on a ground $g = 1$ cell will be at rest for ever. This is taken into account in the rule by modifying equations 2.18 and 2.19 as

$$s_{ul}(t+1) = g_{ul}s_{ul} + (1 - g_{ul})s_{ul}s_{ll}(s_{lr} + (1 - s_{lr})s_{ur}) \quad (2.23)$$
$$s_{ur}(t+1) = g_{ur}s_{ur} + (1 - g_{ur})s_{ur}s_{lr}(s_{ll} + (1 - s_{ll})s_{ul}) \quad (2.24)$$

which means that a particle can leave the upper part of the Margolus block only if it is not a ground cell.

Equations 2.20 and 2.21 are modified in the following way

$$s_{ll}(t+1) = s_{ll} + (1 - s_{ll})[s_{ul}(1 - g_{ul}) + (1 - s_{ul})s_{ur}(1 - g_{ur})s_{lr}] \quad (2.25)$$
$$s_{lr}(t+1) = s_{lr} + (1 - s_{lr})[s_{ur}(1 - g_{ur}) + (1 - s_{ur})s_{ul}(1 - g_{ul})s_{ll}] \quad (2.26)$$

so that an upper particle is made invisible from the lower cells when it stands on the ground.

The behavior of this rule is illustrated in figure 2.18, which represents a simulation of an hour-glass. We can observe that there is an angle of repose of 45 degrees (slope for which a pile is stable and no avalanche occurs). This angle is an obvious consequence of the lattice structure.

Note that more sophisticated models can be considered for describing the flow of granular media. We shall discuss later in this book a model of snow or sand transport by wind in which a different rule is adopted. The cellular automata model developed by A. Károlyi and J. Kertész [70] contains many very interesting additional ingredients, such as dissipation and static friction.

2.2.7   The ant rule

The ant rule is a cellular automaton invented by Chris Langton [71] and Greg Turk which models the behavior of a hypothetical animal (ant) having a very simple algorithm of motion. The ant moves on a square lattice whose sites are either white or gray. When the ant enters a white cell, it turns 90 degrees to the left and paints the cell gray. Similarly, if it enters a gray cell, it paints it white and turns 90 degrees to the right.

It turns out that the motion of this ant exhibits a very complex behavior. Suppose the ant starts in completely white space. After a series of about 500 steps where it essentially keeps returning to its initial position, it enters a chaotic phase during which its motion is unpredictable. Then, after about 10 000 steps of this very irregular motion, the ant suddenly performs a very regular motion which brings it far away from where it started.

Figure 2.19 illustrates these last two phases of motion. The path the ant creates to escape the chaotic initial region has been called a highway [72]. Although this highway is oriented at 45 degrees with respect to the lattice direction, it is traveled by the ant in a way which makes one think of a sewing machine: the pattern is a sequence of 104 steps which repeated indefinitely.

The Langton ant is another example of a cellular automaton whose rule is very simple and yet generates a complex behavior which seems beyond our understanding. Somehow, this fact is typical of the cellular automata approach: although we do know everything about the fundamental laws governing a system (because we set up the rules ourselves!), we are often unable to explain its macroscopic behavior. That is the opposite of the usual scientific procedure: in principle, the physicist has access only (through experiments) to the global properties of a system. From them, he tries to find a universal law. The cellular automata example shows that a fundamental law is quite important from a philosophical point of view, but not sufficient. Complete knowledge of a physical process requires both a microscopic and a macroscopic level of understanding.

To come back to Langton's ant, a general result about this rule is that, during its motion, the ant visits an unbounded region of space, *whatever* the initial space texture is (configuration of gray and white cells). This observation is proved in a theorem due to Bunimovitch and Troubetzkoy.

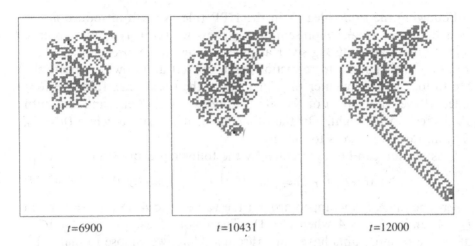

$t=6900$ $t=10431$ $t=12000$

Fig. 2.19. Langton's ant rule. The motion of a single ant starts with a chaotic phase of about 10 000 time steps, followed by the formation of a highway. The figure shows the state of each lattice cell (gray or white) and the ant position (marked by the black dot). In the initial condition all cells are white and the ant is located in the middle of the image.

The proof is as follows: suppose the region the ant visits is bounded. Then, it contains a finite number of cells. Since the number of iterations is infinite, there is a domain of cells that are visited infinitely often. Moreover, due to the rule of motion, a cell is either entered horizontally (we call it an H cell) or vertically (we call it a V cell). Since the ant turns by 90 degrees after each step, an H cell is surrounded by four V cells, and conversely. As a consequence, the H and V cells tile the lattice in a fixed checkerboard pattern. Now, we consider the upper rightmost cell of the domain, that is a cell whose right and upper neighbor are not visited. This cell exists if the trajectory is bounded. If this cell is an H cell (and is so for ever), it has to be entered horizontally from the left and exited vertically downward and, consequently must be gray. However, after the ant has left, the cell is white and there is a contradiction. The same contradiction appears if the cell is a V cell. Therefore, the ant trajectory is not bounded.

Multi-ant motion. Strictly speaking, the motion rule of Langton's ant is not sufficiently complete to specify a cellular automaton model in the sense we defined it. Each cell may, in principle, be occupied by a different ant because any initial configuration of two bits per cell (color and ant) should be acceptable. However, if many ants are simultaneously present, they may want to enter the same site at the same time, from different sides. Langton's rule can be generalized to account for multi-ant situations.

Following the same idea as in the HPP rule, we will introduce $n_i(\vec{r}, t)$ as a Boolean variable representing the presence ($n_i = 1$) or the absence ($n_i = 0$) of an ant entering site $\vec{r}$ at time t along lattice direction $\vec{c}_i$, where $\vec{c}_1$, $\vec{c}_2$, $\vec{c}_3$ and $\vec{c}_4$ stand for direction right, up, left and down, respectively. Up to four ants may enter the same site at the same time, from the four sides. If the color $\mu(\vec{r}, t)$ of the site is gray ($\mu = 0$), *all* entering ants turn 90 degrees to the right. On the other hand, if the site is white ($\mu = 1$), they all turn 90 degrees to the left.

This motion can be represented by the following expression

$$n_i(\vec{r} + \vec{c}_i, t + 1) = \mu n_{i-1}(\vec{r}, t) + (1 - \mu)n_{i+1}(\vec{r}, t) \qquad (2.27)$$

where the index i is wrapped around the values 1 to 4 (i.e $i + 1 = 1$ when $i = 4$, and $i - 1 = 4$ when $i = 1$). The color of each cell is modified after one or more ants have gone through. Here, we choose to make this modification depend on the number of ants present

$$\mu(\vec{r}, t + 1) = \mu(\vec{r}, t) \oplus n_1(\vec{r}, t) \oplus n_2(\vec{r}, t) \oplus n_3(\vec{r}, t) \oplus n_4(\vec{r}, t) \qquad (2.28)$$

where $\oplus$ is the sum modulo 2.

When the initial conditions are such that the cell contains one single ant, rules 2.27 and 2.28 are equivalent to the original Langton's ant motion. When several ants travel simultaneously on the lattice, both cooperative and destructive behavior are observed. First, the erratic motion of several ants favors the formation of a local arrangement of colors allowing the creation of a highway. One has to wait much less time before the first highway. Second, once a highway is created, other ants may use it to travel very fast (they do not have to follow the complicated pattern of the highway builder). In this way, the term "highway" is very appropriate. Third, a destructive effect occurs as the second ant reaches the highway builder. It breaks the pattern and several situations may be observed. For instance, both ants may enter a new chaotic motion, or the highway may be traveled in the other direction (note that the rule is time reversal invariant) and is thus destroyed. Figure 2.20 illustrates the multi-ant behavior.

The problem of an unbounded trajectory occurs again with this generalized motion. The assumption of Bunimovitch–Troubetzkoy's proof no longer holds in this case because a cell may be both an H or a V cell. Indeed, two different ants may enter the same cell, one vertically and the other horizontally.

Actually, the theorem of unbounded motion is wrong in several cases when two ants are present. Periodic motions may occur when the initial positions are well chosen. Again, it comes as a surprise that, for many initial conditions, the two ants follows periodic trajectories.

For instance, when the relative location of the second ant with respect to

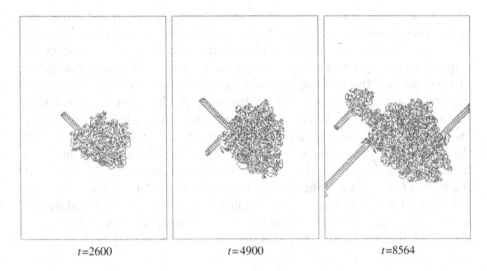

$t=2600$ $t=4900$ $t=8564$

Fig. 2.20. Motion of several Langton's ants. Gray and white indicate the colors
of the cell at the current time. Ant locations are marked by the black dots.
Initially, all cells are white and a few ants are randomly distributed in the central
region, with random directions of motion. The first highway appears much earlier
than when the ant is alone. In addition the highway can be used by other ants
to travel much faster. However, the "highway builder" is usually prevented from
continuing its construction as soon as it is reached by following ants. For instance,
the highway heading north-west after 4900 steps is destroyed. A new highway
emerges later, as we see from the snapshot at time $t = 8564$.

the first one is $(\Delta x, \Delta y) = (2, 3)$, the two ants returns to their initial position
after 478 iterations of the rule (provided they started in a uniformly black
substrate, with the same direction of motion). A very complicated periodic
behavior is observed when $(\Delta x, \Delta y) = (1, 24)$: the two ants start a chaotic-
like motion for several thousands of steps. Then, one ant builds a highway
and escapes from the central region. After a while, the second ant finds
the entrance of the highway and rapidly catches the first one. After the
two ants meet, they start undoing their previous path and return to their
original position. This complete cycle takes about 30 000 iterations.

More generally, it is found empirically that, when $\Delta x + \Delta y$ is odd and
the ants enter the site with the same initial direction, the two-ant motion
is likely to be periodic. However, this is not a rule and the configuration
$(\Delta x, \Delta y) = (1, 0)$ yields an unbounded motion, a diamond pattern of
increasing diameter which is traveled in the same direction by the two
ants.

It turns out that the periodic behavior of a two-ant configuration is
not so surprising. The cellular automata rule 2.27 and 2.28 is reversible

in time, provided that there is never more than one ant at the same site. Time reversal symmetry means that if the direction of motion of all ants are reversed, they will move backwards through their own sequence of steps, with the opposite direction of motion. Therefore, if at some point of their motion the two ants cross each other (on a lattice link, not on the same site), the first ant will go through the past of the second one, and vice versa. They will return to the initial situation (the two ants being exchanged) and build a new pattern, symmetrical to the first one, due to the inversion of the directions of motion. The complete process then cycles for ever. Periodic trajectories are therefore related to the probability that the two ants will, at a some time, cross each other in a suitable way. The conditions for this to happen are fulfilled when the ants sit on a different sublattice (black or white sites on the checkerboard) and exit two adjacent sites against each other. This explain why a periodic motion is likely to occur when $\Delta x + \Delta y$ is odd.

Time reversal invariance.   Time reversal is an important concept in the framework of cellular automata modeling because it is a very fundamental symmetry of the physical world. Checking whether or not a rule is time reversible can be done by considering the microdynamics of the automaton. In our case, suppose the state of the automaton, when the ants enter a site, is the five-component vector $s^{in} = (\mu^{in}, n_1^{in}, n_2^{in}, n_3^{in}, n_4^{in})$. As a result of the color of the site, the directions of motion change and the ants exit the site in other directions and modify the color of the site. Using the updating rule, the output state s^{out} can be computed as the right-hand side of equations 2.27 and 2.28. We shall write this evolution with an operator U such that

$$s^{out} = U s^{in} \tag{2.29}$$

If the dynamics is time reversal invariant, one can change the direction of motion of all ants and the system traces its own past, with the opposite directions of motion. Let us introduce the operator R as the operator which reverses the directions of motion but does not change the color

$$(Rs)_0 = s_0 \qquad (Rs)_i = s_{i-2} \tag{2.30}$$

Time reversal invariance implies that $URs^{out} = Rs^{in}$ since if one reverses the directions of motion and updates the rule, one expects to obtain the initial state with reverse velocities. Since $s^{out} = Us^{in}$, the condition for time reversal invariance reads

$$URU = R \qquad \text{or} \qquad RU = U^{-1}R \tag{2.31}$$

It is easy to check that one-ant configurations satisfy the time reversibility condition. For instance

$$URU \begin{pmatrix} 1 \\ 1 \\ 0 \\ 0 \\ 0 \end{pmatrix} = UR \begin{pmatrix} 0 \\ 0 \\ 1 \\ 0 \\ 0 \end{pmatrix} = U \begin{pmatrix} 0 \\ 0 \\ 0 \\ 0 \\ 1 \end{pmatrix} = \begin{pmatrix} 1 \\ 0 \\ 0 \\ 1 \\ 0 \end{pmatrix} = R \begin{pmatrix} 1 \\ 1 \\ 0 \\ 0 \\ 0 \end{pmatrix} \qquad (2.32)$$

On the other hand, some two-ant configurations do not satisfy the condition, because of the color rule 2.28

$$URU \begin{pmatrix} 1 \\ 1 \\ 1 \\ 0 \\ 0 \end{pmatrix} = UR \begin{pmatrix} 1 \\ 0 \\ 1 \\ 1 \\ 0 \end{pmatrix} = U \begin{pmatrix} 1 \\ 1 \\ 0 \\ 0 \\ 1 \end{pmatrix} = \begin{pmatrix} 1 \\ 1 \\ 1 \\ 0 \\ 0 \end{pmatrix} \neq R \begin{pmatrix} 1 \\ 1 \\ 1 \\ 0 \\ 0 \end{pmatrix} \qquad (2.33)$$

Clearly, time reversal symmetry can be restored by modifying rule 2.28 as follows

$$\mu(t+1) = \begin{cases} \mu(t) & \text{if no ant is present} \\ 1 - \mu(t) & \text{if there is at least one ant} \end{cases} \qquad (2.34)$$

What about real ants? So far we have not considered the behavior of real ants and how they may be connected to this cellular automata. As mentioned before, artificial life is a fascinating domain to which cellular automata models may contribute. The initial purpose of Langton's ant model was apparently to study large-scale regularities in complex systems. However, this rule has some analogies with what happens in the real ant world.

It is well known that real ants leave behind them a substance (the pheromone) which is used by other ants as a stimulus. The idea of modifying the space texture (white or gray) captures an aspect this phenomena. We have seen that the path created by an ant involve a spatial structure that can be used by a companion to travel much more efficiently. It is very tempting to relate this behavior with a famous experiment in which real ants are able to find the shortest path to their food, as the result of a collective effect.

2.2.8 The road traffic rule

Road traffic flow is an every day life problem which can be investigated (at least partially) with cellular automata models [69,73,74]. Cars are treated as point particles moving along a line of sites.

The microdynamic law of motion is taken from Wolfram's rule 184. A car can move only if its destination cell is free. Even if this destination cell will be freed by an occupying car at the time the motion occurs, the first car cannot see it because its horizon of sight is finite and the vehicle stays at rest until the cell in front of it is empty.

This cellular automata rule of motion can be expressed by the following relation

$$n_i(t+1) = n_i^{in}(t)(1 - n_i(t)) + n_i(t)n_i^{out}(t) \qquad (2.35)$$

where $n_i(t)$ is the car occupation number ($n_i = 0$ means a free site, $n_i = 1$ means a vehicle is present at site i). The quantity $n_i^{in}(t)$ represents the state in the cell from which a car could come and $n_i^{out}(t)$ indicates the state of the destination cell. Rule 2.35 means that the next state of cell i is 1 if a car is currently present and the next cell is occupied, or if no car is currently present and a car is arriving.

The number of vehicles $N^{mvt}(t)$ that have moved between iteration t and $t+1$ are those which find n^{out} empty

$$N^{mvt}(t) = \sum_{i=1}^{L} n_i(t)(1 - n_i^{out}(t)) \qquad (2.36)$$

where L is the system size (number of cells). The number of moving cars is also given by the number of empty cells which will receive a new car

$$N^{mvt}(t) = \sum_{i=1}^{L} n_i^{in}(t)(1 - n_i(t)) \qquad (2.37)$$

Therefore, the equation 2.35 of motion can be written as a balance equation

$$\sum_{i=1}^{L} n_i(t+1) - n_i(t) = \sum_{i=1}^{L} [n_i^{in}(t)(1 - n_i(t)) - n_i(t)(1 - n_i^{out}(t))] \qquad (2.38)$$

For a simple one-dimensional ring-like road, the expressions for n^{in} and n^{out} are very simple

$$n_i^{in}(t) = n_{i-1}(t) \qquad n_i^{out}(t) = n_{i+1}(t) \qquad (2.39)$$

and the microdynamics reduces exactly to Wolfram's rule 184.

In order to make the dynamics more interesting, some ingredients can be added to the basic rule. For instance, a car can be blocked at some given site [75] and released with some probability or only at some given time steps, in order to mimic a traffic light or a slowing down due to road construction or any other local events.

Another interesting problem is to generalize the rule to a two-dimensional road network. The question is to define the motion rule of the cars at a

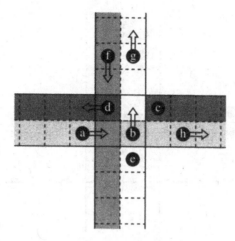

Fig. 2.21. Example of a traffic configuration near a junction. The four central cells represent a rotary which is traveled counterclockwise. The gray levels indicate the different traffic lanes: white is a northbound lane, light gray an eastbound lane, gray a westbound lane and, finally, dark gray is a southbound lane. The dots labeled a,b,c,d,e,f,g and h are cars which will move to the destination cell indicated by the arrows. Cars without an arrow are forbidden to move.

road junction. This can be realized in a simple way if one assumes that a rotary is located at each crossing. Road junctions are formed by central circles around which the traffic moves always in the same direction. A vehicle in a rotary has priority over any entering car.

The implementation we propose for this rule is illustrated in figure 2.21, where a four-corner junction is shown. The four middle cells constitute the rotary. A vehicle on the rotary (b or d) can either rotate counterclockwise or exit. A local flag f is used to decide the motion of a car in a rotary. If $f = 0$, the vehicle (d) exits in the direction allowed by the color of its lane (see figure caption). If $f = 1$, the vehicle moves counterclockwise, like b. The value of the local turn flag f can be updated according to the modeling needs: it can be constant for some amount of time to impose a particular motion at a given junction, completely random, random with some bias to favor a direction of motion, or may change deterministically according to any user-specified rule. For instance, figure 2.22 shows a turn flag configuration which forces cars to move horizontally after the junction.

As before, a vehicle moves only when its destination cell n^{out} is empty. Far from a rotary, the state of the destination cell is determined as the occupation of the down-motion cell. This is also the case for a vehicle turning in the rotary. On the other hand, a car wanting to enter the rotary has to check two cells because it has no priority. For instance, car c

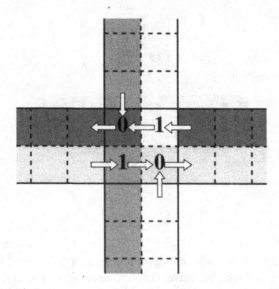

Fig. 2.22. Example of a "traffic-light" configuration. The values indicated on the rotary are the turn flags f. Here, horizontal motion has the priority and vertical traffic has to turn right or left according to a regulation "turn right on red traffic light is permitted".

cannot enter the rotary because b is going to move to the white cell. Car e cannot move either because it sees b (and cannot know whether or not b will actually move). Car a, on the other hand can enter because it sees that d is leaving the rotary and that the gray cell ahead is free.

Similarly, the incoming vehicle n^{in} to a given cell is computed differently inside and outside of the rotary. The light gray cell occupied by car b has two possible inputs: in priority, it is the vehicle from the gray cell at west; if this cell is empty, the input will be the incoming lane, namely the car labeled e.

Figure 2.23 shows two traffic configurations in a Manhattan-like city. The difference between these two situations is the behavior at a rotary. In the first case (a), on each rotary cell, a vehicle has a probability 1/2 to exit. In (b), the turn flag f has an initial random distribution on the rotary. This distribution is fixed for the first 20 iterations and then flips to $f = 1 - f$ for the next 20 steps and so on. In this way, a junction acts as a traffic light which, for some amount of time, allows only a given flow pattern. We observed that the global traffic pattern is different in the two cases: it is much more homogeneous in the random case.

A simple question that can be asked about this traffic model concerns the so-called jamming transition. When the car density is low, all vehicles

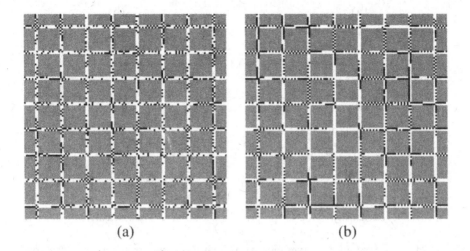

(a)   (b)

Fig. 2.23. Traffic configuration after 600 iterations, for a car density of 30%. Situation (a) corresponds to an equally likely behavior at each rotary junction, whereas image (b) mimics the presence of traffic lights. In the second case, car queues are more likely to form and the global mobility is less than in the first case.

can move because there is a high probability that the next site will be empty. However, if the density of cars is made large enough, there will be an increasing number of vehicles that are blocked by a car occupying the next cell. Actually, it is expected that above a critical car density ρ_c, the average velocity of the vehicles in the road network depends on the total number of cars and decreases as the car density is increased. The traffic congestion can be expressed by measuring how the average velocity $\langle v \rangle$ and the global traffic flow $\rho \langle v \rangle$ depend on the average car density. Precisely, ρ and $\langle v \rangle$ are defined as

$$\rho = N/L \qquad \langle v \rangle = \frac{1}{T} \sum_{\tau=t}^{t+T-1} \frac{N^{mvt}(\tau)}{N}$$

where L is the total length of the road network and $N = \sum_i n_i$ the total number of cars.

We observe in figure 2.24 that above a critical car density (which is quite small when the number of crossings is large), the traffic congestion increases. Below this critical value, the cars move freely. We also observe that the strategy adopted at each crossing affects the global traffic flow.

Figure 2.24 (a) corresponds to a case of "free rotary," where the direction of motion is chosen randomly at each rotary cell. Three curves are represented, corresponding to the length of the road segments between

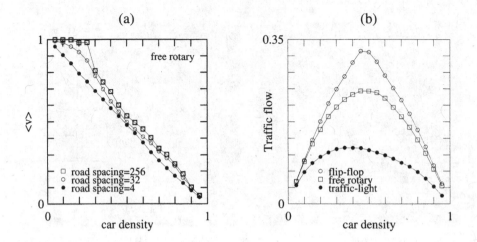

Fig. 2.24. (a) Plot of the average car velocity versus car density in the road network for various road spacings. (b) Plot of the traffic flow $\rho\langle v\rangle$ for the minimal road spacing allowed in the model (four cells) and various strategies of motion.

two successive intersections. For the case of a road network spacing equal to 256, we observe a clear second-order transition from a free moving phase to a situation of slowing down (partial jamming).

In Figure 2.24 (b) the road spacing is fixed at 4 and three different strategies of motion are considered at the rotary: (i)"traffic light" where the behavior at each rotary alternates between two possible paths over a period of 20 iterations; (ii) "free rotary," as in figure (a) and (iii) "flip-flop" which refers to a situation where each car flips the turn flag f as it moves across a rotary cell. In other words, a vehicle imposes the behavior of the next car to come. The initial distribution of f is random. The flip-flop strategy is the most efficient (it permits a higher traffic flow than the others) since it balances the traffic load equally throughout the network.

2.2.9   *The solid body motion rule*

In the previous section, we have presented several cellular automata models where point particles are moving on a lattice (HPP rule, sand rule or traffic rule). Here, we shall present a rule allowing the motion of more complex objects, such as a "solid body" which is composed of many constituent particles and has a coherent structure over large distances.

The problem of solid body motion is quite difficult to address in the framework of cellular automata modeling and, so far, only one-dimensional objects or "strings" can be defined [76]. Strings consist of a chain of pointlike particles which can travel in a three-dimensional space and maintain their size and integrity during their motion. Deformations

Fig. 2.25. A string configuration with five particles.

are allowed (these objects are not rigid), but the particles composing them should not spread out in the entire space.

Here, we will restrict our attention to the motion of these strings in a one-dimensional space and define collision rules between them.

A string can be thought of as a chain of masses linked by springs. More precisely, it is composed of two kinds of particles, say the white ones and the black ones, which alternate along a line. Two successive particles along the chain are either nearest neighbors or separated by one empty cell. Figure 2.25 shows such a chain composed of five particles.

The rule of motion is the following. The time evolution has two phases. First, the black particles are held fixed and the white ones move according to the rule below. Then, the white particles are held fixed and the black ones move with the same prescription.

Let us consider the case where the white particles move. The rule of motion is the following: the new configuration of the string is obtained by interchanging the spacings that separate the white particles from their two adjacent black neighbors at rest.

In other words, the motion of a white particle consists of a reflection with respect to the center of mass of its two black neighbor particles. If $q(t)$ is the position of a white particle on the lattice, and q_+ and q_- are the positions of the right and left adjacent black particles (right and left are defined according to the direction of the x-axis), then the new position $q(t+1)$ of the white particle is given by

$$q(t+1) = q_+ + q_- - q(t) \qquad (2.40)$$

Of course, this rule is only valid for a particle inside the string which has exactly two neighbors. At the extremity, an end particle moves so as to get closer to the string center if its neighbor is two cells apart. On the other hand if its neighbor is located on the next lattice cell, it moves away from the middle of the string. This motion corresponds to a reflection with respect to $q_+ \mp a$, where $a = 3/2$ represents the length of the "spring" linking the particles of the string together. To prevent the particles from moving off the lattice, a has to be integer or half integer.

Thus, if the left end of the string is a white particle located at $q(t)$, its new position will be

$$q(t+1) = 2(q_+ - a) - q(t) \qquad (2.41)$$

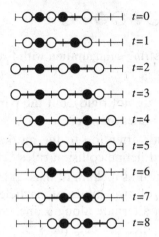

Fig. 2.26.   A complete evolution cycle of a string. From $t = 0$ to $t = 1$ the black particles move and the white ones are stationary. Then, the white particles move while the black ones are held fixed. After $t = 8$ iterations, the configuration is the same as at $t = 0$ and the string has traveled two lattice sites to the right; we thus conclude that its speed is 2/8.

and, similarly, if it is the right end, its new position will be

$$q(t + 1) = 2(q_- + a) - q(t) \qquad (2.42)$$

where q_+ and q_- refer, respectively, to the positions of the particles next to the left end point and next to the right end point.

Figure 2.26 shows an evolution over eight iterations of the string defined in figure 2.25. After these eight iterations, the string configuration is identical to its initial state and new cycle can start. We observe that the string has moved two lattice cells to the right during the basic cycle. Therefore, its velocity is $v = 2/8$.

From a general point of view, mass, momentum and energy can be associated with strings in simple ways. These quantities turn out to be adjustable for a given string, and are conserved during the evolution. Mass is defined as

$$m_1 = \frac{1}{2}m_2 = \ldots = \frac{1}{2}m_{N-1} = m_N = \frac{1}{2} \qquad (2.43)$$

where the particles of a string have been labeled from 1 to N. Note that here, the indices refer to the particles and not to the lattice cells.

The momentum of the string is given by

$$P(t) = \sum_{i=1}^{N} m_i v_i(t) \qquad (2.44)$$

where $v_i(t) = q_i(t+1) - q_i(t)$ is the distance particle i it will travel during the next time step. Finally, the total string energy is

$$E = \frac{1}{2} \sum_{i=1}^{N-1} \left[(x_{i+1} - x_i - a)^2 \right] \tag{2.45}$$

Mass is obviously conserved by the evolution rule because no particle disappears. Energy conservation is also obvious because $x_{i+1} - x_i$ is 1 or 2 and, with $a = 3/2$, one has $(2 - 3/2)^2 = (1 - 3/2)^2 = 1/4$. Thus, energy is proportional to the number of particles of the string. This is no longer the case if the string is allowed to have a transverse motion, too (that is, particles can also move along the y-axis in a two-dimensional system). More generally, energy conservation is due to the fact that the cellular automata rule moves the particles between symmetric positions.

In order to prove that momentum is conserved, it is convenient to write the equations of motion as

$$
\begin{aligned}
q_1(t-1) - 2q_1(t) + q_1(t+1) &= 2(q_2(t) - q_1(t) - a) \\
q_i(t-1) - 2q_i(t) + q_i(t+1) &= q_{i-1}(t) - 2q_i(t) + q_{i+1}(t) \\
q_N(t-1) - 2q_N(t) + q_N(t+1) &= 2(q_{N-1}(t) - q_N(t) + a)
\end{aligned} \tag{2.46}
$$

which follows from the fact that at every other time step, $q_i(t-1) = q_i(t)$ for the kind of particles which are held at rest. The above formulation is manifestly time reversible and has the form of a discrete wave equation.

From 2.44 and 2.46, one has

$$P(t) = \sum_{i=1}^{N} m_i(q_i(t+1) - q_i(t))$$

$$= \frac{1}{2}(q_1(t+1) - q_1(t)) + \sum_{i=2}^{N-1}(q_i(t+1) - q_i(t)) + \frac{1}{2}(q_N(t+1) - q_N(t))$$

$$= \frac{1}{2}[2q_2(t) - q_1(t) - 2a - q_1(t-1)] +$$

$$\sum_{i=2}^{N-1}[q_{i-1}(t) - q_i(t) + q_{i+1}(t) - q_i(t-1)] +$$

$$\frac{1}{2}[2q_{N-1}(t) - q_N(t) - 2a - q_{N-1}(t-1)]$$

$$= \frac{1}{2}[q_1(t) - q_1(t-1)] + \sum_{i=2}^{N-1}(q_i(t) - q_i(t-1)) + \frac{1}{2}(q_N(t) - q_N(t-1))$$

$$= P(t-1) \tag{2.47}$$

which proves that momentum is indeed conserved during string motion.

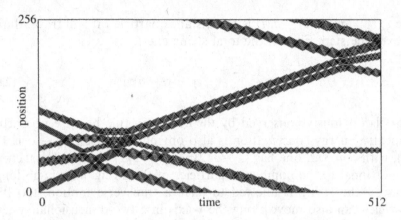

Fig. 2.27. Elastic collisions between five strings of different sizes. The system
is periodic with length 256 lattice sites (vertical axis). This figure shows the
trajectory of each string during 512 successive iterations. The total momentum is
zero.

The possible values of the momentum of a string range between $-M/2$
and $M/2$, where M is the total mass. The smallest possible variation of
momentum for a given string is $\Delta P = \pm 1$.

It is possible to define collisions between different strings so that mass,
momentum and energy is conserved during the interaction. To this end,
we consider only strings which have the same kind of extremities (say,
white particles). Two situations are considered to deal with the conflicting
motion: (i) the left and right ends of two different strings are next to each
other and want to move past each other; (ii) the left and right ends are
one site apart and want to move to the same site.

A way to solve these conflicts is to prevent the particles from moving
when they want to cross each other or to superpose. Momentum is
conserved in this way because, before collision, the two extremities have
opposite velocity (i.e their total momentum vanishes). Due to the collision,
the two particles are held motionless and, therefore, both have zero
momentum. With this mechanism, two colliding strings bounce back by
exchanging a quantum of their momentum. Note that such a collision may
last several time steps before each string has reversed its initial direction
of motion. These types of collisions may be referred to as elastic collisions.
They are illustrated in figure 2.27 in a situation of five interacting strings
moving in a periodic lattice 256 sites long. As we can observe, the motion
of a string is quite complex, due to the collision. In figure 2.28, we show
the position of the middle string over time. No regularity can be identify
during this interval of time which lasts more than 10^5 iterations.

Another type of collision can be defined. In case (ii) mentioned above,
it is also possible to superpose the two white extremities of mass 1/2 and

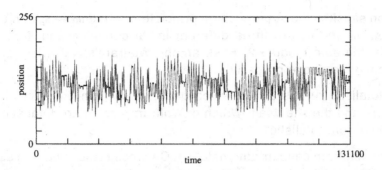

Fig. 2.28. Erratic motion of the middle string of the previous figure, over a large period of time.

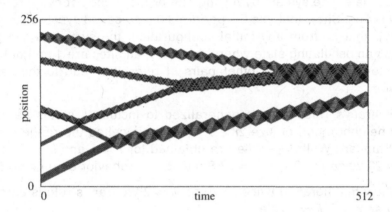

Fig. 2.29. Trajectories of five strings experiencing ballistic aggregation. The first collision does not lead to aggregation because the appropriate conditions are not fulfilled.

make one single white particle of mass 1 surrounded by black particles. As a result, the two strings have stuck to each other. With the same argument as before, it is easy to show that momentum is conserved during this procedure. With this type of collision, the strings perform a ballistic aggregation. Figure 2.29 illustrates this process in a system of five strings.

2.3 Problems

2.1. Some cellular automata rules exhibit the important simplifying feature of additive superposition. A rule is said to be additive if the evolution satisfies the superposition principle $s_0 = t_0 \oplus u_0 \leftrightarrow s_n = t_n \oplus u_n$, where s_0, t_0 and u_0 are three different initial condition and $\oplus$ is, for instance, the sum modulo 2. Find, among Wolfram's 256 rules, those which have the above property.

2.2. A totalistic rule is a rule which depends only on the sum of the neighbors and the site itself. Which of Wolfram's rules are totalistic? Is an additive rule totalistic?

2.3. A nonuniform cellular automata is a CA whose rule is not the same on all lattice sites. A nonuniform cellular automata may be used to perform tasks that a uniform CA (of the same range) cannot achieve. Consider Wolfram's rule 31 on a one-dimensional periodic lattice and "contaminate" the system by having one or more sites obeying rule 81 instead of 31. Show that this CA performs the so-called synchronization task[77]: that is, from any initial configuration, the system converges towards an oscillating state where, at time t, all sites are 1 and at time $t + 1$ they are all 0. Find other pairs of rules that lead to the same behavior.

2.4. Wolfram's rules can be generalized to include k states per cell and a neighborhood of size $2r + 1$, where r is called the range of the rule. Standard Wolfram's rule are obtained for $r = 1$ and $k = 2$. What is rule 20 when $r = 2$ and $k = 2$? Simulate its behavior on a computer.

2.5. Show that iterating twice an $r = 1$, $k = 2$ Wolfram's rule is equivalent to an $r = 2$, $k = 2$ rule.

2.6. Study the stochastic forest fire cellular automata introduced by Drossel and Schwabl which is defined as follows: one starts with a square lattice of size $L \times L$ with periodic boundary conditions. Each site is in one of the three following states: 0 an empty site, T a site occupied by a tree and B a site occupied by a burning tree. The updating rules use the von Neumann neighborhood and read, in terms of the probabilities p, f and g: R1: An empty site sprouts a tree with probability p. R2: A tree catches fire with probability $(1 - g)$, if at least one nearest neighbor tree is burning. R3: A tree catches fire with probability $f(1 - g)$ if no nearest neighbor tree is burning. R4: A burning tree burns down and becomes an empty site.

Identify the different regimes of behavior which are possible. Verify that, when $g = 0, p \ll 1$ and $f \ll p$, a state of self-organized criticality is reached in which clusters of all sizes are burning.

2.7. Consider a multi-lane traffic model: cars move from left to right along a ring, according to Wolfram rule 184. In addition, when a car is blocked by another one in front, it can jump to the nearest lane, provided this move is possible. What kind of neighborhood do you need to implement this rule? Simulate this automaton and measure the flow diagram, assuming some lanes may be slowed down by a random obstacle. Compare the flow with the situation where cars cannot change lane.

2.8. Consider the ant rule of section 2.2.7. Define another strategy to modify the "color" of the cell after several ants have crossed it. Simulate the model and observe the modification with respect to the original rule.

2.9. Assume that each ant of the model of section 2.2.7 has an internal "genome" which specifies how the color of the cell is to be modified after the ant has gone through. For instance, in a genome such as 011001 a 0 means that the color is not modified, whereas 1 means it changes. After each time step, the next digit is read, in a cyclic way. Propose a rule to deal with several ants at a site and simulate the model with ants having various genomes in size and contents.

2.10. Generalize the ant rule by adding more than two states. For instance, each cell can be in any state 0, 1, 2 or 3. After the ant leaves the cell, the state is incremented by 1 mod 4. The ant motion is determine by the state as follows: 0=left turn, 1=right turn 2=straight and 3 reverse direction.

2.11. How long does it takes for the hour glass of section 2.2.6 to empty, as a function of the height of a grain in the upper part? Is time accurately measured by the number of grains that have fallen down?

2.12. Propose a rule to implement the sand model (or a similar "sand" dynamics) without using the Margolus neighborhood. For instance, consider an approach based on the HPP particle motion. Four populations of grain are allowed at each site: grains at rest and those moving vertically in direction south-west, south or south-east. A grain moves preferentially to the south (gravity) unless the destination site is already occupied. When this is the case, it tries to fall left or right, if possible and no conflict can occur. Assuming an exclusion principle for the moving direction (0 or 1 particles), how many rest particles should be allowed on each site in order to be certain of conserving the total number of grains in any situation?

2.13. Simulate a population of robots moving on a 1D lattice. The robots can go left, right or remain stationary. Their motion is determined by a lookup table, which may be different for each robot. This table indicates whether the robot should stop, reverse its direction or continue straight ahead, according to the presence or absence of other robots in the neighborhood. If two (or more) robots arrive on the same site they collide and die (they are removed from the lattice). What is the size of the neighborhood required to implement this rule? What is the size of the lookup tables? Suppose the intelligence of the robot population is measured as the number of robots than can coexists in the long time regime (i.e. when all collisions have taken place). Then, starting from a random initial population (random tables), increase the fitness of the robots by refreshing the population: create offspring obtained from a "mating" of two parents or, more precisely, by mixing the tables of two surviving robots.

2.14. The *Prisoner's Dilemma* is a famous model of a simple social behavior between two persons. The problem is well illutrated by the case of two prisoners who cannot communicate before their trial. If they both cooperate (i.e. do not accuse each other of the crime), they will be convicted to a light punishment. If they both defect (i.e. accuse each other of the crime) they both get a severe punishment. If one of the prisoner defects while the other cooperates, the latter goes to jail and former is freed. Thus, there are four outcomes: CC, CD, DC or DD where C and D stand for Cooperate or Defect and the two letters designate the behavior of each prisoner. There is a different payoff associated with each of these outcomes. Since the prisoners cannot guess in advance what the other will do, it is meaningful to assume that the payoff can be ranked as: DC, CC, DD, CD. That is, the best choice is to defect, *provided* that the other cooperates. A double cooperation is good but cooperating may be quite dangerous in case the other defects. A spatial CA version of this prisoner "game" can be considered. On a two-dimensional lattice each site represents a player with one of the two strategies (C or D). During each iteration of the automaton, each player interacts with its four neighbors (von Neuman neighborhood) and the total payoff of the play is computed. This payoff is cumulated over the iterations and gives a score to each player. After each iteration, the players adopt the strategy of player with the higher score (still in the von Neuman neighborhood). Study the behavior of this system (number of C and D players) for various payoff values. Consider first a situation with DC=9, CC=8, DD=5, CD=3 and cumulate the score modulo 16.

2.15. Consider the string rule of section 2.2.9 but now on a 2D lattice and with a transverse motion instead of the longitudinal one. Assume that successive particles occupy adjacent x-coordinates whereas their y-coordinates differ by $y_i - y_{i-1} \in \{0, \pm 1\}$. Odd particles move at odd time steps and even particles move at even time steps. The motion is along the vertical axis so as to reach the symmetric position with respect to the center of mass of the two standing neighbors. Show that particles will jump 0, 1 or 2 sites up or down. What neighborhood do you need to implement this rule? Simulate the motion of a periodic string. What are the possible speeds you can obtain?

2.16. Find a string rule which combines both the longitudinal and the transverse motion.

2.17. Define a mass and momentum collision mechanism between HPP-like particles and the string model you obtained in the previous problem.

3

Statistical mechanics of lattice gas

3.1 The one-dimensional diffusion automaton

In the previous chapter, we discussed several cellular automata rules which definitely had something to do with the description of physical processes. The question is of course how close these models are to the reality they are supposed to simulate?

In the real world, space and time are not discrete and, in classical physics, the state variables are continuous. Thus, it is crucial to show how a cellular automaton rule is connected to the laws of physics or to the usual quantities describing the phenomena which is modeled. This is particularly important if the cellular automaton is intended to be used as a numerical scheme to solve practical problems.

Lattice gas automata have a large number of potential applications in hydrodynamics and reaction-diffusion processes. The purpose of this chapter is to present the techniques that are used to establish the connection between the macroscopic physics and the microscopic discrete dynamics of lattice gas automata. The problem one has to address is the statistical description of a system of many interacting particles. The methods we shall discuss here are very close, in spirit, to those applied in kinetic theory: the N-body dynamics is described in terms of macroscopic quantities such as the particle density or the velocity field. The derivation of a Boltzmann equation is a main step in this process.

In order to illustrate and understand the main steps leading to a macroscopic description of a cellular automaton model, we first consider a linear rule for which a simple and rigorous calculation can be carried out. From this result we justify the more sophisticated techniques that are necessary when dealing with a nonlinear dynamics. Hydrodynamic models such as the FHP automaton or the reaction-diffusion rules, require such techniques.

3.1.1 A random walk automaton

A random walk describes the motion of a particle which changes its travel direction in a probabilistic way. Many physical situations exhibit a diffusive behavior which is due to random motions at a microscopic level. Here, we shall discuss a probabilistic cellular automaton rule modeling the simultaneous random walk of many particles in one dimension. We will consider the two- and three-dimensional cases in much detail in chapter 4.

We consider a one-dimensional system in which particles may travel left or right. Usually, a random walk is simulated by selecting one particle in the system and moving it at random to the left or to the right, provided that the destination cell is empty. When all particle can move simultaneously, as is the case in a cellular automaton model, this method meets some difficulties: two particles may decide, at the same time to jump to the same unoccupied site. Elaborate communication strategies among the second nearest neighbors are required to solve a double occupancy situation. This solution also means that the mobility of the particles are subject to their density, which is not always desirable.

There is, however, a very simple way to produce a simultaneous random walk compatible with an exclusion principle (i.e no arbitrarily large occupancy). Our evolution rule is inspired by the HPP dynamics described in section 2.2.5. At each site x of the lattice, we define two Boolean variables $n_1(r, t)$ and $n_2(r, t)$. These quantities are occupation numbers indicating whether or not a particle is entering site r at time t in direction e_1 (right) or e_2 (left), respectively.

Random motion is obtained by shuffling the two directions of motion independently at each lattice site and at each time step. In other words, what is traveling in direction e_1 will be exchanged with what is traveling in direction e_2 with a probability $1 - p$. In order to write the microdynamics of such a system, that is the rules governing the behavior of this lattice gas automaton, we need a random bit $\mu(r)$ at each lattice site. We suppose that there is an external mechanism capable of producing, at each iteration, an entire set of statistically independent Boolean variables $\mu(r)$ which are 1 with probability p.

Our goal is to determine the macroscopic behavior of our cellular automaton, that is its behavior in the limit of infinitely short time step τ and small lattice spacing λ. To this end, it is convenient to introduce these space and time units in our rule.

The microdynamics has two phases: shuffling and motion to the left or right nearest neighbor. A particle entering site $r + \lambda$ at time t with a velocity pointing to the right must have been at site r at t. With a probability p, this particle was the one which entered with a velocity pointing to the right and with probability $1 - p$ the one which had a velocity pointing to the left. Therefore, the random walk rule reads

$$
\begin{aligned}
n_1(t+\tau, r+\lambda) &= \mu(t,r)n_1(t,r) + (1-\mu(t,r))n_2(t,r) \\
n_2(t+\tau, r-\lambda) &= (1-\mu(t,r))n_1(t,r) + \mu(t,r)n_2(t,r)
\end{aligned}
\tag{3.1}
$$

As we can see $\mu(r)$ selects at random either $n_1(r,t)$ or $n_2(r,t)$. Note that when both $n_1(r,t)$ $n_2(r,t)$ are one, shuffling has no effect, since the particles are indistinguishable.

3.1.2　*The macroscopic limit*

The first steps necessary to determine the macroscopic behavior of the microdynamics 3.1 is to average it. At a macroscopic scale, the variables n_i no longer appear as Boolean quantities but rather as quantities varying continuously between 0 and 1. Indeed, a macroscopic point of coordinate r designates a microscopic volume comprising many particles. Formally, this averaging $N_i(r,t) = \langle n_i(r,t) \rangle$ is treated as an ensemble average in the sense of statistical mechanics. In practice, it is an average over space, or time, or both, through a coarse-grain procedure.

Since $\mu(r,t)$ is statistically independent of $n_i(r,t)$ (the shuffling mechanism is data blind), the average of $\langle \mu n_i \rangle$ factorizes and we have

$$
\langle \mu(r,t)n_i(r,t) \rangle = \langle \mu(r,t) \rangle \langle n_i(r,t) \rangle = p N_i(r,t)
\tag{3.2}
$$

Therefore, relations 3.1 can be averaged and yield

$$
\begin{aligned}
N_1(t+\tau, r+\lambda) &= p N_1(t,r) + q N_2(t,r) \\
N_2(t+\tau, r-\lambda) &= q N_1(t,r) + p N_2(t,r)
\end{aligned}
\tag{3.3}
$$

where we have defined $q = 1 - p$. These equations are linear and can be solved exactly on a periodic system: any function defined on a periodic lattice of size $L = M\lambda$ is a superposition of plane waves $\exp(ikr)$ with $k = 0, (2\pi/L), ..., (2\pi/L)m, ..., (2\pi/L)(M-1)$. We will come back to this point in more detail in chapter 4 but, for the moment, we will look for solutions to 3.3 of the following form

$$
N(t,r) = A(t)\exp(ikr)
\tag{3.4}
$$

where N and A are the column matrices

$$
N(t,r) = \begin{pmatrix} N_1(t,r) \\ N_2(t,r) \end{pmatrix} \qquad
A(t) = \begin{pmatrix} A_1(t) \\ A_2(t) \end{pmatrix}
\tag{3.5}
$$

$A_1(t)$ and $A_2(t)$ are two functions to be determined. With the relation 3.4, equation 3.3 reads

$$
A(t+\tau) = \begin{pmatrix} p\exp(-ik\lambda) & q\exp(-ik\lambda) \\ q\exp(ik\lambda) & p\exp(ik\lambda) \end{pmatrix} A(t)
\tag{3.6}
$$

We then write

$$A(t) = \exp(i\omega t)A(0) \tag{3.7}$$

and one is left with the condition

$$\begin{pmatrix} p\exp(-ik\lambda) - \exp(i\omega\tau) & q\exp(-ik\lambda) \\ q\exp(ik\lambda) & p\exp(ik\lambda) - \exp(i\omega\tau) \end{pmatrix} A(0) = 0 \tag{3.8}$$

A solution $A(0) \neq 0$ exists provided that the determinant of this matrix is zero, which gives a relation between ω and k (the dispersion relation). This relation is

$$\exp(2i\omega\tau) - 2p\cos(k\lambda)\exp(i\omega\tau) + 2p - 1 = 0 \tag{3.9}$$

and can be solved for $\exp(i\omega\tau)$

$$\exp(i\omega_{\pm}\tau) = p\cos(k\lambda) \pm (1-p)\sqrt{1 - \frac{p^2}{(p-1)^2}\sin^2(k\lambda)} \tag{3.10}$$

A second step when taking the macroscopic limit is to make the lattice spacing λ and the time step τ tend to zero. The quantity giving the time evolution is the factor $\exp i\omega_{\pm}t = \exp(i\omega_{\pm}\tau)^{t/\tau}$. Therefore, one has to compute

$$\lim_{\tau\to0,\lambda\to0}(\exp(i\omega_{\pm}\tau)^{t/\tau}) = \lim_{\tau\to0,\lambda\to0}\left[p\cos(k\lambda) \pm (1-p)\sqrt{1 - \frac{p^2}{(p-1)^2}\sin^2(k\lambda)}\right]^{t/\tau} \tag{3.11}$$

This limit is easy to compute for ω_-, since, when $\lambda \to 0$, the term in the square bracket goes to $2p - 1$. As $0 < p < 1$, $(2p-1)^{t/\tau}$ goes to zero when $\tau \to 0$.

For ω_+, this limit is of the type "1^∞". This indetermination can be removed by taking the logarithm of the expression and differentiating both the numerator and the denominator with respect to τ.

$$\begin{aligned} \lim_{\tau\to0,\lambda\to0} t\frac{\ln(\exp(i\omega_+\tau))}{\tau} &= t \lim_{\tau\to0,\lambda\to0}\frac{\partial_\tau\ln(\exp(i\omega_+\tau))}{\partial_\tau\tau} \\ &= t\lim_{\tau\to0,\lambda\to0}\partial_\tau\ln(\exp(i\omega_+\tau)) \tag{3.12} \end{aligned}$$

To carry out this calculation we need to specify how λ and τ are related when they go to zero. The quantity $\ln(\exp(i\omega_+\tau))$ as given by equation 3.10 is an expression containing λ. Therefore, it is more convenient to write

$$t\lim_{\tau\to0,\lambda\to0}\partial_\tau\ln(\exp(i\omega_+\tau)) = t\lim_{\tau\to0,\lambda\to0}\exp(-i\omega_+\tau)\partial_\lambda\exp(i\omega_+\tau)\frac{d\lambda}{d\tau} \tag{3.13}$$

By differentiating equation 3.10, we obtain that the indetermination left in the calculation is

$$t \lim_{\tau \to 0, \lambda \to 0} \partial_\tau \ln(\exp(i\omega_+\tau)) = t \lim_{\lambda \to 0} \left[-pk \sin k\lambda - \frac{kp^2}{(1-p)} \sin k\lambda \right] \left(\frac{d\lambda}{d\tau}\right) \quad (3.14)$$

because the other terms gives factors of 1 in the limit $\lambda \to 0$.

There are two natural choices to compute $(d\lambda/d\tau)$ in the macroscopic limit:

(a) $\lambda/\tau = v = \text{const} \neq 0$ which reflects a situation where the propagation speed v of the particles remains finite. Therefore $d\lambda = vd\tau$ and the limit 3.14 gives 0. Consequently, we have

$$[\exp(i\omega_+\tau)]^{t/\tau} \to 1 \quad (3.15)$$

and the system appears frozen since $A(t) = A(0)$.

(b) A more interesting case is obtained when $\lambda^2/\tau \to \text{const} \neq 0$. In this limit, the time step goes faster to 0 than the lattice spacing. We have $2\lambda d\lambda = \text{const } d\tau$, that is

$$\frac{d\lambda}{d\tau} = \left(\frac{1}{2\lambda}\right)\frac{\lambda^2}{\tau} \quad (3.16)$$

The singularity in limit 3.14 is $(\sin k\lambda/\lambda) \to k$ and we obtain

$$[\exp(i\omega_+\tau)]^{t/\tau} \to \exp\left(-tk^2 \frac{p}{2(1-p)} \frac{\lambda^2}{\tau}\right) \quad (3.17)$$

Therefore, in this limit, there is a quantity $\rho(r,t)$ which evolves as

$$\rho(r,t) = \sum_k \exp(ikr) \exp(-tDk^2) \quad (3.18)$$

where D (the diffusion coefficient) is defined as

$$D = \frac{p}{2(1-p)} \frac{\lambda^2}{\tau} \quad (3.19)$$

It turns out that ρ is precisely a solution of the one-dimensional diffusion equation

$$\partial_t \rho = D\partial_r^2 \rho \quad (3.20)$$

Therefore, the simultaneous random walk rule we have defined here simulates the behavior of a diffusive process in the continuous limit $(\lambda^2/\tau) = \text{const}$. This is an expected result, of course, since diffusion corresponds to a microscopic random motion. However, this very simple example illustrates that a simplified microscopic dynamics may very well capture the essential features of a macroscopic reality.

3.1.3 The Chapman–Enskog expansion

The derivation given in the previous section is possible because equation 3.3 is linear. This is usually not the case and no explicit solution can be expected to the lattice Boltzmann dynamics. Instead, we shall try to obtain the differential equations describing, in the continuous limit, the behavior of the macroscopic quantities such as the density of particle and so on.

The Chapman–Enskog expansion technique is commonly used in statistical mechanics to derive the macroscopic laws governing the relevant physical quantities when the Boltzmann equation is known. The Boltzmann equation is an equation for the quantity $f(r,v,t)$ where $f(r,v,t)dr\,dv$ gives, at time t, the average number of particles in the region dr around r and having a velocity between v and $v+dv$.

The quantity N_i defined above plays the role of the Boltzmann density function f, except that the velocities are labeled with a discrete index i instead of a continuous variable v. Equation 3.3 is then the lattice Boltzmann equation of random walk microdynamics.

The local density of particle is the integral of f over the velocity variable. In our case the density ρ reads

$$\rho(r,t) = \sum_{i=1}^{2} N_i(r,t) \tag{3.21}$$

The idea of the Chapman–Enskog expansion is the following: to a first approximation, one expects that $N_1 = N_2 = \rho/2$. Then, there are small corrections that are found with perturbation analysis. More precisely, we shall write

$$N_i = N_i^{(0)} + \epsilon N_i^{(1)} + \epsilon^2 N_i^{(2)} + ... \tag{3.22}$$

where ϵ is a small parameter and the $N_i^{(\ell)}$s are functions of r and t to be determined.

The first step of the Chapman–Enskog expansion is to write equation 3.3 as

$$N_1(t+\tau, r+\lambda) - N_1(r,t) = (p-1)(N_1(t,r) - N_2(t,r))$$
$$N_2(t+\tau, r-\lambda) - N_2(r,t) = (p-1)(N_2(t,r) - N_1(t,r)) \tag{3.23}$$

By summing these two equations, we obtain

$$N_1(t+\tau, r+\lambda) + N_2(t+\tau, r-\lambda) - \rho(r,t) = 0 \tag{3.24}$$

The fact that the right-hand side vanishes is not by accident. It reflects the fact that the number of particles is locally conserved when the directions are shuffled: the number of particles entering site r at time t are exiting at time $t + \tau$. For this reason, equation 3.24 is called the *continuity* equation. This equation will play a crucial role in what follows.

The next step is to consider a Taylor expansion of the left-hand side of equation 3.23. One has

$$N_i(t+\tau, r+\lambda c_i) - N_i(r,t) = \left[\tau \partial_t + \frac{\tau^2}{2}\partial_t^2 + \lambda c_i \partial_r + \frac{\lambda^2}{2}c_i^2\partial_r^2 + \tau\lambda c_i \partial_t \partial_r \right] N_i(t,r)$$
(3.25)

where we have defined $c_1 = -c_2 = 1$ and neglected third-order terms in the expansion.

From the calculation we performed in section 3.1.2, we know that there is only one way to take the continuous limit $\lambda \to 0$ and $\tau \to 0$, namely that $\tau \sim \lambda^2$. As a consequence, λ and τ are not of the same order of magnitude. For the sake of convenience, we shall write

$$\lambda \sim \epsilon\lambda \qquad \tau \sim \epsilon^2\tau$$
(3.26)

We can now compare, order by order, the two sides of equation 3.23. Using relations 3.22 and 3.25, we have, for the order $O(\epsilon^0)$

$$N_1^{(0)} = N_2^{(0)}$$
(3.27)

because the left-hand side is of order ϵ, at least. The equation has no unique solution. The key to the Chapman–Enskog expansion is to impose the condition that the density ρ is completely expressed in terms of the $N_i^{(0)}$s

$$\rho = \sum_{i=1}^{2} N_i^{(0)}$$
(3.28)

Consequently, we have

$$N_1^{(0)} = N_2^{(0)} = \frac{1}{2}\rho$$
(3.29)

and the condition that

$$\sum_{i=1}^{2} N_i^{(\ell)} = 0 \qquad \text{if } \ell \geq 1$$
(3.30)

The next order $O(\epsilon)$ of the Boltzmann equation is given by taking the term $\lambda c_i \partial_r N_i^{(0)}$ of the Taylor expansion and the term $N_i^{(1)}$ in the right-hand side of equation 3.23. Since $N_i^{(0)} = \rho/2$, we obtain

$$\frac{\lambda}{2}\begin{pmatrix} 1 \\ -1 \end{pmatrix}\partial_r\rho = (p-1)\begin{pmatrix} 1 & -1 \\ -1 & 1 \end{pmatrix}\begin{pmatrix} N_1^{(1)} \\ N_2^{(1)} \end{pmatrix}$$
(3.31)

This equation has a solution because the vector $(1, -1)$ is an eigenvector of the right-hand side matrix

$$\begin{pmatrix} 1 & -1 \\ -1 & 1 \end{pmatrix} \begin{pmatrix} 1 \\ -1 \end{pmatrix} = 2 \begin{pmatrix} 1 \\ -1 \end{pmatrix} \tag{3.32}$$

Since this matrix is not invertible, it is not expected that an equation such as equation 3.31 always has a solution. However, the magic of the Chapman–Enskog expansion is that, precisely, the left-hand side of the equation is in the image of the matrix. We will return to this problem in more detail in section 3.2. For the present time, it important to notice that the reason why the matrix is not invertible is particle conservation.

The solution of equation 3.31 is then straightforward

$$N_i^{(1)} = \frac{\lambda}{4(p-1)} c_i \partial_r \rho \tag{3.33}$$

The equation governing the evolution of ρ is given by the continuity equation 3.24. The order $O(\epsilon^1)$ reads

$$O(\epsilon^1) : \quad \sum_{i=1}^{2} \lambda c_i \partial_r N_i^{(0)} = 0 \tag{3.34}$$

which is obviously satisfied by our solution. The next order is

$$O(\epsilon^2) : \quad \sum_{i=1}^{2} \left[\tau \partial_t N_i^{(0)} + \lambda c_i \partial_r N_i^{(1)} + \frac{\lambda^2}{2} c_i^2 \partial_r^2 N_i^{(0)} \right] = 0 \tag{3.35}$$

Using the solution for $N_i^{(0)}$ and $N_i^{(1)}$, we obtain the equation

$$\partial_\tau \rho + \frac{\lambda^2}{\tau} \left(\frac{1}{2(p-1)} + \frac{1}{2} \right) \partial_r^2 \rho = 0 \tag{3.36}$$

and, finally

$$\partial_t \rho = D \partial_r^2 \rho \tag{3.37}$$

which is the expected diffusion equation with the diffusion constant

$$D = \frac{\lambda^2}{\tau} \frac{p}{2(1-p)} \tag{3.38}$$

The derivation here is particularly simple because the microdynamics of the random walk cellular automaton is simple. For more complex rules, such as the FHP rule we shall discuss in section 3.2, the principle of the expansion will be identical but involves more complex steps that were trivially verified here. For instance, the relation between λ and τ is not known. In general, phenomena at various scales may be present. In addition to diffusive behavior, convective phenomena may appear at a scale where $\lambda \sim \tau$. In order to differentiate between these various orders of magnitude, a multiscale has to be used in conjunction with the Chapman–Enskog method.

3.1.4 Spurious invariants

Conserved quantities or invariants generally play a very important role in physics and, in particular, in cellular automata modeling. Local conservation laws at the level of the automaton rule are still visible at the macroscopic scale of observation. The diffusion equation obtained in the previous section is a continuity equation

$$\partial_t \rho + \text{div} J = 0 \tag{3.39}$$

where J is the particle current whose expression is obtained by a Taylor expansion of equation 3.24 and retaining only terms of order ϵ

$$
\begin{aligned}
J &= \frac{\lambda}{\tau}(N_1^{(1)} - N_2^{(1)}) + \frac{\lambda^2}{2\tau}\partial_r(N_1^{(0)} + N_2^{(0)}) \\
&= -D\partial_r \rho \tag{3.40}
\end{aligned}
$$

The expression for J depends on the microscopic nature of the particle motion, but the continuity equation itself is only a consequence of the conservation of particle numbers.

Are there any other conservation laws in our microdynamics that we have not extracted with the Chapman–Enskog expansion and which could play a role in a numerical simulation? The answer is yes: there is a spurious invariant in the random walk cellular automaton rule presented here. To find it, let us return to the dispersion relation 3.10:

$$\exp(2i\omega\tau) - 2p\cos(k\lambda)\exp(i\omega\tau) + 2p - 1 = 0 \tag{3.41}$$

A quantity will be conserved if $\| \exp(i\omega\tau) \| = 1$. The condition $\exp(i\omega\tau) = 1$ implies

$$\cos(k\lambda) = 1 \tag{3.42}$$

which is satisfied for $k\lambda = 2n\pi$. Since k only takes the values

$$k = 0, (2\pi/L), ..., (2\pi/L)m, ..., (2\pi/L)(M-1)$$

the solution of equation 3.42 is $k = 0$.

If one expresses the density ρ as

$$\rho(r,t) = \sum_k \rho_k e^{i(\omega t + kr)} \quad \text{with} \quad \rho_k = \sum_{r \in lattice} \rho(r,0)e^{-ikr} \tag{3.43}$$

one sees that the physical quantity associated with the wave number k is the total number of particle $\rho_0 = \sum_r \rho(r) = N$.

Equation 3.10 has another solution for $\exp(i\omega\tau) = -1$ which is

$$\cos(k\lambda) = -1 \qquad (3.44)$$

Within the possible range of values of k, this implies that $k\lambda = \pi$. Therefore, in our dynamics, there is a quantity which changes its sign at each iteration. The quantity associated with $k\lambda = \pi$ is

$$\rho_{\pi/\lambda} = \sum_{r \in lattice} \rho(r, 0) e^{-i\pi(r/\lambda)} \qquad (3.45)$$

Thus,

$$\rho_{\pi/\lambda} = N_{\text{even}} - N_{\text{odd}} \qquad (3.46)$$

where N_{even} and N_{odd} denote the total number of particles on even and odd lattice sites, respectively. Our result is easily explained by the following fact: all particles entering an even site at time t will move to an odd site at time $t + 1$, and conversely. Therefore, there is no interaction, at any time, between particles that are not located on the same odd or even sublattice. Particle conservation holds independently for each sublattice and our system actually contains two non-interacting subsystems. This feature is known as the checkerboard invariant.

This spurious invariant may have undesirable effects in reaction-diffusion processes, such as the annihilation reaction $A + A \to 0$ in which two particles entering the same site are removed with a probability p_{reaction}. If the odd and even sublattices are not well distinguished, the amount of A particles will eventually reach a constant concentration because no reaction can take place between particles on different sublattices.

3.2 The FHP model

The FHP rule is a model of a two-dimensional fluid which was introduced by Frisch, Hasslacher and Pomeau [32], in 1986. We show here how the fully discrete microscopic dynamics maps onto the macroscopic behavior of hydrodynamics. Due to the nonlinearity of the rule, the calculation will be much more complex than in the case of the one-dimensional diffusion rule discussed in the previous section.

3.2.1   The collision rule

The FHP model describes the motion of particles traveling in a discrete space and colliding with each other, very much in the same spirit as the HPP lattice gas discussed in section 2.2.5. The main difference is that, for isotropy reasons that will become clear below, the lattice is hexagonal (i.e. each site has six neighbors, as shown in figure 3.1).

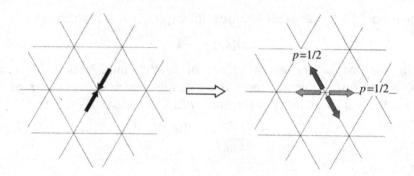

Fig. 3.1. Two-body collision in the FHP model. On the right part of the figure, the two possible outcomes of the collision are shown in dark and light gray, respectively. They both occur with probability 1/2.

The FHP model is an abstraction, at a microscopic scale, of a fluid. It is expected to contain all the salient features of a real fluid. It is well known that the continuity and Navier–Stokes equations of hydrodynamics express the local conservation of mass and momentum in a fluid. The detailed nature of the microscopic interactions does not affect the form of these equations but only the values of the coefficients (such as the viscosity) appearing in them. Therefore, the basic ingredients one has to include in the microdynamics of the FHP model are the conservation of particles and momentum after each updating step. In addition, some symmetries are required so that, in the macroscopic limit where time and space can be considered as continuous variables, the system is isotropic.

As in the case of the HPP model, the microdynamics of FHP is given in terms of Boolean variables describing the occupation numbers at each site of the lattice and at each time step (i.e. the presence or the absence of a fluid particle). The FHP particles move in discrete time steps, with a velocity of constant modulus, pointing along one of the six directions of the lattice. The dynamics is such that no more than one particle enters the same site at the same time with the same velocity. This restriction is the exclusion principle; it ensures that six Boolean variables at each lattice site are always enough to represent the microdynamics.

In the absence of collisions, the particles would move in straight lines, along the direction specified by their velocity vector. The velocity modulus is such that, in a time step, each particle travels one lattice spacing and reaches a nearest-neighbor site.

Interactions take place among particles entering the same site at the same time and result in a new local distribution of particle velocities. In order to conserve the number of particles and the momentum during each interaction, only a few configurations lead to a non-trivial collision

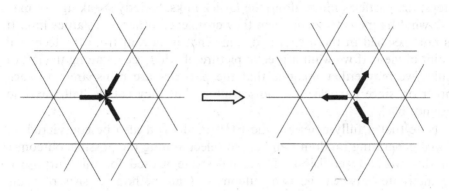

Fig. 3.2. Three-body collision in the FHP model.

(i.e a collision in which the directions of motion have changed). For instance, when exactly two particles enter the same site with opposite velocities, both of them are deflected by 60 degrees so that the output of the collision is still a zero momentum configuration with two particles (see figure 3.1). As shown in figure 3.1, the deflection can occur to the right or to the left, indifferently. For symmetry reasons, the two possibilities are chosen randomly, with equal probability.

Another type of collision is considered: when exactly three particles collide with an angle of 120 degrees between each other, they bounce back to where they come from (so that the momentum after the collision is zero, as it was before the collision). Figure 3.2 illustrates this rule. Several variants of the FHP model exist in the literature [3], including some with rest particles (models FHP-II and FHP-III).

For the simplest case we are considering here, all interactions come from the two collision processes described above. For all other configurations (i.e those which are not obtained by rotations of the situations given in figures 3.1 and 3.2) no collision occurs and the particles go through as if they were transparent to each other.

Both two- and three-body collisions are necessary to avoid extra conservation laws. The two-particle collision removes a pair of particles with a zero total momentum and moves it to another lattice direction. Therefore, it conserves momentum along each line of the lattice. On the other hand, three-body interactions deflect particles by 180 degrees and cause the net momentum of each lattice line to change. However, three-body collisions conserve the number of particles within each lattice line.

In the FHP rule, particles interact only at the lattice sites. It is tempting to think of this dynamics as snapshots, at integer clock cycles, of a continuous time evolution on a lattice. Between two consecutive time steps, the particles move along the lattice links. Strictly speaking, we may ask what happens to them when they cross each other on a lattice link. It is not specified in the model and somewhat irrelevant from the technical point of view. If we want a precise picture of what occurs along the lattice links, we may either imagine that the particles are transparent to each other or, since particles are indiscernible, that they collide half way and bounce back.

Note that a fully deterministic FHP model can also be considered, to avoid generating random numbers to select among the possible outcomes of the two-body collision. The usual strategy used for this purpose is to alternate between the two outcomes. One method consists of using time parity (i.e at even time steps, a $n_i n_{i+3}$ collision results in a $n_{i+1} n_{i+4}$ configuration and, at odd time steps, to a $n_{i+5} n_{i+2}$ configuration). Another technique is to add a seventh bit to the automaton state, which acts as a control bit. When this bit is 1, the $n_{i+1} n_{i+4}$ configuration is chosen and the bit is flipped to 0. Conversely, when the seventh bit is 0, the $n_{i+5} n_{i+2}$ output is selected and the control bit is flipped to 1.

A deterministic FHP dynamics is interesting also because it guarantees time reversibility invariance, which is a fundamental symmetry of microscopic physics. However, in the rest of this section, we shall consider the probabilistic FHP rule initially described because no macroscopic difference is expected to show up when one dynamics is used in place of the other.

3.2.2   The microdynamics

The full microdynamics of the FHP model can be expressed by evolution equations for the occupation numbers: we introduce $n_i(\vec{r}, t)$ as the number of particles (which can be either 0 or 1) entering site $\vec{r}$ at time t with a velocity pointing along direction $\vec{c}_i$, where $i = 1, 2, ..., 6$ labels the six lattice directions. The unit vectors $\vec{c}_i$ are shown in figure 3.3.

We also define the time step as τ and the lattice spacing as λ. Thus, the six possible velocities $\vec{v}_i$ of the particles are related to their directions of motion by

$$\vec{v}_i = \frac{\lambda}{\tau} \vec{c}_i \tag{3.47}$$

Without interactions between particles, the evolution equations for the n_i would be given by

$$n_i(\vec{r} + \lambda \vec{c}_i, t + \tau) = n_i(\vec{r}, t) \tag{3.48}$$

which expresses that a particle entering site $\vec{r}$ with velocity along $\vec{c}_i$ will continue in a straight line so that, at the next time step, it will enter site

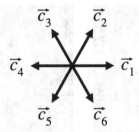

Fig. 3.3. The direction of motion $\vec{c}_i$.

$\vec{r} + \lambda \vec{c}_i$ with the same direction of motion. However, due to collisions, a particle can be removed from its original direction or another one can be deflected into direction $\vec{c}_i$.

For instance, if only n_i and n_{i+3} are 1 at site $\vec{r}$, a collision occurs and the particle traveling with velocity $\vec{v}_i$ will then move with either velocity $\vec{v}_{i-1}$ or $\vec{v}_{i+1}$ (note that the operations on index i are taken to be modulo 6). The quantity

$$D_i = n_i n_{i+3}(1 - n_{i+1})(1 - n_{i+2})(1 - n_{i+4})(1 - n_{i+5}) \qquad (3.49)$$

indicates, when $D_i = 1$ that such a collision will take place. Therefore,

$$n_i - D_i \qquad (3.50)$$

is the number of particles left in direction $\vec{c}_i$ due to a two-particle collision along this direction.

Now, when $n_i = 0$, a new particle can appear in direction $\vec{c}_i$, as the result of a collision between n_{i+1} and n_{i+4} or a collision between n_{i-1} and n_{i+2}. It is convenient to introduce a random Boolean variable $q(\vec{r}, t)$ which decides whether the particles are deflected to the right ($q = 1$) or to the left ($q = 0$) when a two-body collision takes place. Therefore, the number of particle created in direction $\vec{c}_i$ is

$$q D_{i-1} + (1 - q) D_{i+1} \qquad (3.51)$$

Particles can also be deflected into (or removed from) direction $\vec{c}_i$ because of a three-body collision. The quantity which expresses the occurrence of a three-body collision with particles n_i, n_{i+2} and n_{i+4} is

$$T_i = n_i n_{i+2} n_{i+4}(1 - n_{i+1})(1 - n_{i+3})(1 - n_{i+5}) \qquad (3.52)$$

As before, the result of a three-body collision is to modify the number of particles in direction $\vec{c}_i$ as

$$n_i - T_i + T_{i+3} \qquad (3.53)$$

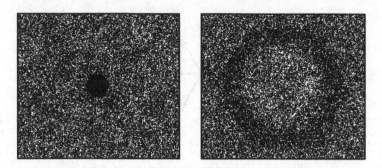

Fig. 3.4. Development of a sound wave in a FHP gas, due to particle overconcentration in the middle of the system.

Thus, according to our collision rules, the microdynamics of the FHP model reads

$$
\begin{aligned}
n_i(\vec{r} + \lambda\vec{c}_i, t + \tau) \;=\; & n_i(\vec{r}, t) \\
& -D_i + qD_{i-1} + (1 - q)D_{i+1} \\
& -T_i + T_{i+3}
\end{aligned}
\tag{3.54}
$$

where the right-hand side is computed at position $\vec{r}$ and time t. These equations are easy to code in a computer and yield a fast and exact implementation of the model. As an example, figure 3.4 illustrates a sound wave in the FHP gas at rest. Although the initial concentration is analogous to the situation of figure 2.13, we observe here a much more isotropic behavior.

3.2.3   *From microdynamics to macrodynamics*

The macroscopic variables.   The physical quantities of interest are not so much the Boolean variables n_i but macroscopic quantities or average values, such as, for instance, the average density of particles and the average velocity field at each point of the system. These quantities are defined from the ensemble average $N_i(\vec{r}, t) = \langle n_i(\vec{r}, t) \rangle$ of the microscopic occupation variables. $N_i(\vec{r}, t)$ is also the probability of having a particle entering site $\vec{r}$, at time t, with velocity $\vec{v}_i = (\lambda/\tau)\vec{c}_i$.

Following the usual definition of statistical mechanics, the local density of particles is the sum of the average number of particles traveling along each direction $\vec{c}_i$

$$
\rho(\vec{r}, t) = \sum_{i=1}^{6} N_i(\vec{r}, t)
\tag{3.55}
$$

Similarly, the particle current, which is the density ρ times the velocity field $\vec{u}$, is expressed as

$$\rho(\vec{r},t)\vec{u}(\vec{r},t) = \sum_{i=1}^{6} \vec{v}_i N_i(\vec{r},t) \qquad (3.56)$$

Another quantity which plays an important role in the following derivation is the momentum tensor Π, which is defined as

$$\Pi_{\alpha\beta} = \sum_{i=1}^{6} v_{i\alpha}v_{i\beta}N_i(\vec{r},t) \qquad (3.57)$$

where the greek indices α and β label the two spatial components of the vectors. The quantity Π represents the flux of the α-component of momentum transported along the β-axis. This term will contain the pressure contribution and the effects of viscosity.

The multiscale and Chapman–Enskog expansion. It is important to show that our discrete world is, at some appropriate scale of observation, governed by admissible equations: the physical conservation laws and the symmetry of the space are to be present and the discreteness of the lattice should not show up. The connection between the microscopic Boolean dynamics and the macroscopic, continuous world has to be established in order to assess the validity of the model.

The starting point to obtain the macroscopic behavior of the FHP automaton is to derive an equation for the N_is. Averaging the microdynamics 3.54 yields

$$N_i(\vec{r} + \lambda\vec{c}_i, t + \tau) - N_i(\vec{r},t) = \langle \Omega_i \rangle \qquad (3.58)$$

where

$$\langle \Omega_i \rangle = -\langle T_i \rangle + \langle T_{i+3} \rangle - \langle D_i \rangle + \frac{1}{2}\langle D_{i-1} \rangle + \frac{1}{2}\langle D_{i+1} \rangle \qquad (3.59)$$

The factor $1/2$ comes from the fact that $\langle q \rangle = 1/2$ because, as shown in figure 3.1, both collisions are chosen with equal probability.

Equation 3.58 is still discrete in space and time. The N_is vary between 0 and 1 and, at a scale $L \gg \lambda$, $T \gg \tau$, one can expect that they are smooth functions of the space and time coordinates. Therefore, equation 3.58 can be Taylor expanded up to second order to give

$$\tau\partial_t N_i + \lambda(\vec{c}_i \cdot \nabla)N_i + \frac{\tau^2}{2}\partial_t^2 N_i + \frac{\lambda^2}{2}(\vec{c}_i \cdot \nabla)^2 N_i + \lambda\tau(\vec{c}_i \cdot \nabla)\partial_t N_i = \langle \Omega_i \rangle \qquad (3.60)$$

The macroscopic limit of the FHP dynamics will require the solution of this equation. However, under the present form, there is little hope of solving it. Several approximations will be needed.

The first one involves the Chapman–Enskog expansion presented in section 3.1.3. We shall write

$$N_i = N_i^{(0)} + \epsilon N_i^{(1)} + \epsilon^2 N_i^{(2)} + ... \tag{3.61}$$

where ϵ is a small parameter. As in section 3.1.3, there will be here again some freedom in the choice of this expansion. In order to determine the $N_i^{(\ell)}$ uniquely and consistently, it is necessary to impose the condition that the macroscopic quantities ρ and $\rho \vec{u}$ are entirely given by the zero order of the expansion

$$\rho = \sum_{i=1}^{6} N_i^{(0)} \qquad \rho \vec{u} = \sum_{i=1}^{6} \vec{v}_i N_i^{(0)} \tag{3.62}$$

and, therefore

$$\sum_{i=1}^{6} N_i^{(\ell)} = 0 \qquad \sum_{i=1}^{6} \vec{v}_i N_i^{(\ell)} = 0, \qquad \text{for } \ell \geq 1 \tag{3.63}$$

These relations will appear as a natural choice when we come to the solution of equation 3.60.

In the case of FHP microdynamics, as opposed to the random walk, we do not know how λ and τ are related in the continuous limit. On physical grounds, we expect to observe phenomena on several time scales. For instance, the ratio λ/τ is the velocity of the particles and should not vanish in the continuous limit. In addition, dissipative phenomena (such as viscosity) will be present too, and with a coefficient of the order λ^2/τ.

The difficulty is to capture, in equation 3.60, terms of the same magnitude. We could formally introduce, as in section 3.1.3, quantities λ_1, τ_1 and τ_2 such that $\lambda = \epsilon \lambda_1$ and $\tau = \epsilon \tau_1 + \epsilon^2 \tau_2$. It is, however, more appropriate to consider two macroscopic time scales T_1 and T_2 satisfying

$$\frac{\tau}{T_1} = O(\epsilon) \qquad \frac{\tau}{T_2} = O(\epsilon^2) \tag{3.64}$$

and one macroscopic length scale

$$\frac{\lambda}{L_1} = O(\epsilon) \tag{3.65}$$

We then introduce two time variables t_1 and t_2 such that

$$t = \frac{t_1}{\epsilon} + \frac{t_2}{\epsilon^2} \tag{3.66}$$

As a consequence, the time derivative ∂_t becomes

$$\partial_t = \epsilon \partial_{t_1} + \epsilon^2 \partial_{t_2} \tag{3.67}$$

Similarly, a macroscopic space variable $\vec{r}_1$ is introduced. It is related to $\vec{r}$ through $\vec{r} = \vec{r}_1/\epsilon$, and the space derivative reads

$$\frac{\partial}{\partial r_\alpha} = \epsilon \partial_{1\alpha} \tag{3.68}$$

where $\partial_{1\alpha}$ is defined as the derivative with respect to the α-component of $\vec{r}_1$.

Except for the right-hand side, $\langle \Omega_i \rangle$, which is not yet expressed in terms of the N_i, we are now in a position to identify the various orders in ϵ appearing in the left-hand side of equation 3.60.

The balance equation. We are interested in finding the equation of motion of the particle density ρ and the particle current $\rho \vec{u}$. Without solving equation 3.60 and without developing the term $\langle \Omega_i \rangle$, the general structure of these equations can be derived.

Indeed, Ω_i, which is called the collision term, has some generic properties

$$\sum_{i=1}^{6} \Omega_i = 0 \qquad \sum_{i=1}^{6} \vec{v}_i \Omega_i = 0 \tag{3.69}$$

The first equation holds because clearly

$$\sum_{i=1}^{6} T_i = \sum_{i=1}^{6} T_{i+3} \tag{3.70}$$

$$\sum_{i=1}^{6} D_i = \sum_{i=1}^{6} D_{i-1} = \sum_{i=1}^{6} D_{i+1} \tag{3.71}$$

The second equation in 3.69 is true because, due to their definition

$$D_i = D_{i+3} \qquad \vec{c}_i = -\vec{c}_{i+3} \tag{3.72}$$

and

$$T_i = T_{i+2} = T_{i+4} \qquad \vec{c}_i + \vec{c}_{i+2} + \vec{c}_{i+4} = 0 \tag{3.73}$$

Thus, after summation, all the terms cancel. Equation 3.69 is not true by accident. It follows from the local conservation of particle numbers and momentum during a collision.

Of course, equation 3.69 is also satisfied for the average $\langle \Omega_i \rangle$. Thus, when equation 3.60 is summed over i, or multiplied by $\vec{c}_i$ and then summed, the right-hand side vanishes. This property provides the general form of the equation of motion for ρ and $\rho \vec{u}$.

Using equations 3.61, 3.67 and 3.68, we obtain for the order ϵ of equation 3.60

$$\sum_{i=1}^{6}\left[\tau\partial_{t_1}N_i^{(0)} + \lambda c_{i\beta}\partial_{1\beta}N_i^{(0)}\right] = 0 \tag{3.74}$$

and

$$\sum_{i=1}^{6}\left[\tau\partial_{t_1}c_{i\alpha}N_i^{(0)} + \lambda c_{i\alpha}c_{i\beta}\partial_{1\beta}N_i^{(0)}\right] = 0 \tag{3.75}$$

where we have adopted the convention that repeated Greek indices imply a summation over the space coordinates

$$c_{i\beta}\partial_{1\beta} \equiv c_{ix}\partial_{1x} + c_{iy}\partial_{1y} \tag{3.76}$$

We can perform the summation over i, using equations 3.62. We obtain

$$O(\epsilon) : \qquad \partial_{t_1}\rho + \mathrm{div}_1\rho\vec{u} = 0 \tag{3.77}$$

and

$$O(\epsilon) : \qquad \partial_{t_1}\rho u_\alpha + \partial_{1\beta}\Pi_{\alpha\beta}^{(0)} = 0 \tag{3.78}$$

where $\Pi_{\alpha\beta}^{(0)}$ is the zero-order Chapman–Enskog approximation of the momentum tensor defined in 3.57. The reader will recognize in equation 3.77 the usual continuity equation, reflecting particle conservation. Equation 3.78 expresses momentum conservation and corresponds, as we shall see, to the Euler equation of hydrodynamics, in which dissipative effects are absent.

The same calculation can be repeated for the order $O(\epsilon^2)$. Remembering relations 3.63, we find

$$\partial_{t_2}\rho + \frac{\tau}{2}\partial_{t_1}^2\rho + \frac{\tau}{2}\partial_{1\alpha}\partial_{1\beta}\Pi_{\alpha\beta}^{(0)} + \tau\partial_{t_1}\partial_{1\alpha}\rho u_\alpha = 0 \tag{3.79}$$

and

$$\partial_{t_2}\rho u_\alpha + \partial_{1\beta}\Pi_{\alpha\beta}^{(1)} + \frac{\tau}{2}\partial_{t_1}^2\rho u_a + \frac{\tau}{2}\partial_{1\beta}\partial_{1\gamma}S_{\alpha\beta\gamma}^{(0)} + \tau\partial_{t_1}\partial_{1\beta}\Pi_{\alpha\beta}^{(0)} = 0 \tag{3.80}$$

where S is a third-order tensor defined as

$$S_{\alpha\beta\gamma} = \sum_{i=1}^{6} v_{i\alpha}v_{i\beta}v_{i\gamma}N_i \tag{3.81}$$

These last two equations can be simplified using relations 3.77 and 3.78. Let us first consider the case of equation 3.79. One has

$$\frac{\tau}{2}\partial_{t_1}^2\rho = -\frac{\tau}{2}\partial_{t_1}\partial_{1\alpha}\rho u_a \tag{3.82}$$

and, therefore

$$\frac{\tau}{2}\partial_{t_1}^2\rho + \frac{\tau}{2}\partial_{1\alpha}\partial_{1\beta}\Pi_{\alpha\beta}^{(0)} + \tau\partial_{t_1}\partial_{1\alpha}\rho u_\alpha = \frac{\tau}{2}\partial_{1\alpha}\left[\partial_{t_1}\rho u_a + \partial_{1\beta}\Pi_{\alpha\beta}^{(0)}\right] = 0 \tag{3.83}$$

Thus, equation 3.79 reduces to

$$\partial_{t_2}\rho u_\alpha = 0 \tag{3.84}$$

which means that at the time scale T_2, the density variations are negligible. Similarly, since

$$\frac{\tau}{2}\partial_{t_1}^2 \rho u_a = -\frac{\tau}{2}\partial_{t_1}\partial_{1\beta}\Pi_{\alpha\beta}^{(0)} \tag{3.85}$$

equation 3.80 becomes

$$\partial_{t_2}\rho u_\alpha + \partial_{1\beta}\left[\Pi_{\alpha\beta}^{(1)} + \frac{\tau}{2}\left(\partial_{t_1}\Pi_{\alpha\beta}^{(0)} + \partial_{1\gamma}S_{\alpha\beta\gamma}^{(0)}\right)\right] = 0 \tag{3.86}$$

This last equation contains the dissipative contributions to the Euler equation 3.78. The first contribution is $\Pi_{\alpha\beta}^{(1)}$ which is the dissipative part of the momentum tensor. The second part, namely $\frac{\tau}{2}\left(\partial_{t_1}\Pi_{\alpha\beta}^{(0)} + \partial_{1\gamma}S_{\alpha\beta\gamma}^{(0)}\right)$ comes from the second-order terms of the Taylor expansion of the discrete Boltzmann equation. These terms account for the discreteness of the lattice and have no counterpart in standard hydrodynamics. As we shall see, they lead to the so-called *lattice viscosity*.

The order ϵ and ϵ^2 can be grouped together to give the general equations governing our system. Summing equations 3.86 and 3.78 with the appropriate power of ϵ as factor gives

$$\partial_t\rho u_\alpha + \frac{\partial}{\partial r_\beta}\left[\Pi_{\alpha\beta} + \frac{\tau}{2}\left(\epsilon\partial_{t_1}\Pi_{\alpha\beta}^{(0)} + \frac{\partial}{\partial r_\gamma}S_{\alpha\beta\gamma}^{(0)}\right)\right] = 0 \tag{3.87}$$

where we have used $\partial_t = \epsilon\partial_{t_1} + \epsilon^2\partial_{t_2}$ and $\frac{\partial}{\partial r_\alpha} = \epsilon\partial_{1\alpha}$. Similarly, equations 3.77 and 3.84 yield

$$\partial_t\rho + \mathrm{div}\rho\vec{u} = 0 \tag{3.88}$$

which is the standard continuity equation. Equation 3.87 corresponds to the Navier–Stokes equation. With the present form, it is not very useful because the tensors Π and S are not given in terms of the quantities ρ and $\vec{u}$. To go further, we will have to solve equation 3.60 and find expressions for $N_i^{(0)}$ and $N_i^{(1)}$ as functions of ρ and $\vec{u}$. However, for the time being, it is important to notice that the derivation of the continuity equation 3.88 and the Navier–Stokes equation 3.87 are based on very general considerations, namely mass and momentum conservation which ensure that $\sum\Omega_i = \sum\vec{v}_i\Omega_i = 0$. The specific collision rules of the FHP model do not affect the structure of these balance equations. However, the details of the collision rule will play a role for the explicit expression of Π and S.

The Boltzmann equation. In order to solve equation 3.60 we first need to express the right-hand side $\langle \Omega_i \rangle$ in a suitable form. Unfortunately, there is no exact way to relate $\langle \Omega_i \rangle$ with the N_is because, the collison term is nonlinear and, in general, the average of a product is not equal to the product of the average. However, if the quantities n_i can be considered as independent random variables, factorization is possible.

It is certainly an approximation to assume that $\langle n_i n_j \rangle = \langle n_i \rangle \langle n_j \rangle = N_i N_j$ when $i \neq j$ because there are correlations among the particles (note that when $i = j$, one has $n_i^2 = n_i$, due to the Boolean nature of n_i and $\langle n_i^2 \rangle = \langle n_i \rangle$). These correlations may play a crucial role in the dynamics, as we shall see in chapter 5, for a simple reaction-diffusion system. However, in many situations, it turns out that, due to the numerous collisions the particles experience, some form of molecular chaos takes place in the system and it can be assumed that different occupation numbers n_i and n_j are statistically independent. This is the so-called Boltzmann hypothesis leading to a closed equation for the $\langle n_i \rangle$s.

It is quite difficult to prove that the Boltzmann hypothesis holds or not in a given system. It is, however, a technical step which is necessary to perform in order to progress with our derivation of the macroscopic behavior of the lattice gas automaton. In section 5.8, we shall present another method which does not require factorization. Unfortunately, this method is limited to rather simple cellular automata dynamics, due to technical difficulties.

The above discussion points out an important aspect of cellular automata modeling: a numerical simulation of, say, the FHP model does not imply any simplification such as the Boltzmann hypothesis. The dynamics contains all degrees of freedom of the system and the computer provides an exact N-body simulation (although the interactions are much simpler than in real systems). Therefore, spontaneous fluctuations and many-body correlation naturally build up as the dynamics goes on. Cellular automata thus offer a very interesting way to study correlations in nonequilibrium systems and to take into account the effect of the fluctuations. We shall come back to these question in chapter 5.

For the moment, let us assume that the Boltzmann hypothesis is true for the FHP model. Therefore,

$$\langle \Omega_i(n_1, n_2, n_3, n_4, n_5, n_6) \rangle = \Omega_i(N_1, N_2, N_3, N_4, N_5, N_6) \qquad (3.89)$$

and equation 3.60 reads

$$\tau \partial_t N_i + \lambda(\vec{c}_i \cdot \nabla)N_i + \frac{\tau^2}{2}\partial_t^2 N_i + \frac{\lambda^2}{2}(\vec{c}_i \cdot \nabla)^2 N_i + \lambda\tau(\vec{c}_i \cdot \nabla)\partial_t N_i = \Omega_i(N) \quad (3.90)$$

Using the explicit expression for Ω_i in terms of T_i and D_i (see equations 3.49

and 3.52), we obtain

$$
\begin{aligned}
\Omega_i(N) = \; & -N_i N_{i+2} N_{i+4}(1 - N_{i+1})(1 - N_{i+3})(1 - N_{i+5}) \\
& +N_{i+1} N_{i+3} N_{i+5}(1 - N_i)(1 - N_{i+2})(1 - N_{i+4}) \\
& -N_i N_{i+3}(1 - N_{i+1})(1 - N_{i+2})(1 - N_{i+4})(1 - N_{i+5}) \\
& +\frac{1}{2} N_{i+1} N_{i+4}(1 - N_i)(1 - N_{i+2})(1 - N_{i+3})(1 - N_{i+5}) \\
& +\frac{1}{2} N_{i+2} N_{i+5}(1 - N_i)(1 - N_{i+1})(1 - N_{i+3})(1 - N_{i+4}) \quad (3.91)
\end{aligned}
$$

Equation 3.90 is called the Boltzmann equation.

The Chapman–Enskog expansion revisited. Equation 3.90 can be solved using the Chapman–Enskog method. Using equations 3.61, 3.67 and 3.68 for the expressions of N_i, ∂_t and $(\partial/\partial r_\alpha)$ in terms of ϵ, one obtains an equation for each order $O(\epsilon^\ell)$. The right-hand side of equation 3.91 can be expanded as

$$
\Omega_i(N) = \Omega_i(N^{(0)}) + \epsilon \sum_{j=1}^{6} \left(\frac{\partial \Omega_i(N^{(0)})}{\partial N_j} \right) N_j^{(1)} + O(\epsilon^2) \quad (3.92)
$$

The left-hand side of 3.91 is of order ϵ^1, at least, whereas the right-hand side has a contribution of order $O(\epsilon^0)$. The two lowest-order equations are simply (see equations 3.74 and 3.75).

$$
O(\epsilon^0) : \qquad \Omega_i(N^{(0)}) = 0 \quad (3.93)
$$

and

$$
O(\epsilon^1) : \qquad \partial_{t_1} N_i^{(0)} + \partial_{1\alpha} v_{i\alpha} N_i^{(0)} = \frac{1}{\tau} \sum_{j=1}^{6} \left(\frac{\partial \Omega_i(N^{(0)})}{\partial N_j} \right) N_j^{(1)} \quad (3.94)
$$

The first equation determines the $N_i^{(0)}$s. Once they are known, they can be substituted into the second equation in order to obtain a solution for the $N_i^{(1)}$. Unfortunately, this procedure is not as simple as it first looks, because the matrix $(\partial\Omega/\partial N)$ (whose elements are $\partial\Omega_i/\partial N_j$) is not invertible. Indeed, due to the conservation laws, one has

$$
\sum_i \left(\frac{\partial \Omega_i}{\partial N_j} \right) = \frac{\partial}{\partial N_j} \sum_i \Omega_i = 0 \quad (3.95)
$$

and similarly

$$
\sum_i c_{i\alpha} \left(\frac{\partial \Omega_i}{\partial N_j} \right) = 0 \quad (3.96)
$$

which shows that the columns of the matrix $(\partial\Omega/\partial N)$ are linear combinations of each other and that the determinant is zero. The above two equations can also be written as

$$\left(\frac{\partial\Omega}{\partial N}\right)^T E_0 = \left(\frac{\partial\Omega}{\partial N}\right)^T E_1 = \left(\frac{\partial\Omega}{\partial N}\right)^T E_2 = 0 \qquad (3.97)$$

where the superscript T indicates a matrix transpose. The quantities E_0, E_1 and E_2 are called the *collisional invariants* and are vectors of $\mathbb{R}^6$ defined as

$$\begin{aligned}
E_0 &= (1,1,1,1,1,1) \\
E_1 &= (c_{11}, c_{21}, c_{31}, c_{41}, c_{51}, c_{61}) \\
E_2 &= (c_{21}, c_{22}, c_{32}, c_{42}, c_{52}, c_{62})
\end{aligned} \qquad (3.98)$$

where $c_{i\alpha}$ denotes the spatial component α of lattice direction vector $\vec{c}_i$. The reason the E_k are called collisional invariants is because they describe the conserved quantities of the dynamics, namely

$$N \cdot E_0 = \rho \qquad vN \cdot E_1 = \rho u_1 \qquad vN \cdot E_2 = \rho u_2 \qquad (3.99)$$

where $\cdot$ denotes the scalar product in $\mathbb{R}^6$ and $v = \lambda/\tau$.

In order for equation 3.94 to have a solution, it is necessary that $\partial_{t_1} N_i^{(0)} + \partial_{1\alpha} v_{i\alpha} N_i^{(0)}$ be in the image space of $(\partial\Omega/\partial N)$. It is well known from linear algebra that the image of a matrix is orthogonal (in the sense of the scalar product) to the kernel of its transpose

$$\mathrm{Im}\left(\frac{\partial\Omega}{\partial N}\right) = \left[\mathrm{Ker}\left(\frac{\partial\Omega}{\partial N}\right)^T\right]^{\perp} \qquad (3.100)$$

Therefore, the solubility condition of equation 3.94 requires that $\partial_{t_1} N_i^{(0)} + \partial_{1\alpha} v_{i\alpha} N_i^{(0)}$ be orthogonal to E_0, E_1 and E_2. But that is precisely what equations 3.77 and 3.78 express. Therefore, a solution $N_i^{(1)}$ is guaranteed to exists provided that $\partial_{t_1} N_i^{(0)}$ is suitably given in terms of the balance equation. As we shall see, the appropriate choice will be to write

$$\partial_{t_1} N_i^{(0)} = \frac{\partial N_i^{(0)}}{\partial\rho}\partial_{t_1}\rho + \frac{\partial N_i^{(0)}}{\partial\rho u_\alpha}\partial_{t_1}\rho u_\alpha \qquad (3.101)$$

and express $\partial_{t_1}\rho$ and $\partial_{t_1}\rho\tilde{u}_\alpha$ as given by equations 3.77 and 3.78.

Finally, note that when a solution to equation 3.94 exists, it is not unique. As we said in section 3.2.3 we also impose the condition that

$$\sum_{i=1}^{6} N_i^{(\ell)} = 0 \qquad \sum_{i=1}^{6} \tilde{v}_i N_i^{(\ell)} = 0, \qquad \text{for } \ell \geq 1 \qquad (3.102)$$

This amounts to asking that the solution $N^{(1)}$ is also orthogonal to the collisional invariants and belongs to $\mathrm{Im}(\partial\Omega/\partial N)$.

The next subsections will be devoted to the explicit solution of equations 3.93 and 3.94.

The local equilibrium solution. The solutions $N_i^{(0)}$ which make the collision term Ω vanish are known as the local equilibrium solutions. Physically, they correspond to a situation where the rate of each type of collision equilibrates. Since the collision time τ is much smaller than the observation times T_1 and T_2, it is reasonable to expect, to a first approximation, that an equilibrium is reached locally.

However, to find the $N_i^{(0)}$s such that $\Omega_i(N^{(0)}) = 0$, one needs an extra assumption. The quantity Ω_i is the sum of several contributions, as shown by equation 3.59. The quantity $T_i(N)$ gives the probability of a three-body collision with one particle traveling along direction $\vec{c}_i$. In a local equilibrium situation, the probability of the other three-body collision T_{i+3} has no reason to be different from T_i. Similarly, the probabilities of all two-body collisions should be identical. Therefore, we can cancel $\Omega_i(N^{(0)})$ as follows

$$T_i(N^{(0)}) = T_{i+3}(N^{(0)}) \tag{3.103}$$

$$D_i(N^{(0)}) = D_{i+1}(N^{(0)}) = D_{i-1}(N^{(0)}) \tag{3.104}$$

Since these relations should be satisfied for all i, equation 3.104 is equivalent to simply $D_i(N^{(0)}) = D_{i+1}(N^{(0)})$. The first condition ($T_i = T_{i+3}$) reads

$$\frac{N_i^{(0)}}{(1 - N_i^{(0)})} \frac{N_{i+2}^{(0)}}{(1 - N_{i+2}^{(0)})} \frac{N_{i+4}^{(0)}}{(1 - N_{i+4}^{(0)})} = \frac{N_{i+1}^{(0)}}{(1 - N_{i+1}^{(0)})} \frac{N_{i+3}^{(0)}}{(1 - N_{i+3}^{(0)})} \frac{N_{i+5}^{(0)}}{(1 - N_{i+5}^{(0)})} \tag{3.105}$$

We take the logarithm of this equation and introduce the notation

$$M_i = \frac{N_i^{(0)}}{(1 - N_i^{(0)})} \tag{3.106}$$

so that our condition becomes

$$\log(M_i) + \log(M_{i+2}) + \log(M_{i+4}) - \log(M_{i+1}) - \log(M_{i+3}) - \log(M_{i+5}) = 0 \tag{3.107}$$

Note that this relation is the same for all i and we shall write it as

$$\log M_1 - \log M_2 + \log M_3 - \log M_4 + \log M_5 - \log M_6 = 0 \tag{3.108}$$

The second set of conditions (equation 3.104) reads

$$\log M_i + \log M_{i+3} = \log M_{i+1} + \log M_{i+4} \tag{3.109}$$

which gives two independent relations

$$\log M_1 - \log M_2 + \log M_4 - \log M_5 = 0 \tag{3.110}$$

for $i = 1$ and

$$\log M_2 - \log M_3 + \log M_5 - \log M_6 = 0 \qquad (3.111)$$

for $i = 2$. The other values of i yield relations that are a linear combination of 3.110 and 3.111. Instead of equations 3.110 and 3.111, it is convenient to replace them by their sum and difference

$$\log M_1 - \log M_3 + \log M_4 - \log M_6 = 0 \qquad (3.112)$$

and

$$\log M_1 - 2\log M_2 + \log M_3 + \log M_4 - 2\log M_5 + \log M_6 = 0 \qquad (3.113)$$

In order to combine equations 3.108, 3.112 and 3.113, it is convenient to use the collisional invariants defined in equation 3.98

$$
\begin{aligned}
E_0 &= (1,1,1,1,1,1) \\
E_1 &= (c_{11}, c_{21}, c_{31}, c_{41}, c_{51}, c_{61}) \\
E_2 &= (c_{21}, c_{22}, c_{32}, c_{42}, c_{52}, c_{62})
\end{aligned}
\qquad (3.114)
$$

as well as the quantities

$$
\begin{aligned}
E_3 &= (1,-1,1,-1,1,-1) \\
E_4 &= (1,0,-1,1,0,-1) \\
E_5 &= (1,-2,1,1,-2,1)
\end{aligned}
\qquad (3.115)
$$

where $c_{i\alpha}$ denotes the spatial component α of lattice direction vector $\vec{c}_i$.

The six vectors $E_0, ..., E_5$ form an orthogonal basis of $\mathbb{R}^6$ because the $\vec{c}_i$s have the following properties

$$\vec{c}_i = -c_{i+3} \qquad \vec{c}_i + \vec{c}_{i+2} + \vec{c}_{i+3} = 0 \qquad (3.116)$$

and

$$\sum_{i=1}^{6} c_{i\alpha} c_{i\beta} = 3\delta_{\alpha\beta} \qquad (3.117)$$

The first two relations are obvious from figure 3.3. The last one can be demonstrated as follows. First, we remark that it is equivalent to the relation

$$\frac{1}{3} \sum_i \vec{c}_i \vec{c}_i = \mathbf{1} \qquad (3.118)$$

where $\mathbf{1}$ is the 2×2 unity matrix and $\vec{c}_i \vec{c}_i$ is the matrix whose components are $c_{i\alpha} c_{i\beta}$. Then, it is easy to check that if one multiplies the matrix $\sum_i \vec{c}_i \vec{c}_i$ by any of the $\vec{c}_k$, one obtains $3\vec{c}_k$ because $\vec{c}_k \cdot \vec{c}_{k+1} = 1/2$, $\vec{c}_k \cdot \vec{c}_{k+2} = -1/2$ and $\vec{c}_k + \vec{c}_{k+2} + \vec{c}_{k+4} = 0$. With the vectors E_n, the local equilibrium conditions 3.108, 3.112 and 3.113 can be written as

$$\log M \cdot E_3 = 0 \qquad \log M \cdot E_4 = 0 \qquad \log M \cdot E_5 = 0 \qquad (3.119)$$

where $\log M$ is a notation for $(\log M_1, \log M_2, \log M_3, \log M_4, \log M_5, \log M_6)$ and also represents the scalar product in $\mathbb{R}^6$. Since the E_ns form a basis of the space, $\log M$ is any linear combination of the collisional invariants E_0, E_1 and E_2

$$\log M = aE_0 + b_1E_1 + b_2E_2 \tag{3.120}$$

or, by using the definition of E_0, E_1, E_2 and the notation $\vec{b} = (b_1, b_2)$

$$\log M_i = a + \vec{b} \cdot \vec{c}_i \tag{3.121}$$

We can obtain the solution for $N^{(0)}$ using 3.106 and 3.121

$$N_i^{(0)} = \frac{1}{1 + \exp(-a - \vec{b} \cdot \vec{c}_i)} \tag{3.122}$$

This expression has the form of a Fermi–Dirac distribution. This is a consequence of the exclusion principle we have imposed in the cellular automaton rule (no more than one particle per site and direction).

The Euler equation. The quantities a and $\vec{b}$ in 3.122 are functions of the density ρ and the velocity field $\vec{u}$ and are determined according to equations 3.62. In order to carry out this calculation, we consider a Taylor expansion of $N_i^{(0)}$, up to second order in the velocity field $\vec{u}$.

Although straightforward, this calculation is rather tedious. We write $N_i^{(0)}$ as

$$N_i^{(0)} = f(X_i) \equiv \frac{1}{1 + \exp(-X_i)} \tag{3.123}$$

where the six-dimensional vector $X = (X_1, X_2, ..., X_6)$ is defined as

$$X = a(\rho, \vec{u})E_0 + b_1(\rho, \vec{u})E_1 + b_2(\rho, \vec{u})E_2 \tag{3.124}$$

The expansion of $N_i^{(0)}$ around $\vec{u} = 0$ yields

$$N_i^{(0)} = f(X_i(\vec{u} = 0)) + u_\alpha \frac{\partial}{\partial u_\alpha} f(X_i(\vec{u} = 0)) + \frac{1}{2} u_\alpha u_\beta \frac{\partial^2}{\partial u_\alpha \partial u_\beta} f(X_i(\vec{u} = 0)) \tag{3.125}$$

We have

$$\frac{\partial}{\partial u_\alpha} = \frac{\partial X_i}{\partial u_\alpha} \frac{\partial}{\partial X_i} \tag{3.126}$$

and

$$\frac{\partial^2}{\partial u_\alpha \partial u_\beta} = \frac{\partial^2 X_i}{\partial u_\alpha \partial u_\beta} \frac{\partial}{\partial X_i} + \frac{\partial X_i}{\partial u_\alpha} \frac{\partial X_i}{\partial u_\beta} \frac{\partial^2}{\partial X_i^2} \tag{3.127}$$

When the fluid has no macroscopic motion, that is when $\vec{u} = 0$, all the $N_i^{(0)}$ should be identical, for symmetry reasons. Therefore, $X_i(\vec{u} = 0)$ does not depend on i and we write

$$X_i(\rho, \check{u} = 0) = x(\rho) \tag{3.128}$$

Thus, equation 3.125 becomes

$$N_i^{(0)} = f(x) + u_\alpha \left(\frac{\partial X_i}{\partial u_\alpha}\right)_{\check{u}=0} f'(x) + \frac{1}{2} u_\alpha u_\beta \left[\frac{\partial^2 X_i}{\partial u_\alpha \partial u_\beta} f'(x) + \frac{\partial X_i}{\partial u_\alpha} \frac{\partial X_i}{\partial u_\beta} f''(x)\right]_{\check{u}=0} \tag{3.129}$$

The unknowns in this relation are determined according to 3.99, which states that

$$\rho = \sum_i E_{0i} \cdot N_i^{(0)} \qquad \rho u_1 = v \sum_i E_{1i} \cdot N_i^{(0)} \qquad \rho u_2 = v \sum_i E_{2i} \cdot N_i^{(0)} \tag{3.130}$$

First we observe that

$$\sum_i E_{0i} \frac{\partial X_i}{\partial u_\alpha} = 6 \frac{\partial a}{\partial u_\alpha} \qquad \sum_i E_{1i} \frac{\partial X_i}{\partial u_\alpha} = 3 \frac{\partial b_1}{\partial u_\alpha} \qquad \sum_i E_{2i} \frac{\partial X_i}{\partial u_\alpha} = 3 \frac{\partial b_2}{\partial u_\alpha} \tag{3.131}$$

because

$$E_0 \cdot E_0 = 6 \qquad E_1 \cdot E_1 = E_2 \cdot E_2 = 3 \tag{3.132}$$

and $E_k \cdot E_l = 0$ if $k \neq l$. Similarly,

$$\sum_i E_{0i} \frac{\partial X_i}{\partial u_\alpha} \frac{\partial X_i}{\partial u_\beta} = 6 \frac{\partial a}{\partial u_\alpha} \frac{\partial a}{\partial u_\beta} + 3 \frac{\partial b_1}{\partial u_\alpha} \frac{\partial b_1}{\partial u_\beta} + 3 \frac{\partial b_2}{\partial u_\alpha} \frac{\partial b_2}{\partial u_\beta} \tag{3.133}$$

$$\sum_i E_{1i} \frac{\partial X_i}{\partial u_\alpha} \frac{\partial X_i}{\partial u_\beta} = 3 \frac{\partial a}{\partial u_\alpha} \frac{\partial b_1}{\partial u_\beta} + 3 \frac{\partial b_1}{\partial u_\alpha} \frac{\partial a}{\partial u_\beta} \tag{3.134}$$

and

$$\sum_i E_{2i} \frac{\partial X_i}{\partial u_\alpha} \frac{\partial X_i}{\partial u_\beta} = 3 \frac{\partial a}{\partial u_\alpha} \frac{\partial b_2}{\partial u_\beta} + 3 \frac{\partial b_2}{\partial u_\alpha} \frac{\partial a}{\partial u_\beta} \tag{3.135}$$

because

$$\sum_i E_{0i} E_{0i} E_{ki} = \sum_i E_{ki} = \begin{cases} 6 & \text{if } k = 0 \\ 0 & \text{otherwise} \end{cases} \tag{3.136}$$

$$\sum_i E_{0i} E_{ki} E_{li} = 3\delta_{kl} \qquad \text{if } k, l \neq 0 \tag{3.137}$$

and

$$\sum_i E_{ki} E_{li} E_{mi} = 0 \qquad \text{if } k, l, m \neq 0 \tag{3.138}$$

These results follow from the geometric properties of the lattice directions $\check{c}_i$ and, in particular, from 3.117.

The condition $\rho = E_0 \cdot N^{(0)}$ now reads

$$
\begin{aligned}
\rho = {} & 6f(x) + 6u_\alpha a_\alpha f'(x) + 3u_\alpha u_\beta a_{\alpha\beta} f'(x) \\
& + \frac{1}{2} u_\alpha u_\beta \left[6a_\alpha a_\beta + 3b_{1\alpha} b_{1\beta} + 3b_{2\alpha} b_{2\beta} \right] f''(x)
\end{aligned}
\tag{3.139}
$$

where we have introduced the notation

$$
a_\alpha \equiv \left(\frac{\partial a}{\partial u_\alpha} \right)_{\vec{u}=0} \qquad a_{\alpha\beta} \equiv \left(\frac{\partial^2 a}{\partial u_\alpha \partial u_\beta} \right)_{\vec{u}=0}
\tag{3.140}
$$

and similarly for b_1 and b_2. Since 3.139 must be satisfied for all values of ρ and $\vec{u}$, one concludes that

$$
6f(x) = \rho \qquad a_\alpha = 0
\tag{3.141}
$$

and

$$
a_{\alpha\beta} f'(x) + \frac{1}{2} \left(b_{1\alpha} b_{1\beta} + b_{2\alpha} b_{2\beta} \right) f''(x) = 0
\tag{3.142}
$$

Now, since $E_1 \cdot N^{(0)} = \rho(u_1/v)$, we also have

$$
\rho \frac{u_1}{v} = 3u_\alpha b_{1\alpha} f'(x) + \frac{1}{2} u_\alpha u_\beta \left[b_{1\alpha\beta} f'(x) + (3a_\alpha b_{1\beta} + 3b_{1\alpha} a_\beta) f''(x) \right]
\tag{3.143}
$$

Consequently,

$$
3b_{11} f'(x) = \frac{\rho}{v} \qquad b_{12} = 0 \qquad b_{1\alpha\beta} = 0
\tag{3.144}
$$

For symmetry reasons, the final condition $E_2 \cdot N^{(0)} = \rho(u_2/v)$ gives

$$
3b_{22} f'(x) = \frac{\rho}{v} \qquad b_{21} = 0 \qquad b_{2\alpha\beta} = 0
\tag{3.145}
$$

Finally, equation 3.142 reduces to

$$
a_{11} f'(x) + \frac{1}{2} b_{11}^2 f''(x) = 0 \qquad a_{22} f'(x) + \frac{1}{2} b_{22}^2 f''(x) = 0
\tag{3.146}
$$

and the only non-zero quantities of our expansion are

$$
f(x) = \frac{\rho}{6} \qquad b_{11} = b_{22} = \frac{\rho}{3v f'(x)}
\tag{3.147}
$$

and

$$
a_{11} = a_{22} = -\frac{\rho^2}{18v^2} \frac{f''(x)}{(f'(x))^3}
\tag{3.148}
$$

We can now substitute these results into the expression 3.125 for $N_i^{(0)}$

$$
\begin{aligned}
N_i^{(0)} = {} & \frac{\rho}{6} + \frac{\rho}{3v} \vec{c}_i \cdot \vec{u} - \frac{\rho^2}{36} \frac{u^2}{v^2} \frac{f''(x)}{(f'(x))^2} \\
& + \frac{\rho^2}{18} \frac{f''(x)}{(f'(x))^2} c_{i\alpha} c_{i\beta} \frac{u_\alpha}{v} \frac{u_\beta}{v}
\end{aligned}
\tag{3.149}
$$

where $u^2 = u_\alpha u_\alpha = u_1^2 + u_2^2$.

Since f is a Fermi–Dirac distribution, one also has

$$f' = f(1 - f) \qquad f'' = f(1 - f)(1 - 2f) \tag{3.150}$$

and, with $f(x) = (\rho/6)$,

$$\frac{\rho^2}{18} \frac{f''(x)}{(f'(x))^2} = \frac{2\rho(3 - \rho)}{3(6 - \rho)} \tag{3.151}$$

We now have the final expression for the local equilibrium solution of our problem

$$N_i^{(0)} = \frac{\rho}{6} + \frac{\rho}{3v}\check{c}_i \cdot \check{u} - \frac{1}{2}\rho G(\rho)\frac{u^2}{v^2} + \rho G(\rho)c_{i\alpha}c_{i\beta}\frac{u_\alpha}{v}\frac{u_\beta}{v} \tag{3.152}$$

with the function G defined as

$$G(\rho) = \frac{2}{3}\frac{(3 - \rho)}{(6 - \rho)} \tag{3.153}$$

The Euler equation is the first approximation describing the fluid motion. It neglects corrections to local equilibrium, in particular those which produce viscosity. The Euler equation is obtained from equation 3.78 with the expression 3.57 for the momentum tensor. The calculation of $\Pi_{\alpha\beta}^{(0)}$ consists of multiplying equation 3.152 by $v_{i\alpha}v_{i\beta}$ and summing over i. We recall that $\sum_i v_{i\alpha} = \sum_i v_{i\alpha}v_{i\beta}v_{i\gamma} = 0$, because $\check{c}_i = -\check{c}_{i+3}$ and $\sum_i v_{i\alpha}v_{i\beta} = 3v^2\delta_{\alpha\beta}$.

We also need to evaluate fourth-order quantities in the $\check{v}_i$. An important relation is that

$$\sum_{i=1}^{6} c_{i\alpha}c_{i\beta}c_{i\gamma}c_{i\delta} = \frac{3}{4}(\delta_{\alpha\beta}\delta_{\gamma\delta} + \delta_{\alpha\gamma}\delta_{\beta\delta} + \delta_{\alpha\delta}\delta_{\beta\gamma}) \tag{3.154}$$

The above relation is satisfied for any particular orientation of the six vectors $\check{c}_i$. The fourth-order tensor $\sum_{i=1}^{6} c_{i\alpha}c_{i\beta}c_{i\gamma}c_{i\delta}$ is isotropic and this is the reason why a hexagonal lattice is considered for a hydrodynamical model, rather than a square lattice. Isotropic fourth-order tensors made of the lattice directions play a crucial role in the lattice gas models of hydrodynamics because of the convective nonlinear term $\check{u} \cdot \vec{\nabla}\check{u}$ in the Navier–Stokes equation. The isotropy requirement constraints the possible lattices which can be used. This is a problem in three dimensions where such an isotropic lattice does not exist. One has to consider a four-dimensional lattice and then project in three-dimensional space. As a result, one obtains two different speeds of particles and some directions with two particles per links.

The result of equation 3.154 can be obtained by general symmetry consideration [33] but a direct derivation is also possible. For this purpose, we write

$$\check{c}_k = (c_{k1}, c_{k2}) = (\cos(k\theta + \phi), \sin(k\theta + \phi)) \tag{3.155}$$

where

$$\theta = \frac{2\pi}{6} \quad \text{and} \quad \phi \in [0, 2\pi] \tag{3.156}$$

The quantities to be computed in relation 3.154 are actually

$$\sum_k c_{k1}^4, \quad \sum_k c_{k2}^4, \quad \sum_k c_{k1}^2 c_{k2}^2, \quad \sum_k c_{k1}^3 c_{k2}, \quad \text{and} \quad \sum_k c_{k1} c_{k2}^3 \tag{3.157}$$

Using $\sum_{k=0}^{n-1} r^k = (r^n - 1)/(r - 1)$ for $r \neq 1$ and $\exp(i\theta) = \cos\theta + i\sin\theta$, one has

$$\sum_{k=1}^{6} (c_{k1} + ic_{k2})^4 = \sum_{k=0}^{5} \exp(4ik\theta)\exp(4i\phi)$$

$$= \exp(4i\phi)\frac{\exp(24i\theta) - 1}{\exp(4i\theta) - 1} = 0 \tag{3.158}$$

$$\sum_{k=1}^{6} (c_{k1} + ic_{k2})^3(c_{k1} - ic_{k2}) = \sum_{k=0}^{5} \exp(2ik\theta)\exp(2i\phi)$$

$$= \exp(2i\phi)\frac{\exp(12i\theta) - 1}{\exp(2i\theta) - 1} = 0 \tag{3.159}$$

and

$$\sum_{k=1}^{6} (c_{k1} + ic_{k2})^2(c_{k1} - ic_{k2})^2 = \sum_{k=0}^{5} 1 = 6 \tag{3.160}$$

On the other hand, a directed calculation of the left-hand side of relations 3.158–3.160 yields·

$$\sum_{k=1}^{6} (c_{k1} + ic_{k2})^4 = \sum_{k=1}^{6} \left[(c_{k1}^4 + c_{k2}^4 - 6c_{k1}^2 c_{k2}^2) + 2i(c_{k1}^3 c_{k2} - c_{k1} c_{k2}^3) \right] \tag{3.161}$$

$$\sum_{k=1}^{6} (c_{k1} + ic_{k2})^3(c_{k1} - ic_{k2}) = \sum_{k=1}^{6} \left[(c_{k1}^4 - c_{k2}^4) + 2i(c_{k1}^3 c_{k2} + c_{k1} c_{k2}^3) \right] \tag{3.162}$$

and

$$\sum_{k=1}^{6} (c_{k1} + ic_{k2})^2(c_{k1} - ic_{k2})^2 = \sum_{k=1}^{6} (c_{k1}^4 + c_{k2}^4 + 2c_{k1}^2 c_{k2}^2) \tag{3.163}$$

Comparing equations 3.161–3.163 with 3.158–3.160 yields

$$\sum_k c_{k1}^2 c_{k2}^2 = \frac{3}{4} \quad \sum_k c_{k1}^4 = \sum_k c_{k2}^4 = \frac{9}{4} \quad \sum_k c_{k1}^3 c_{k2} = \sum_k c_{k1} c_{k2}^3 = 0 \tag{3.164}$$

which, in a more compact form, is equivalent to relation 3.154, namely

$$\sum_{i=1}^{6} c_{i\alpha} c_{i\beta} c_{i\gamma} c_{i\delta} = \frac{3}{4}(\delta_{\alpha\beta}\delta_{\gamma\delta} + \delta_{\alpha\gamma}\delta_{\beta\delta} + \delta_{\alpha\delta}\delta_{\beta\gamma}) \tag{3.165}$$

We now return to the calculation of $\Pi^{(0)}$. Using equation 3.154 and equation 3.152, we have

$$\begin{aligned}
\Pi^{(0)}_{\alpha\beta} &= \sum_i N_i^{(0)} v_{i\alpha} v_{i\beta} \\
&= \frac{\rho}{2}v^2 \delta_{\alpha\beta} - \frac{3\rho}{2} G(\rho) u^2 \delta_{\alpha\beta} \\
&\quad + \frac{3\rho}{4} G(\rho) \left(u^2 \delta_{\alpha\beta} + 2u_\alpha u_\beta \right) \\
&= \left(\frac{v^2}{2}\rho - \frac{\rho}{2} g(\rho) u^2 \right) \delta_{\alpha\beta} + \rho g(\rho) u_\alpha u_\beta \tag{3.166}
\end{aligned}$$

where

$$g(\rho) \equiv \frac{3}{2} G(\rho) = \frac{\rho - 3}{\rho - 6} \tag{3.167}$$

As we can see, this result is independent of the lattice directions $\vec{c}_i$, because of the isotropy of the tensor 3.154.

The quantity $\left(\frac{v^2}{2}\rho - \frac{\rho}{2} g(\rho) u^2 \right) \delta_{\alpha\beta}$ is called the pressure term and $\rho g(\rho) u_\alpha u_\beta$ the convective part of the momentum tensor. Our microscopic dynamics gives an explicit expression for the pressure

$$p = \left(\frac{v^2}{2}\rho - \frac{\rho}{2} g(\rho) u^2 \right) \tag{3.168}$$

The term $\frac{v^2}{2}\rho$ corresponds to a perfect gas contribution, at fixed temperature. In the FHP model, the temperature is not defined. In addition, the balance equation for the kinetic energy is identical to the mass conservation equation, since all particles have the same velocities. Temperature has been introduced in multispeed lattice gas models, through the equipartition theorem [37,78,79].

We are now in a position to write down the Euler approximation (i.e. at scale T_1 and L_1) describing the behavior of our discrete fluid.

It is a standard approximation of hydrodynamics [80] that, at low Mach number (i.e for $\vec{u}$ much smaller than the speed of sound), the density ρ can be considered as a constant $\rho = \rho_0$, except in the pressure term.

Within this limit and using the expression for $\Pi^{(0)}$, equation 3.78 becomes

$$\partial_{t_1} \vec{u} + g(\rho)(\vec{u} \cdot \nabla_1)\vec{u} = -\frac{1}{\rho} \nabla_1 p \tag{3.169}$$

Compared with the standard Euler equation, the above relation differs because of the $g(\rho)$ factor in front of the convective term $(\vec{u} \cdot \nabla)\vec{u}$. The fact that $g(\rho) \neq 1$ is of course a failure of the model to properly describe a fluid flow. This factor comes from the non-Galilean invariance of the cellular automata dynamics and implies that $\partial_t u_\alpha + g(\rho)(\vec{u} \cdot \nabla)\vec{u}$ is no longer the total derivative $(d\vec{u}/dt)$.

However, at low Mach, $g(\rho)$ can be assumed constant and, by renormalizing the time t properly, one can transform ∂_{t_1} into $g(\rho)\partial'_t$ and get rid of the $g(\rho)$ factor in the left-hand side of equation 3.169.

The speed of sound. Within the Euler approximation (that is neglecting the damping due to viscous effects), we can compute the speed of sound of the FHP model. For this purpose, we consider a fluid at rest ($\vec{u} = 0$) with a given density ρ_{eq}. Small fluctuations $\Delta\rho$ and $\Delta\vec{u}$ will obey a wave equation with a propagation speed which will be the speed of sound. We have, according to 3.77 and 3.78

$$\partial_{t_1}\rho + \mathrm{div}_1\rho\vec{u} = 0 \tag{3.170}$$

and

$$\partial_{t_1}\rho u_\alpha + \partial_{1\beta}\Pi^{(0)}_{\alpha\beta} = 0 \tag{3.171}$$

where, to first order in $\vec{u}$, $\Pi^{(0)}_{\alpha\beta} = (v^2/2)\rho\delta_{\alpha\beta}$ (see equation 3.166). Since ρ_{eq} is constant, we have, to first order in $\Delta\rho$ and $\Delta\vec{u}$

$$\partial_{t_1}\Delta\rho + \rho_{eq}\mathrm{div}_1\Delta\vec{u} = 0 \tag{3.172}$$

and

$$\rho_{eq}\partial_{t_1}u_\alpha + \frac{v^2}{2}\partial_{1\alpha}\Delta\rho = 0 \tag{3.173}$$

Combining these two equation yields

$$\partial^2_{t_1}\Delta\rho - \frac{v^2}{2}\nabla^2_1\Delta\rho = 0 \tag{3.174}$$

This relation has the form of a wave equation with a speed

$$c^2_s = \frac{v^2}{2} \tag{3.175}$$

Thus, we can write

$$\Pi^{(0)}_{\alpha\beta} = c^2_s\rho\delta_{\alpha\beta} \tag{3.176}$$

The lattice viscosity. We shall now proceed to the explicit calculation of the terms involved in the Navier–Stokes equation derived in 3.87, namely

$$\partial_t \rho u_\alpha + \frac{\partial}{\partial r_\beta}\left[\Pi_{\alpha\beta}^{(0)} + \epsilon\Pi_{\alpha\beta}^{(1)} + \frac{\tau}{2}\left(\epsilon\partial_{t_1}\Pi_{\alpha\beta}^{(0)} + \frac{\partial}{\partial r_\gamma}S_{\alpha\beta\gamma}^{(0)}\right)\right] = 0 \qquad (3.177)$$

The new terms with respect to the Euler approximation of the previous section are of two kinds: those involving the zeroth order of the Chapman–Enskog expansion and $\Pi_{\alpha\beta}^{(1)}$ which implies computing $N_i^{(1)}$ from equation 3.94. For symmetry reasons, these new contributions are expected to be odd functions of the velocity flow $\vec{u}$. Therefore, we consider only the lowest-order terms in the expansion of $N_i^{(0)}$ and $\Pi_{\alpha\beta}^{(0)}$.

We first investigate the two contributions

$$\frac{\tau}{2}\frac{\partial}{\partial r_\beta}\epsilon\partial_{t_1}\Pi_{\alpha\beta}^{(0)} \quad\text{and}\quad \frac{\tau}{2}\frac{\partial^2}{\partial r_\beta\partial r_\gamma}S_{\alpha\beta\gamma}^{(0)} \qquad (3.178)$$

From equation 3.166 we approximate

$$\Pi_{\alpha\beta}^{(0)} = \frac{v^2}{2}\rho\delta_{\alpha\beta} + O(u^2) \qquad (3.179)$$

Since $\partial_{t_1}\rho = -\mathrm{div}_1\rho\vec{u}$ (by equation 3.77), one has $\epsilon\partial_{t_1}\rho = -\mathrm{div}\rho\vec{u}$ and

$$\begin{aligned}\frac{\tau}{2}\frac{\partial}{\partial r_\beta}\epsilon\partial_{t_1}\Pi_{\alpha\beta}^{(0)} &= \frac{\tau v^2}{4}\frac{\partial}{\partial r_\beta}\epsilon\partial_{t_1}\rho\delta_{\alpha\beta} \\ &= -\frac{\tau v^2}{4}\frac{\partial\mathrm{div}\rho\vec{u}}{\partial r_\alpha}\end{aligned} \qquad (3.180)$$

To compute the term involving

$$S_{\alpha\beta\gamma}^{(0)} = \sum_{i=1}^{6} v_{i\alpha}v_{i\beta}v_{i\gamma}N_i^{(0)}, \qquad (3.181)$$

we first notice that the only contribution to $N_i^{(0)}$ given by equation 3.152 will be

$$N_i^{(0)} = \frac{\rho}{3v^2}\vec{v}_i \cdot \vec{u} \qquad (3.182)$$

because the other terms contain an odd number of $\vec{v}_i$. Thus, we have, using 3.154,

$$\begin{aligned}S_{\alpha\beta\gamma}^{(0)} &= \frac{v^2\rho}{3}\frac{3}{4}(\delta_{\alpha\beta}\delta_{\gamma\delta} + \delta_{\alpha\gamma}\delta_{\beta\delta} + \delta_{\alpha\delta}\delta_{\beta\gamma})u_\delta \\ &= \frac{v^2}{4}\rho(\delta_{\alpha\beta}u_\gamma + \delta_{\alpha\gamma}u_\beta + \delta_{\beta\gamma}u_\alpha)\end{aligned}$$

Therefore

$$\frac{\tau}{2}\frac{\partial^2}{\partial r_\beta r_\gamma}S^{(0)}_{\alpha\beta\gamma} = \frac{\tau v^2}{8}\nabla^2\rho u_\alpha + \frac{\tau v^2}{4}\frac{\partial}{\partial r_\alpha}\text{div}\rho\vec{u} \qquad (3.183)$$

Substituting the results 3.180 and 3.183 into the Navier–Stokes equation 3.177 yields

$$\partial_t\rho u_\alpha + \frac{\partial}{\partial r_\beta}\Pi^{(0)}_{\alpha\beta} = -\frac{\partial}{\partial r_\beta}\epsilon\Pi^{(1)}_{\alpha\beta} - \frac{\tau v^2}{8}\nabla^2\rho u_\alpha \qquad (3.184)$$

We observe that the contribution of the two terms we just computed has the form of a viscous effect $\nu_\text{lattice}\nabla^2\rho\vec{u}$, where

$$\nu_\text{lattice} = -\frac{\tau v^2}{8} \qquad (3.185)$$

where ν_lattice is a *negative* viscosity. As mentioned previously, the origin of this contribution is due to the discreteness of the lattice ($S^{(0)}_{\alpha\beta\gamma}$ and $\partial_{t_1}\Pi^{(0)}_{\alpha\beta}$ comes from the Taylor expansion). For this reason, this term is referred to as a lattice contribution to the viscosity. The fact that it is negative is of no consequence because, the contribution $-\frac{\partial}{\partial r_\beta}\epsilon\Pi^{(1)}_{\alpha\beta}$ which we still have to calculate will be positive and larger than the present one.

The collisional viscosity. The usual contribution to viscosity is due to the collision between the fluid particles. This contribution is captured by the term $\frac{\partial}{\partial r_\beta}\epsilon\Pi^{(1)}_{\alpha\beta}$ in equation 3.184. In order to compute it, we first have to solve equation 3.94 for $N^{(1)}_i$. To lowest order in the velocity flow $\vec{u}$, we have

$$\frac{1}{\tau}\sum_{j=1}^{6}\left(\frac{\partial\Omega_i(N^{(0)})}{\partial N_j}\right)N^{(1)}_j = \partial_{t_1}N^{(0)}_i + \partial_{1\alpha}v_{i\alpha}N^{(0)}_i$$

$$= -\frac{\partial N^{(0)}_i}{\partial\rho}\text{div}_1\rho\vec{u} - \frac{\partial N^{(0)}_i}{\partial\rho u_\alpha}\partial_{1\beta}\Pi^{(0)}_{\alpha\beta} + \partial_{1\alpha}v_{i\alpha}N^{(0)}_i$$

$$(3.186)$$

where we have expressed the time derivative of $N^{(0)}_i$ in terms of the derivatives with respect to ρ and ρu_α, as mentioned in equation 3.101

$$\partial_{t_1}N^{(0)}_i = \frac{\partial N^{(0)}_i}{\partial\rho}\partial_{t_1}\rho + \frac{\partial N^{(0)}_i}{\partial\rho u_\alpha}\partial_{t_1}\rho u_\alpha \qquad (3.187)$$

and used equations 3.77 and 3.78 to express $\partial_{t_1}\rho$ and $\partial_{t_1}\rho u_\alpha$. These substitutions will ensure that the right-hand side of equation 3.186 will be in the image of $(\partial\Omega/\partial N)$.

As we did for the lattice viscosity, we shall consider only the first order in the velocity flow $\vec{u}$. The omitted terms are expected to be of the order $O(u^3)$. From the expressions 3.152 and 3.166, we have for the lowest order in $\vec{u}$

$$N_i^{(0)} = \frac{\rho}{6} + \frac{\vec{c}_{i\alpha}}{3v}\rho u_\alpha \quad \text{and} \quad \Pi_{\alpha\beta}^{(0)} = \frac{v^2}{2}\rho\delta_{\alpha\beta} \qquad (3.188)$$

Thus

$$\frac{\partial N_i^{(0)}}{\partial \rho} = \frac{1}{6} \quad \text{and} \quad \frac{\partial N_i^{(0)}}{\partial \rho u_\alpha} = \frac{c_{i\alpha}}{3v} \qquad (3.189)$$

and we can rewrite 3.186 as

$$
\begin{aligned}
\frac{1}{\tau}\sum_{j=1}^{6}\left(\frac{\partial\Omega_i(N^{(0)})}{\partial N_j}\right)N_j^{(1)} &= -\frac{\partial N_i^{(0)}}{\partial\rho}\mathrm{div}_1\rho\vec{u} - \frac{\partial N_i^{(0)}}{\partial\rho u_\alpha}\partial_{1\beta}\Pi_{\alpha\beta}^{(0)} + \partial_{1\alpha}v_{i\alpha}N_i^{(0)} \\
&= -\frac{1}{6}\mathrm{div}_1\rho\vec{u} - \frac{v_{i\alpha}}{6}\partial_{1\alpha}\rho + \frac{v_{i\alpha}}{6}\partial_{1\alpha}\rho + \frac{c_{i\beta}}{3}\partial_{1\beta}c_{i\alpha}\rho u_\alpha \\
&= -\frac{1}{6}\mathrm{div}_1\rho\vec{u} + \frac{1}{3}c_{i\alpha}c_{i\beta}\partial_{1\beta}\rho u_\alpha \\
&= \frac{1}{3}(c_{i\alpha}c_{i\beta} - \frac{1}{2}\delta_{\alpha\beta})\partial_{1\beta}\rho u_\alpha \qquad (3.190)
\end{aligned}
$$

From this result, it is now clear that equation 3.186 will have a solution: the six-dimensional vectors $Q_{\alpha\beta}$ of components $Q_{\alpha\beta i}$ given by $(c_{i\alpha}c_{i\beta} - \frac{1}{2}\delta_{\alpha\beta})$ are orthogonal to the collisional invariant E_0, E_1 and E_2 defined in 3.98 (this is in particular due to the fact that $\sum_i c_{i\alpha}c_{i\beta} = 3\delta_{\alpha\beta}$). Since E_0, E_1 and E_2 are in the kernel of $(\partial\Omega/\partial N)^T$, then $Q_{\alpha\beta}$ is in the image space of $(\partial\Omega/\partial N)$ (see equation 3.100).

We can now consider the left-hand side of equation 3.186. We compute it for $\vec{u} = 0$ and introduce the notation

$$s = \frac{\rho}{6} \qquad (3.191)$$

From the explicit expression of Ω_i given by equation 3.91 we obtain

$$
\begin{aligned}
\left(\frac{\partial\Omega_1}{\partial N_1}\right)_{\vec{u}=0} &= -s^2(1-s)^3 - s^3(1-s)^2 - s(1-s)^4 \\
&\quad -\frac{1}{2}s^2(1-s)^3 - \frac{1}{2}s^2(1-s)^3 \\
&= -s(1-s)^2\left[s(1-s) + s^2 + (1-s)^2 + s(1-s)\right] \\
&= -s(1-s)^2 \qquad (3.192)
\end{aligned}
$$

Similarly, we compute the other terms and we obtain

$$
\left(\frac{\partial \Omega}{\partial N}\right)_{\tilde{u}=0} = \frac{1}{2}s(1-s)^2
$$

$$
\times \begin{pmatrix}
-2 & s+1 & -3s+1 & 4s-2 & -3s+1 & s+1 \\
s+1 & -2 & s+1 & -3s+1 & 4s-2 & -3s+1 \\
-3s+1 & s+1 & -2 & s+1 & -3s+1 & 4s-2 \\
4s-2 & -3s+1 & s+1 & -2 & s+1 & -3s+1 \\
-3s+1 & 4s-2 & -3s+1 & s+1 & -2 & s+1 \\
s+1 & -3s+1 & 4s-2 & -3s+1 & s+1 & -2
\end{pmatrix}
$$

$$(3.193)$$

We have already mentioned that the vectors $Q_{\alpha\beta}$ are in $\mathrm{Im}\left(\frac{\partial\Omega}{\partial N}\right)$. We observe here that they also are eigenvectors. To see that, let us consider the product of the first row of $(\partial\Omega/\partial N)$ with $Q_{\alpha\beta}$. Using

$$
Q_{\alpha\beta i} = (c_{i\alpha}c_{i\beta} - \frac{1}{2}\delta_{\alpha\beta}) \tag{3.194}
$$

one has

$$
\frac{2}{s(1-s)^2}\sum_i \frac{\partial\Omega_1^{(0)}}{\partial N_i}Q_{\alpha\beta i} = \sum_i (c_{i\alpha}c_{i\beta}) - 3c_{1\alpha}c_{1\beta} - 3c_{4\alpha}c_{4\beta}
$$

$$
+ s[\sum_i (c_{i\alpha}c_{i\beta}) - 4c_{3\alpha}c_{3\beta} + 2c_{4\alpha}c_{4\beta} - 4c_{5\alpha}c_{5\beta}]
$$

$$
= 3\delta_{\alpha\beta} - 6c_{1\alpha}c_{1\beta}
$$

$$
+ s[3\delta_{\alpha\beta} - 4(c_{1\alpha}c_{1\beta} + c_{3\alpha}c_{3\beta} + c_{5\alpha}c_{5\beta}) + 6c_{1\alpha}c_{1\beta}]
$$

$$
= -6(1-s)(c_{1\alpha}c_{1\beta} - \frac{1}{2}\delta_{\alpha\beta}) \tag{3.195}
$$

where we have again used 3.117. Since the rows of $(\partial\Omega/\partial N)$ are circular permutations of each other, the same calculation can be repeated and we find that

$$
\left(\frac{\partial\Omega}{\partial N}\right)Q_{\alpha\beta} = -3s(1-s)^3 Q_{\alpha\beta} \tag{3.196}
$$

This yields immediately the solution for $N^{(1)}$ as a multiple of $Q_{\alpha\beta}$. Since $Q_{\alpha\beta}$ is orthogonal to the collisional invariants, $N^{(1)}$ will clearly satisfy the extra conditions 3.63. Thus we have

$$
N_i^{(1)} = -\frac{2\tau}{3\rho(1-s)^3}(c_{i\alpha}c_{i\beta} - \frac{1}{2}\delta_{\alpha\beta})\partial_{1\beta}\rho u_\alpha \tag{3.197}
$$

We may now compute the correction $\epsilon\Pi^{(1)}$ to the momentum tensor. Since

$\epsilon \partial_{1\beta} = \partial_\beta$, we obtain

$$
\begin{aligned}
\epsilon \Pi_{\alpha\beta}^{(1)} &= \epsilon \sum_i N_i^{(1)} v_{i\alpha} v_{i\beta} \\
&= \frac{2\tau v^2}{3\rho(1-s)^3} \left[\frac{3}{2} \mathrm{div} \rho \vec{u} \delta_{\alpha\beta} - \frac{3}{4} (\delta_{\alpha\beta}\delta_{\gamma\delta} + \delta_{\alpha\gamma}\delta_{\beta\delta} + \delta_{\alpha\delta}\delta_{\beta\gamma}) \partial_\gamma \rho u_\delta \right] \\
&= \frac{\tau v^2}{2\rho(1-s)^3} \left[\mathrm{div}\rho\vec{u}\delta_{\alpha\beta} - (\partial_\alpha \rho u_\beta + \partial_\beta \rho u_\alpha) \right]
\end{aligned}
\tag{3.198}
$$

We can now rewrite explicitly, to first order in ϵ and second order in the velocity flow $\vec{u}$, the Navier-Stokes equation obtained in 3.184. Using expression 3.166 for $\Pi_{\alpha\beta}^{(0)}$, we have

$$
\partial_t \rho u_\alpha + \partial_\beta (\rho g(\rho) u_\alpha u_\beta) = -\nabla p - \partial_\beta \left[\frac{\tau v^2}{2\rho(1-s)^3} (\delta_{\alpha\beta}\mathrm{div}\rho\vec{u} - (\partial_\alpha u_\beta + \partial_\beta u_\alpha)) \right]
$$
$$
- \frac{\tau v^2}{8} \nabla^2 \rho u_\alpha
\tag{3.199}
$$

where the pressure p is given by relation 3.168.

Again, in the limit of low Mach number, the density can be assumed to be a constant, except in the pressure term [80]. From the continuity equation 3.88, we then get $\mathrm{div}\rho\vec{u} = 0$ and $\Pi^{(1)}$ contribution reduces to

$$
\begin{aligned}
\frac{1}{\rho} \partial_\beta \epsilon \Pi_{\alpha\beta}^{(1)} &= -\frac{1}{\rho} \frac{\tau v^2}{2\rho(1-s)^3} \left[\partial_\alpha \partial_\beta \rho u_\beta + \rho \partial_\beta^2 u_\alpha \right] \\
&= -\frac{\tau v^2}{2\rho(1-s)^3} \nabla^2 u_\alpha \\
&= -v_{\mathrm{coll}} \nabla^2 u_\alpha
\end{aligned}
\tag{3.200}
$$

Finally, within this approximation equation 3.199 can be cast into

$$
\partial_t \vec{u} + g(\rho)(\vec{u} \cdot \nabla)\vec{u} = -\frac{1}{\rho}\nabla p + v\nabla^2 \vec{u}
\tag{3.201}
$$

The quantity v is the kinematic viscosity of our discrete fluid, whose expression is composed of the lattice and collisional viscosities

$$
v = \tau v^2 \left(\frac{1}{2\rho(1-\frac{\rho}{6})^3} - \frac{1}{8} \right)
\tag{3.202}
$$

We observe that the viscosity of the FHP model depends strongly on the density and may become arbitrarily large for the limiting values $\rho = 0$ and $\rho = 6$. Its minimal value is obtained for $\rho = 3/2$

Up to the factor $g(\rho) = (3 - \rho)/(6 - \rho)$, equation 3.201 is the standard Navier–Stokes equation. This term $g(\rho) \neq 1$ is an indication of the non-Galilean invariance of the model. However, as we explained previously, since we assumed that $\rho \simeq$ const, the factor $g(\rho)$ can be absorbed in a renormalization of the time and our lattice dynamics is described by the usual hydrodynamic equation.

Several models have been proposed to restore the Galilean invariance [42,44,81], by adding more particles on the lattice with different velocities. We shall return to this problem in the framework of the lattice Boltzmann models which are a generalization of the fully discrete dynamics of the FHP model.

In this section we have derived the macroscopic behavior of the FHP model using the Chapman–Enskog and a multiscale expansion. Whereas the form of equation 3.201 depends little on the type of collision the particles experience, the expression 3.202 for the viscosity is very sensitive to the collision processes. In a lattice gas dynamics, the viscosity is intrinsic to the model and is not an adjustable parameter. The collision rules can be modified in order to change the viscosity. Other techniques than the full calculation we have presented here can be used to obtain the viscosity. M. Hénon [82,83] has developed a method in which a perturbed shear flow can be considered to compute the viscosity.

Mean free path. It is interesting to note that there is a proportionality relation between the collisional viscosity (see 3.202)

$$\nu_{\text{coll}} = \tau v^2 \left(\frac{1}{2\rho(1 - \frac{\rho}{6})^3} \right)$$

and the mean free path.

The mean free path is the average distance a particle travels between two collisions. If we denote by p_k the probability of having a collision after exactly k steps, the mean free path is defined as

$$\lambda_{\text{mfp}} = \lambda \sum_{k=1}^{\infty} k p_k$$

The probability of having a collision after k steps is the probability of having no collision during the first $k - 1$ iterations and then a collision. Thus, we write

$$p_k = (1 - p_{\text{coll}})^{k-1} p_{\text{coll}}$$

where p_{coll} is the probability for a particle to undergo a collision.

Thus, the mean free path becomes

$$\lambda_{\text{mfp}} = \lambda p_{\text{coll}} \sum_{k=1}^{\infty} k(1 - p_{\text{coll}})^{k-1} = \frac{\lambda}{p_{\text{coll}}}$$

This quantity can be computed from the microdynamics

$$n_i(t + \tau, \vec{r} + \lambda \vec{c}_i) = n_i(t, \vec{r}) + \Omega_i(n)$$

A particle traveling with velocity $\vec{v}_i$ (i.e. $n_i(t, \vec{r}) = 1$) undergoes a collision if $\Omega_i = -1$ (i.e. $n_i(t + \tau, \vec{r} + \lambda \vec{c}_i) = 0$). From equation 3.91, we obtain

$$\begin{aligned}\Omega_i(n_i = 1) &= -(1 - n_{i+1})(1 - n_{i+5})[n_{i+2}n_{i+4}(1 - n_{i+3}) \\ &\quad + n_{i+3}(1 - n_{i+2})(1 - n_{i+4})]\end{aligned}$$

The probability p_{coll} of having a collision is then the probability that Ω_i takes the value -1. Since $\Omega_i(n_i = 1)$ can be either 0 or -1,

$$p_{coll} = \text{Prob}(\Omega_i = -1) = -(-1 \cdot \text{Prob}(\Omega_i = -1) + 0 \cdot \text{Prob}(\Omega_i = 0) = -\langle \Omega_i \rangle$$

Consequently,

$$\lambda_{mfp} = -\frac{1}{\langle \Omega_i(n_i = 1) \rangle}$$

In the Boltzmann approximation and at equilibrium ($\vec{u} = 0$, $N_i \equiv s = (\rho/6)$), we obtain

$$-\langle \Omega_i \rangle = (1 - s)^2 \left[s^2(1 - s) + s(1 - s)^2 \right] = \frac{\rho}{6} \left(1 - \frac{\rho}{6} \right)^3$$

Therefore,

$$\lambda_{mfp} = \frac{\lambda}{\frac{\rho}{6} \left(1 - \frac{\rho}{6} \right)^3}$$

and

$$v_{coll} = \frac{v}{12} \lambda_{mfp}$$

This result is important because it shows that in order to have a small viscosity, one has to decrease the mean free path. This is achieved by increasing the collision probability. In conclusion, the more collisions we have, the smaller the viscosity is.

Numerical viscometer. The viscosity can also be measured directly on a simulation of the lattice gas automaton. The method developed by Kadanoff et al. [84] is to build a numerical viscometer by considering two Poiseuille flows [80], opposite to each other. The system is periodic in both x- and y-directions. Opposite uniform forces are applied to the upper and lower half of the system by injecting momentum with a given probability at each lattice site. Practically, this is achieved by reorganizing the local directions of motion of the particle so as to modify the momentum in the x-direction. Momentum is forced in the positive direction in the upper part of the system and forced in the negative direction in the lower part.

From the parabolic velocity profile (Poiseuille flow), one can deduce the viscosity.

It turns out [84] that, as the system size gets larger, this viscosity increases logarithmically. Such a logarithmic divergence is expected for a two-dimensional fluid, from general considerations of statistical mechanics. This observation seems in contradiction with the result 3.202. The point is that the transport coefficients, such as the viscosity given by 3.202, are renormalized by hydrodynamic contributions (when going beyond the Boltzmann approximation) whose effects are divergent with the system size in two dimensions.

This problem is related to the existence of long time tails in the velocity autocorrelation function in a fluid. It was first observed numerically by Alder and Wrainwright in 1968 that the velocity autocorrelation in hydrodynamics decays as $t^{-d/2}$, where d is the dimensionality of the system and t the time. This long time tail is also well observed in lattice gas models [85,86].

The connection with the divergence of the viscosity is the following. It is well known from statistical mechanics [87] that transport coefficients (such as the viscosity) can be computed as an integral over time of suitable autocorrelation functions. This is the Green-Kubo formalism. In two dimensions, the velocity autocorrelation goes for long times as t^{-1} and its integral diverges, as an indication that some pathology is present in a two-dimensional fluid.

3.2.4 The collision matrix and semi-detailed balance

In section 3.2.2, we described the collision between FHP particles in terms of two- and three-body collisions (see equation 3.54).

A more general formulation is often considered, in which no explicit collision is given and which is generic of any lattice gas model. The idea of this formulation is to introduce a *collision matrix* $A(n, n')$ which gives the *probability* that an input state n is transformed into an output state n' in a collision process (note that here n' designate the state just before the propagation step takes place). For instance, in the plain FHP model, figure 3.1 shows that the probability of the transition

$$(n_1, n_2, n_3, n_4, n_5, n_6) = (0, 1, 0, 0, 1, 0) \rightarrow (1, 0, 0, 1, 0, 0) = (n_1', n_2', n_3', n_4', n_5', n_6')$$

is

$$A(n, n') = \frac{1}{2}$$

Clearly the whole collision process can be defined by giving the the full collision matrix A. In the example of the FHP model, there are 64 input states and 64 possible output states and thus, A is a 64×64 matrix.

In general, A is a sparse matrix: $A(n, n')$ is zero for most pairs (n, n'). This is the case in particular when the states n and n' do not have the same momtentum or do not contain the same number of particles. Since $A(n, n')$ is a probability one has obviously

$$\sum_{n'} A(n, n') = 1 \qquad (3.203)$$

for any n.

In some models, the collision matrix is symmetrix: $A(n, n') = A(n', n)$ and one says that *detailed balance* holds. This is the case with the HPP and plain FHP models. However, in general, this is not true. We may for instance modify the two-body collision rule in the FHP model so that, with probability 1, one has

$$n = (0, 1, 0, 0, 1, 0) \quad \rightarrow \quad (1, 0, 0, 1, 0, 0) = n'$$
$$n' = (1, 0, 0, 1, 0, 0) \quad \rightarrow \quad (0, 0, 1, 0, 0, 1) = n''$$
$$n'' = (0, 0, 1, 0, 0, 1) \quad \rightarrow \quad (0, 1, 0, 0, 1, 0) = n$$

Then $1 = A(n, n') \neq A(n', n) = 0$ and detailed balance is violated.

A weaker property is the so-called *semi-detailed balance* which only requires the symmetric of 3.203, namely that

$$\sum_{n} A(n, m) = 1$$

for any m. Obviously, detailed balance implies semi-detailed balance.

Semi-detailed balance is obeyed by most lattice gas models. In particular, by the above modified two-body FHP collision rule. However, if the interaction was such that, for instance,

$$n' \rightarrow n \qquad n'' \rightarrow n \qquad n \rightarrow n$$

then semi-detailed balance would not be satisfied. Note that this situation also implies that time reversal is not obeyed.

When all input states appear with equal probability, semi-detailed balance guarantees that all output states are also equally likely. This property is important because it is necessary to show the existence of an H-theorem in lattice gases and prove rigorously [88] that the dynamics of a system at equilibrium relaxes to the completely factorized, Fermi–Dirac universal distribution 3.122.

The lattice Boltzman equation. Using the collision matrix A, the associated lattice Boltzmann equation is easily written. In the Boltzmann approximation (factorization assumption), if N_i is the probability that

$n_i = 1$, the probability $P(n)$ of an input state n can be expressed as

$$P(n) = \prod_j N_j^{n_j} (1 - N_j)^{(1-n_j)} \qquad (3.204)$$

Hence

$$\sum_n P(n)A(n, n')$$

is the probability of output state n' and the average value $N_i' = \langle n_i' \rangle$ is

$$N_i' = \sum_{n,n'} n_i' P(n)A(n, n')$$

Since N_i can also be written as $\sum_n n_i P(n)$ and $\sum_{n'} A(n, n') = 1$, we have

$$N_i' - N_i = \sum_{n,n'} (n_i' - n_i)P(n)A(n, n')$$

Using expression 3.204 for $P(n)$ and remembering that $N_i' = N_i(\vec{r} + \lambda c_i, t + \tau)$, we finally obtain the following lattice Boltzmann equation

$$N_i(\vec{r} + \lambda c_i, t + \tau) - N_i(\vec{r}, t) = \sum_{n,n'} (n_i' - n_i)A(n, n') \prod_j N_j^{n_j} (1 - N_j)^{(1-n_j)} \quad (3.205)$$

3.2.5 The FHP-III model

Several important problems in hydrodynamics occur at high Reynolds numbers $R = (\ell u / v)$, where ℓ is the characteristic scale of the flow, u its characteristic velocity and v the kinematic viscosity [80]. The Reynolds number appears when the Navier–Stokes equation is scaled according to the characteristic sizes of the physical flow. It indicates that the flows of different fluids are similar when they have a similar Reynolds number. This similarity of flows explains why, in principle, lattice gas automata can be used to simulate any real fluid.

In the case of the FHP model, the factor $g(\rho)$ has been absorbed in a rescaling of time, which results in a rescaling of the viscosity $v(\rho) \rightarrow (v(\rho)/g(\rho))$. The Reynolds number can then be rewritten as

$$R = \ell M R_*(\rho) \qquad (3.206)$$

where $M = (u/c_s)$ is the the Mach number (c_s is the speed of sound derived in section 3.2.3) and $R_*(\rho)$ is defined as

$$R_*(\rho) = \frac{c_s g(\rho)}{v(\rho)} \qquad (3.207)$$

and is the reduced Reynolds number which contains the intrinsic properties of the lattice gas dynamics [88]. The maximum value of R_* that can

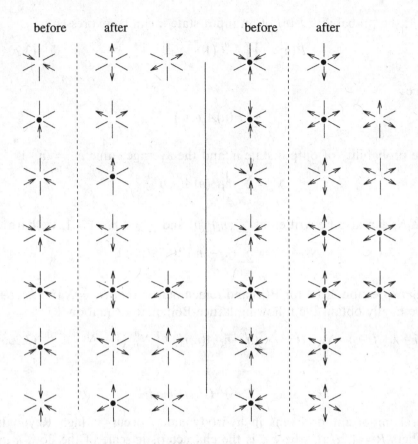

Fig. 3.5. The collision rule of the FHP-III model. Before and after collision configurations are shown. When two possible outputs are given, each of them is selected with equal probability.

be obtained with the FHP model is

$$R_*^{\max} = 0.387 \qquad \text{for} \qquad \rho = 1.12 \qquad (3.208)$$

Considering a Mach number of $M = 0.3$, which is a standard choice for assuming incompressibility, and a typical length scale of $\ell = 1000$ lattice size, one obtains a Reynolds number of about $R = 130$.

A way to reach larger Reynolds numbers without increasing the lattice size is to have the smallest possible viscosity [88]. For this purpose, it is necessary to work with another set of collision rules.

The FHP-III model [3] is a variation of the FHP dynamics which is significantly less viscous. Its particularity resides in the presence of rest particles (i.e. up to seven particles per lattice site). In addition, the number of configurations leading to a collision is increased with respect to the original FHP automaton.

Figure 3.5 illustrates the collision rules. We observe that while mass and momentum are exactly conserved in a collision, kinetic energy is not. A collision between a rest particle and a moving one results in two moving particles and there is production of kinetic energy. However, the inverse collision is expected to occur with the same probability and, therefore, energy is conserved in a statistical way.

The viscosity of the FHP-III model is found to be [3]

$$v = \tau v^2 \left(\frac{1}{4\rho(1 - \frac{\rho}{7})^3(1 - \frac{4\rho}{49})} - \frac{1}{8} \right) \tag{3.209}$$

which is smaller than the viscosity of the FHP model. Therefore, increasing the number of collisions results in decreasing viscosity. For this model, the maximum value of R_* is [88]

$$R_*^{\max} = 2.22 \quad \text{for} \quad \rho = 2 \tag{3.210}$$

Although the FHP-III model has a better physical behavior, it is much more complex than the FHP dynamics. From the point of view of numerical efficiency, it is highly desirable to implement the collision rule as a lookup table in which all the 2^7 possible collisions are computed in advance. Otherwise the microdynamics requires an enormous number of terms. As an example of the complexity of the collision term, we give below the algebraic expression for the FHP-III collision rules for N_0, the population of rest particle

$$
\begin{aligned}
N_0(t + \tau, r) - N_0 = & - N_0[((N_1\overline{N_2} + N_2\overline{N_1})\overline{N_3} + N_3\overline{N_1}\ \overline{N_2})\overline{N_4}\ \overline{N_5}\ \overline{N_6} \\
& + \overline{N_1}\ \overline{N_2}\ \overline{N_3}\ ((N_4\overline{N_5} + N_5\overline{N_4})\overline{N_6} + N_6\overline{N_4}\ \overline{N_5})] \\
& + [(N_1N_3\overline{N_2}\ \overline{N_4} + N_2N_4\overline{N_1}\ \overline{N_3})\overline{N_5}\ \overline{N_6} \\
& + (N_3N_5\overline{N_4}\ \overline{N_6} + N_4N_6\overline{N_3}\ \overline{N_5})\overline{N_1}\ \overline{N_2} \\
& + (N_1N_5\overline{N_2}\ \overline{N_6} + N_2N_6\overline{N_1}\ \overline{N_5})\overline{N_3}\ \overline{N_4}](\overline{N_0} - N_0) \\
& + \frac{1}{2}[N_6\overline{N_2}\ (N_1N_4\overline{N_3}\ \overline{N_5} + N_3N_5\overline{N_1}\ \overline{N_4}) \\
& + N_4\overline{N_6}\ (N_2N_5\overline{N_1}\ \overline{N_3} + N_1N_3\overline{N_2}\ \overline{N_5}) \\
& + N_2\overline{N_4}\ (N_3N_6\overline{N_1}\ \overline{N_5} + N_1N_5\overline{N_3}\ \overline{N_6})](\overline{N_0} - N_0) \\
& + \frac{1}{2}[N_1\overline{N_5}\ (N_2N_4\overline{N_3}\ \overline{N_6} + N_3N_6\overline{N_2}\ \overline{N_4}) \\
& + N_5\overline{N_3}\ (N_2N_6\overline{N_1}\ \overline{N_4} + N_1N_4\overline{N_2}\ \overline{N_6}) \\
& + N_3\overline{N_1}\ (N_4N_6\overline{N_2}\ \overline{N_5} + N_2N_5\overline{N_4}\ \overline{N_6})](\overline{N_0} - N_0) \\
& + [N_4(N_5(N_2N_6\overline{N_1}\ \overline{N_3} + N_1N_3\overline{N_2}\ \overline{N_6}) \\
& + N_2N_3N_6\overline{N_1}\ \overline{N_5}) + N_1(N_2(N_3N_5\overline{N_4}\ \overline{N_6} \\
& + N_4N_6\overline{N_3}\ \overline{N_5}) + N_3N_5N_6\overline{N_2}\ \overline{N_4})](\overline{N_0} - N_0)
\end{aligned}
$$

$$+ [N_6(N_5(N_4(N_3(N_2\overline{N_1} + N_1\overline{N_2}) + N_1N_2\overline{N_3})$$
$$+N_1N_2N_3\overline{N_4}) + N_1N_2N_3N_4\overline{N_5})$$
$$+N_1N_2N_3N_4N_5\overline{N_6}]\overline{N_0} \qquad\qquad (3.211)$$

where N_i, $i = 1, ..., 6$ label the moving particles. The right-hand side of this equation is computed at time t and position $\vec{r}$. The factors $1/2$ account for the two possible outcomes arising for some collisions. Quantity q is a stochastic Boolean variable at each lattice site. The evolution law of the moving particles N_i, $i \neq 0$ is even more complex.

3.2.6   Examples of fluid flows

In this section, we show, as an illustration, two examples of fluid flows past an obstacle. Figure 3.6 show the streaklines lines resulting from the presence of a plate in the way of a fluid flowing from right to left. The flow pattern is produced by letting particles in suspension follow the fluid streaklines (note that streaklines are different from streamlines in the sense that the latter are defined as the path suspensions would follow if one could "freeze" the velocity field at a given time).

The rule we have used to produce the motion of suspensions is the following: each of the particles cumulates the local velocity flow, averaged over a 32×32 neighborhood of sites and 10 consecutive time steps. When it has gathered enough momentum along the x- or y-axis, it moves by one lattice site in the corresponding direction by consuming an amount Δ_x or Δ_y of its momentum.

Particles in suspension are injected along a vertical line on the right part of the system, with regular spacing. One observes the formation of eddies, resulting from a von Karman flow instability. Note that the fluid particles are not shown in this figure, but only the suspensions.

The cellular automaton fluid we have considered here does not exactly obey the FHP rule. It is a variant which enhance three- and four-body collisions: all possible collisions with three or four fluid particles entering a site are taken into account, as is the case for the FHP-III model. However, here, there is no rest particle. In order to maximize the frequency of collisions, the simulation has been performed at an average density of $\rho = 3$ (i.e. three particles per site, on average).

The initial fluid speed is produced (and maintained) in a standard way, by accelerating the particles in the first quarter of the system. This is typically obtained by reversing the direction of motion of a particle traveling rightward with a given probability.

In figure 3.7, we show the flow produced past a vertical plate. The parameters of the simulation are identical to those of figure 3.6, except that the system size is 512×128 and, consequently, the Reynolds number

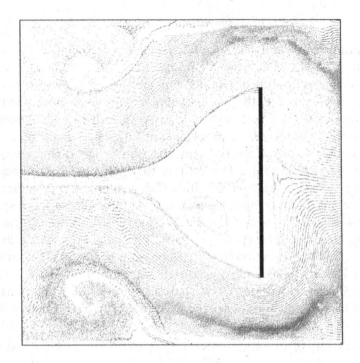

Fig. 3.6. Streaklines shown by particles in suspension in a FHP-like fluid. The system size is 1024 × 1024 and periodic in the vertical direction.

Fig. 3.7. Streaklines shown by particles in suspension in a FHP-like fluid. The system size is 512 × 128 and periodic in the vertical direction.

is smaller. Oscillations are seen along the streaklines, indicating the onset of an instability.

Cellular automata fluids require quite large systems to exhibit non-trivial flow patterns. This is due to their relatively high viscosity which limits the Reynolds number that can be attained. On the other hand, we can see that the symmetry can be spontaneously broken due to the intrinsic fluctuations in the fluid automaton.

3.2.7   *Three-dimensional lattice gas models*

Hydrodynamics is usually a three-dimensional problem and it is clearly important to generalize the two-dimensional FHP fluid to three dimensions. Although there is no difficulty in defining collision rules between particles moving on a 3-d lattice, there are severe isotropy problems. It turns out that no regular three-dimensional lattice exists with enough symmetry to guarantee that fourth-order tensors such as (3.154) are isotropic.

The solution to this problem has been proposed by d'Humières, Lallemand and Frisch [36,88]. The idea is to consider a *four*-dimensional lattice, the so-called the FCHC lattice (face-centered-hyper-cubic) which is isotropic and then to project the motion to three-dimensional space.

In the 4-d FCHC model, there are 24 directions of motion, corresponding to the lines to connect a site to its nearest neighbors. These links are of length $\sqrt{2}$. The projection in three dimension results in six double links of length 1 and 12 single links of length $\sqrt{2}$. After projection, we obtain a multispeed model with up to two particles along some directions.

The 24 lattice directions $\vec{c}_i$ are the following [89]. First the double links are

$$\vec{c}_1 = (\ 1,\ 0,\ 0) \qquad \vec{c}_2 = (-1,\ 0,\ 0) \qquad \vec{c}_3 = (-1,\ 0,\ 0)$$
$$\vec{c}_4 = (\ 1,\ 0,\ 0) \qquad \vec{c}_5 = (\ 0,\ 1,\ 0) \qquad \vec{c}_6 = (\ 0,-1,\ 0)$$
$$\vec{c}_7 = (\ 0,-1,\ 0) \qquad \vec{c}_8 = (\ 0,\ 1,\ 0) \qquad \vec{c}_9 = (\ 0,\ 0,\ 1)$$
$$\vec{c}_{10} = (\ 0,\ 0,-1) \qquad \vec{c}_{11} = (\ 0,\ 0,-1) \qquad \vec{c}_{12} = (\ 0,\ 0,\ 1)$$

and, second, the single links

$$\vec{c}_{13} = (\ 1,\ 1,\ 0) \qquad \vec{c}_{14} = (-1,-1,\ 0) \qquad \vec{c}_{15} = (\ 1,-1,\ 0)$$
$$\vec{c}_{16} = (-1,\ 1,\ 0) \qquad \vec{c}_{17} = (\ 1,\ 0,\ 1) \qquad \vec{c}_{18} = (-1,\ 0,-1)$$
$$\vec{c}_{19} = (\ 1,\ 0,-1) \qquad \vec{c}_{20} = (-1,\ 0,\ 1) \qquad \vec{c}_{21} = (\ 0,\ 1,\ 1)$$
$$\vec{c}_{22} = (\ 0,-1,-1) \qquad \vec{c}_{23} = (\ 0,\ 1,-1) \qquad \vec{c}_{24} = (\ 0,-1,\ 1)$$

Figure 3.8 illustrates this projection in three-dimensional space. We refer the reader to [88] for a detailed discussion of the collision rules.

3.3   Thermal lattice gas automata

3.3.1   *Multispeed models*

In the HPP, FHP and related models discussed in the previous sections we did not pay much attention to energy conservation during the collision process. The reason is that these models are devised to describe athermal systems, without a notion of temperature.

Here we would like to present some LGA models which may account for thermal properties in a fluid. The key ingredient is to build a model having a velocity distibution instead of one single speed. By velocity

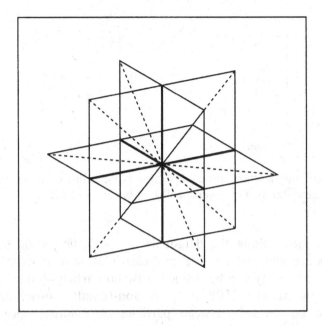

Fig. 3.8. The possible particle velocities of the FCHC lattice gas model of a fluid. Bold links represent the direction in which two particles may travel simultaneously.

distibution we mean here that the *modulus* of the particle speed may vary in addition to its direction.

With several velocity populations, kinetic energy can be defined in a non-trivial way and a local temperature associated to the fluid through the equipartition theorem.

Since we work on a lattice and would like to keep the dynamics to its maximum simplicity, we will not consider a large range of different velocities for the particles. Typically we shall have "hot" (or fast) particles that travel to second nearest neighbor sites in one time step, normal, speed-one particles and, finally "cold" particles at rest. These models are usually termed *multispeed* LGA.

The main difficulty with a multispeed model is to define appropriate collision rules. We would like to conserve mass, momentum and enery independently and we expect that collisions result in a change of the velocity distribution. The constraint that particles are moving on a lattice reduces very much the number of possible collisions and the velocity set should be chosen carefully.

The first multispeed model of this type was proposed by Lallemand and d'Humières [37] in 1986. It is defined on a square lattice and comprises particles of the same mass $m = 1$ but speed 0, 1 and $\sqrt{2}$. The

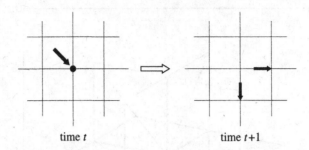

time t time $t+1$

Fig. 3.9.   Mass, momentum and energy conserving collision in the three-speed square lattice automaton. The reverse collision is also possible.

fast particles move along the diagonals of the lattice while the speed-one particles are identical to HPP particles. Collisions conserving mass, momentum and energy can be defined between particles having the same speed using the regular HPP rule. A non-trivially conserving energy collision can be defined between particles of different speed. It is illustrated in figure 3.9 and involves one fast, speed-$\sqrt{2}$ particle colliding with one rest particles to produce two speed-one particles. Of course, the reverse collision, in which two speed-one particles collide at right angle and give one rest particle and one fast particle, is also taken into account.

Energy is defined as pure kinetic energy

$$E = m\frac{v^2}{2}$$

Rest particles have zero kinetic energy, speed-one particles have energy $1/2$ and fast particles have energy 2. With this definition, it is easy to check that the collision described in figure 3.9 conserves mass, momentum and energy. It is also interesting to note that this collision rule modifies the velocity distribution: a pair of particles with speed 0 and $\sqrt{2}$ are transformed into two particles of speed one. Thus rest particles couple the population of speed-one and speed-$\sqrt{2}$ particles.

Of course, since this model is defined on a squre lattice we expect that isotropy problems will show up with fourth-order tensors. Thus this model is not appropriate to describe a fluid flow but, on the other hand, may well simulate a fluid at rest, submitted to a temperature gradient. An interesting application is the study of nonequilibrium fluctuations in such a system [90].

In section 3.3.3, we shall present other multispeed LGA that are defined on a hexagonal lattice and can simulate isotropic flows.

3.3.2 Thermo-hydrodynamical equations

When energy is taken into account, a fluid is decribed by the so-called thermohydrodynamic equations [87,91]. They are the Navier–Stokes equation complemented with a new equation for the local *energy density* or, using thermodynamics transformations, an equation for the local temperature at each point of the fluid.

In this section we shall briefly sketch how these equations can be obtained from a multispeed LGA dynamics, without giving a complete, nor a general derivation.

The microdynamics of a multispeed model can be written using the usual formalism discussed in this chapter. We denote by $n_i \in \{0,1\}$ the presence or absence of a particle traveling with velocity $\vec{v}_i$

$$n_i(\vec{r} + \tau\vec{v}_i, t + \tau) - n_i(\vec{r}, t) = \Omega_i(n(\vec{r}, t))$$

The right-hand side term Ω_i is the collision term which now obeys an extra conservation law for energy

$$\sum_i m_i \frac{\vec{v}_i^2}{2} \Omega_i = 0$$

Of course the usual mass and momentum conservation are still required

$$\sum_i m_i \Omega_i = \sum_i m_i \vec{v}_i \Omega_i = 0$$

Clearly, in a one-speed model, energy and mass conservation are identical.

The precise expression for Ω depends on the model, the type of collision and the lattice topology. But the three above properties must be satisfied in order to describe a thermal cellular automata fluid.

The standard procedure can be repeated to obtained the macrosopic equations related to the LGA microdynamics: ensemble average, Bóltzmann factorization and multiscale Chapman–Enskog expansion. The fluid density ρ and velocity field $\vec{u}$ are defined as previously

$$\rho = \sum_i m_i N_i \qquad \rho\vec{u} = \sum_i m_i \vec{v}_i N_i$$

where $N_i = \langle n_i \rangle$ is the averaged occupation number.

A new physical quantity is introduced: the local energy per particle $\tilde{e}$

$$\rho\tilde{e} = \sum_i m \frac{\vec{v}_i^2}{2} N_i$$

and the corresponding internal energy e can be defined as

$$\rho e = \sum_i m \frac{(\vec{v}_i - \vec{u})^2}{2} N_i = \rho\tilde{e} - \rho\frac{\vec{u}^2}{2}$$

A temperature T can be introduced in this model, assuming a local version of the equipartition theorem [87] is valid. Thus we define, for a two-dimensional system

$$e = kT$$

where k plays the role of the Boltzmann constant. Taking

$$k = \frac{1}{2}\left(\frac{\lambda}{\tau}\right)^2$$

turns out to be a natural choice for k in a lattice gas system.

Another important quantity is the heat current $\vec{J}_Q$

$$\vec{J}_Q = \sum_i \frac{(\vec{v}_i - \vec{u})^2}{2}(\vec{v}_i - \vec{u})N_i$$

It is a rather tedious task to obtain the explicit thermohydrodynamic equations governing ρ, $\vec{u}$, e, starting from a given model and we will not develop further this derivation. In general, we expect a loss of Galilean invariance but, in some models [81], tricks may be used to remove this problem.

As an illustration of the properties of a thermal model, it is also interesting to consider the simpler case of a fluid at rest, in a stationary state and submitted to a temperature gradient. In this case we typically expect the following result [78]

$$\vec{J}_Q = -\kappa \nabla T$$

where $\kappa = \kappa(\rho, T)$ is the thermal conductivity which is built-in in the collision rule.

3.3.3   *Thermal FHP lattice gases*

Multispeed models on a hexagaonal lattice which generalize the FHP dynamics to include thermal properties have also been proposed. However, due to six-fold symmetry, they are slightly more complicated than the thermal HPP dynamics.

A three-speed FHP model is illustrated in figure 3.10. Particles may have speed 0, 1 or 2, in lattice units. We see that the collision which couples the three populations involves a speed-two and a rest particle to produce two speed-one particles (or conversely). But to conserve mass, momentum and energy the different velocity populations must have different masses. Rest particles have mass 3/2 and zero kinetic energy, speed-one particles have mass 1 and energy 1/2 and, finally, speed-two particles have mass 1/2 and energy 1.

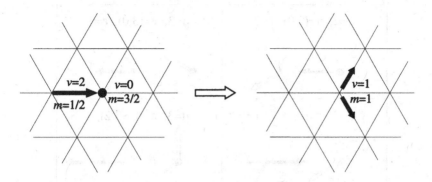

Fig. 3.10. Mass, momentum and energy conserving collision rule in a thermal FHP model. The reverse collision is also possible. Note that particles with a different speed have a different mass.

In order to define a multispeed dynamics on a hexagonal lattice with all equal masses it is necessary to include a fourth velocity population. The model proposed by Grosfils *et al.* contains up to 19 particles per site (one rest particle plus three times six moving particles [79]). In addition to speed 0, 1, and 2, a population of speed $\sqrt{3}$ is included. These particles travel along the lines bisecting the lattice directions and, in one time step, they jump from one site to one of the six next-nearest neighbors.

In this model, a particle has kinetic energy 0, 1/2, 3/2 or 2. The collision rules are defined by selecting at random the output velocity configuration among all configurations having the same mass, momentum and energy as the input configuration. This approach requires one to classify all situations in a table, according to their particle number, momentum and energy. Figure 3.11 shows some of these classes and some configurations belonging to it.

This four-speed FHP model has been shown to reproduce correctly spontaneous fluctuations by observing the behavior of the dynamic structure factor which is expected to exhibit three peaks for a real fluid [79].

3.4 The staggered invariants

As we have seen previously, quantities that are conserved by the microdynamics play a crucial role in the derivation of the macroscopic equations governing the cellular automaton.

In the FHP model, the basic ingredients we have put in the dynamics are the local conservation of mass and momentum, which are prescribed by the laws of physics.

In principle, it is not desirable to have any other conserved quantities in the dynamics. It turns out that extra conservation laws are frequent

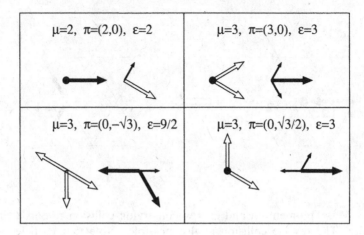

Fig. 3.11. Example of particle configurations with same mass μ, momentum π and energy ϵ in the four-speed FHP model. Particles with speed 0, 1, $\sqrt{3}$ and 2 are represented by a dot, a thin arrow, an open arrow and a large, black arrow, respectively.

in lattice gas models. We have already observed that the HPP model described in section 2.2.5 conserved momentum along each lattice line and each lattice column. We have mentioned that both two- and three-body collisions are necessary in the FHP rule to avoid spurious conservations.

These kinds of undesirable conservations are related to the choice of the collision rule and generally show up when performing the Chapman–Enskog expansion. However, in many cellular automata models, there are other types of spurious conserved quantities, which are an artifact of the discrete nature of the underlying lattice and have no physical counterpart. They are the so-called *staggered invariants*, first discovered by McNamara and Zanetti [92]. They have received this name because they usually depend on the parity of space and time.

For instance, we have already noticed in section 3.1.4 the existence of a spurious invariant directly related to the checkerboard structure of the square lattice

$$N_{\text{even}}(t) - N_{\text{odd}}(t) = (-1)^{t/\tau}[N_{\text{even}}(0) - N_{\text{odd}}(0)] \qquad (3.212)$$

The checkerboard invariant generalizes to almost all lattice gas automata. In the FHP model, this unphysical conserved quantity is called the *staggered momentum*.

Consider for instance figure 3.12, and more particularly the horizontal line of sites labeled "+". The particles sitting on this line at time t will either stay on this line if they travel horizontally or move to one of the horizontal lines labeled "-", above or below. Now, suppose we compute

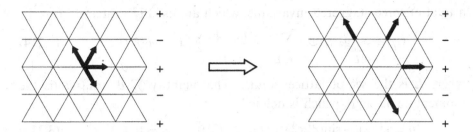

Fig. 3.12. Illustration of the staggered momentum in the FHP dynamics: The vertical component of momentum due to particles sitting on a horizontal line labeled "+" is transferred, at the next iteration, to the two lines labeled "-" above and below.

the total momentum along the y-axis (vertical direction) carried by each particles. The particles moving horizontally have, of course, no vertical component of momentum and do not contribute. Therefore, all the *vertical* momentum of the "+" line will be distributed, at time $t + \tau$, on the two horizontal "-" lines.

If we continue to label the horizontal lines by alternating "+" and "-" we obtain similarly that all the vertical momentum of a "-" line at t is transferred to the surrounding "+" lines at the next step. We can build a globally conserved quantity by summing over all "+" or "-" lines

$$\sum_{\vec{r} \in L_+} P_y(t + \tau, \vec{r}) = \sum_{\vec{r} \in L_-} P_y(t, \vec{r}) \qquad \sum_{\vec{r} \in L_-} P_y(t + \tau, \vec{r}) = \sum_{\vec{r} \in L_+} P_y(t, \vec{r}) \quad (3.213)$$

where L_+ and L_- denote the two sublattices obtained with our line labeling and P_y represent the y-component of momentum. If we add the two equations, we obtain no more than the expected physical conservation of momentum. But, by subtracting them, we obtain the typical form of a staggered invariant, namely

$$\sum_{\vec{r} \in L_+} P_y(t + \tau, \vec{r}) - \sum_{\vec{r} \in L_-} P_y(t + \tau, \vec{r}) = - \left(\sum_{\vec{r} \in L_+} P_y(t, \vec{r}) - \sum_{\vec{r} \in L_-} P_y(t, \vec{r}) \right)$$

$$= (-1)^{(t/\tau)} \left(\sum_{\vec{r} \in L_+} P_y(0, \vec{r}) - \sum_{\vec{r} \in L_-} P_y(0, \vec{r}) \right)$$

Here, we see that the staggered momentum results from the detailed way in which the momentum is conserved on the lattice.

Of course, the same construction we consider for the vertical component of the momentum can be repeated for the lattice lines oriented 60 and 120 degrees with respect to the horizontal direction. As a result, we obtain a total of three staggered invariants, which are usually expressed as

$$G_{\vec{\theta}} \equiv \sum_{\vec{r} \in L} g_{\vec{\theta}}(t, \vec{r}) \equiv \sum_{\vec{r} \in L} (-1)^{(t/\tau) + \vec{\theta} \cdot \vec{r}} \sum_i \vec{\theta} \cdot \vec{v}_i n_i(\vec{r}, t) \qquad (3.214)$$

where L is the set of lattice points. The quantity $\sum_i \vec{\theta} \cdot \vec{v}_i n_i(\vec{r}, t)$ is the momentum along $\vec{\theta}$, which is defined as

$$\vec{\theta} = (1/\lambda)(-\sin(n(\pi/3)), \cos(n(\pi/3))) \qquad n = 0, 1, 2 \qquad (3.215)$$

For each value of n, $\vec{\theta}$ is orthogonal to one of the three families of line orientations. The scalar product $\vec{\theta} \cdot \vec{r}$ specifies one sublattice division because it take the same integer value (even or odd) on a line perpendicular to $\vec{\theta}$.

The local staggered invariants, such as $g_{\vec{\theta}}$ often show a diffusive behavior but sometimes may exhibit propagating modes, as shown by Ernst and Brito.

Staggered invariants may play an important role in the physical behavior of a lattice gas model because they couple to the *nonlinear* terms in the Navier–Stokes equations [84,92–94]. For instance the $\vec{u} \cdot \nabla \vec{u}$ term will be accompanied by an analogous term containing the average staggered momentum ·density $g_{\vec{\theta}}$. However, no direct evidence of pernicious effects have been detected in fluid flow simulations, at least qualitatively.

The influence of a staggered invariant on the physical behavior of a cellular automata fluid can be minimized by properly choosing the initial condition of the fluid particles [84]. It is indeed always possible to prepare an initial state in which the total staggered momentum $G_{\vec{\theta}}$ vanishes exactly.

In addition, it is possible to chose the initial positions of the particles so that the staggered momentum coincides with the regular momentum. For this purpose, they are only placed on the lattice of spacing 2λ shown in figure 3.13. The other sites are left empty. These positions correspond to choosing the intersection of the L_+ lines for the three orientations. Along the bold lines of figure 3.13, one has

$$(-1)^{\vec{\theta} \cdot \vec{r}} = 1 \qquad (3.216)$$

and, at time $t = 0$, $G_{\vec{\theta}}$ reads

$$G_{\vec{\theta}} = \vec{\theta} \cdot \sum_{r \in L} \sum_i \vec{v}_i n_i(\vec{r}, 0) = \vec{\theta} \cdot \vec{P}_{\text{tot}} \qquad (3.217)$$

where $\vec{P}_{\text{tot}}$ denotes the total momentum of the system. Since both $G_{\vec{\theta}}$ and $\vec{P}_{\text{tot}}$ are invariants, the fact that they are equal at the initial time guarantees that they are always equal.

Usually, there are many staggered invariants in a lattice gas automaton [95] and there is apparently no general way to find all of them. One

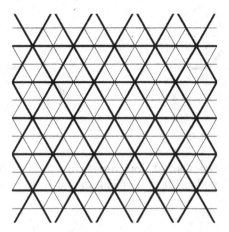

Fig. 3.13. The bold lines represents the sublattice for which the total staggered momentum reduces to the usual momentum.

often distinguishes between *dynamic* and *geometric* staggered invariants. The former are staggered in both space and time (like the staggered momentum or the checkerboard invariant), whereas the latter are staggered only in space.

As an example of a geometrical staggered invariant, consider the case of a two-speed HPP model [78]. This model consists of two types of interacting HPP particles: the slow particles obey the same dynamics as discussed in section 2.2.5; the fast particles are also subject to the usual HPP collisions but travel along the diagonals of the square lattice (which forms a square lattice of spacing $\sqrt{2}\lambda$). The slow and fast particles have a different kinetic energy. Interactions between the two species which conserves mass, momentum and energy can be devised. This system may represent either a binary mixture or a gas with a temperature. In the latter case, temperature is defined through the equipartition theorem. Figure 3.14 shows the result of a typical collision between a fast and a slow particle.

A simple geometric staggered invariant exists for the fast particles. Since they move along the diagonals of the square lattice, there are two families of fast particles: those moving on the "white" sites of the checkerboard and those moving on the "black" sites. Therefore, their number ($N_{\text{even}}^{\text{fast}}$ and $N_{\text{odd}}^{\text{fast}}$) is conserved independently of each other. The sum $N_{\text{odd}}^{\text{fast}} + N_{\text{even}}^{\text{fast}}$ expresses the expected mass conservation of the fast particles and the difference

$$N_{\text{odd}}^{\text{fast}} - N_{\text{even}}^{\text{fast}} = \sum_{\vec{r} \in L} (-1)^{\vec{r} \cdot \vec{\theta}} \sum_{i=1,4} n_i^{\text{fast}}(\vec{r}, t) \tag{3.218}$$

is a geometric invariant, where $\vec{\theta} = (1/\lambda)(1,1)$ characterizes the checkerboard division of the lattice.

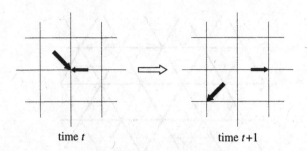

Fig. 3.14. Collision between a fast (long arrow) and a slow (short arrow) particle in the two-speed HPP model. Mass, momentum and energy are conserved during this interaction process.

Of course, the fast particles also yield the usual checkerboard dynamical invariant,

$$\sum_{\vec{r}\in L}(-1)^{t/\tau+\vec{r}\cdot\vec{\theta}}\sum_{i=1,4}n_i^{\text{fast}}(\vec{r},t) \qquad (3.219)$$

which is now characterized by $\vec{\theta}=(1/\lambda)(1,0)$ or $\vec{\theta}=(1/\lambda)(0,1)$ and means that a fast particle entering a vertical (respectively, horizontal) line at time t will clearly enter, at the next step, one of the adjacent lines, at left or right (up and down, respectively).

3.5   Lattice Boltzmann models

3.5.1   Introduction

Cellular automata fluids, such as those discussed so far in this chapter, represent idealized N-body systems. Their time evolution can be performed exactly on a computer, without many of the approximations usually done when computing numerically the motion of a fluid. In particular, there is no need, in a lattice gas simulation to assume some factorization of the many-body correlation functions into a product of one-particle density functions.

Of course, the cellular automata model may be inadequate to represent a real situation but it includes naturally the intrinsic fluctuations present in any system composed of many particles. This features is out of reach of most tractable numerical techniques. In many physical situations, spontaneous fluctuations and many-particle correlations can be safely ignored. This is, however, not the case if one is interested in long time tails present in the velocity autocorrelation (see section 3.2.3). We will also consider in the next chapter other systems for which correlations or local fluctuations cannot be neglected and where the N-body nature of the cellular automata approach is necessary.

On the other hand, a cellular automata simulation is very noisy (because it deals with Boolean quantities). In order to obtain the macroscopic behavior of system (like streaklines in the flow past an obstacle), one has to average the state of each cell over a rather large patch of cells around it (for instance a 32×32 square) and over several consecutive time steps. This slows down very much the simulation speed and requires large systems. Therefore, the benefits of the cellular automata approach over more traditional numerical techniques becomes blurred [38].

In addition, due to its Boolean nature, cellular automata models offer little flexibility to finely adjust external parameters. Tuning is done through probabilities, which is not always the most efficient way.

When correlations can be neglected and the Boltzmann molecular chaos hypothesis is valid, it may be much more effective to directly simulate on the computer the lattice Boltzmann equation

$$N_i(\vec{r} + \lambda \vec{c}_i, t + \tau) = N_i(\vec{r}, t) + \Omega_i(N) \tag{3.220}$$

with Ω_i given, for instance, by (3.91) or (3.211). It is more advantageous to average the microdynamics *before* simulating it rather than after. The quantities of interest N_i are no longer Boolean variables but probabilities of presence which are continuous variables ranging in the interval $[0, 1]$. This approach reflects the fact that, at a macroscopic level of observation, the discrete nature of the fluid particles vanishes.

A direct simulation of the lattice Boltzmann dynamics was first considered by McNamara and Zanetti [51]. It considerably decreases the statistical noise that plagues cellular automata models and considerably reduces the computational requirements. As mentioned previously, the main drawback is that this approach neglects fluctuations and many-body correlation functions.

Strictly speaking, one can argue that a dynamics based on a set of continuous variables is no longer a CA model because a real number requires an infinite amount of information to be stored in a computer. However, in practice, real numbers are approximated using a finite number of bits (typically 32 or 64 in modern computers) and, therefore, the lattice Boltzmann approach is also a cellular automata rule such that the state of each cell is a large (but finite) number of bits. Of course, the use of a lookup table is not possible and a direct calculation of the rule should be repeated for each site and each iteration.

The lattice Boltzmann method or lattice Boltzmann equation (often referred to as LBM and LBE) have been widely used for simulating various fluid flows [96] and is believed to be a very serious candidate to overcome traditional numerical techniques of computational fluid dynamics. Their microscopic levels of description provide a natural interpretation of the numerical scheme and permit intuitive generalizations to complex flow problems (two-phase flow [97,45,46], magnetohydrodynamics [98], flow in porous media [40,41] or thermohydrodynamics [99]).

In the following chapters, we shall consider some applications of the lattice Boltzmann method to fluid problems (snow transport and deposition by wind) and also to other domains such as pattern formation in reaction-diffusion systems and wave propagation in heterogeneous media.

In lattice Boltzmann fluids, the most natural way to define the collision term Ω_i is to average the microdymanics of a given underlying cellular automata fluid and factorize it into a product of average, as we did previously to get the Boltzmann approximation. We then obtain an expression like those derived in 3.91 or 3.211. It turns out that the collision term given by 3.211 (and its counterpart for moving particles) requires more than one thousand floating point operations at each lattice site and time step. Even on a massively parallel computer, in which every cell is computed simultaneously, this is usually not acceptable.

The first solution to this problem is to consider the same approximation as we used with the Chapman–Enskog expansion when deriving the macroscopic behavior of the FHP fluid. The idea is to linearize the collision term around its local equilibrium solution. This approach has been proposed by Higuera and coworkers [52] and considerably reduces the complexity of the operations involved.

Following the same idea, a further simplification can be considered [100]: the collision term need not be related to an existing cellular automata microdynamics, as long as particle and momentum are conserved. In its simplest form, the lattice Boltzmann dynamics can be written as a relaxation equation [101]

$$f_i(\vec{r} + \lambda \vec{c}_i, t + \tau) - f_i(\vec{r}, t) = \frac{1}{\xi} \left(f_i^{(0)}(\vec{r}, t) - f_i(\vec{r}, t) \right) \qquad (3.221)$$

or, equivalently as

$$f_i(\vec{r} + \lambda \vec{c}_i, t + \tau) = \frac{1}{\xi} f_i^{(0)}(\vec{r}, t) + \left(1 - \frac{1}{\xi} \right) f_i(\vec{r}, t) \qquad (3.222)$$

where the local equilibrium solution $f_i^{(0)}$ is now a given function of the density $\rho = \sum f_i$ and velocity flow $\rho \vec{u} = \sum f_i \vec{v}_i$. The function $f_i^{(0)}$ can be *chosen* so as to produce a given behavior, first to guarantee mass and momentum conservation but also to yield the appropriate local equilibrium probability distribution. In particular, the lack of Galilean invariance that plagues cellular automata fluid can be cured, as well the spurious velocity contribution appearing in the expression 3.168 of the pressure term. In a more general context, $f_i^{(0)}$ could include other physical features, such as a local temperature. Equation 3.221 is often referred to as the Lattice BGK (or LBGK) method [102] (the abbreviation BGK stands for Bhatnager, Gross and Krook [103] who first considered a collision term with a single relaxation time, in 1954).

Equation 3.221 has been studied by several authors [104], due to its ability to deal with high Reynolds number flows. However, one difficulty of this approach is the numerical instabilities which may develop in the simulation, under some circumstances.

Finally, let us mention, in a similar context, the *multiparticle* lattice gas which is an intermediate approach (see section 7.3) between a pure automata and a real-valued description and allows an arbitrary number of particles in each direction [105]. This technique is not yet very well explored and could possibly alleviate some of the instability problems and restore the many-body correlations.

3.5.2 A simple two-dimensional lattice Boltzmann fluid

In this section, we present a detailed derivation of a lattice Boltzmann dynamics of a two-dimensional fluid. A natural starting point would be to consider again the hexagonal lattice used for the FHP model. However, in most computers (and especially on parallel computers), the data structure representing a hexagonal lattice is more difficult to work with than a regular square lattice which is naturally implemented as an array of data. Although simple coordinate transforms can be devised to map the hexagonal lattice onto a square lattice (see section 7.2), boundary conditions should be treated in a separate way.

The choice of a hexagonal lattice stems from isotropic requirements. Actually, the FCHC model discussed in section 3.2.7 provides an alternative since it can be projected onto a two-dimensional square lattice.

Figure 3.8 suggests that the resulting lattice is composed of eight velocities $\vec{v}_i$ with $\vec{v}_1$, $\vec{v}_3$, $\vec{v}_5$ and $\vec{v}_7$ collinear with the horizontal and vertical directions, while $\vec{v}_2$, $\vec{v}_4$, $\vec{v}_6$ and $\vec{v}_8$ correspond to the diagonals of the lattice. These eight velocities are shown in figure 3.15.

Clearly, one has, for odd i,

$$\|\vec{v}_{i+1}\| = \sqrt{2}\|\vec{v}_i\| = \sqrt{2}\frac{\lambda}{\tau} \tag{3.223}$$

where, as usual, λ and τ are the lattice spacing and the time step, respectively.

Since the square lattice results from the projection of the 24-particle FCHC model, all velocities do not have the same weight. Actually, four particles give the same projection along the main lattice directions. This fact can be taken into account by assigning a mass $m_i = 4$ to the horizontal

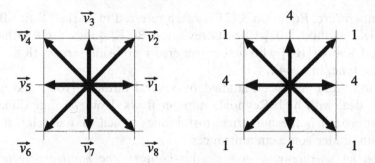

Fig. 3.15. The eight velocities in the lattice Boltzmann model of a two-dimensional fluid (on the left) and the mass associated to each of these directions (on the right).

and vertical directions and a mass $m_i = 1$ for the diagonal motions, as illustrated in figure 3.15.

The macroscopic quantities, such as the local density ρ or the velocity flow $\vec{u}$ will be defined as

$$\rho = \sum_{i=1}^{8} m_i f_i \qquad \rho\vec{u} = \sum_{i=1}^{8} m_i f_i \vec{v}_i \qquad (3.224)$$

Isotropy. From the derivation we carried out for the FHP model, we learned that isotropy is satisfied because the right-hand sides of equations 3.117 and 3.154 do not depend on a particular choice of the coordinate system. In our case, since all velocity vectors do not have the same modulus and all particles the same mass, the isotropy condition will take the form

$$\sum_{i=1}^{8} m_i v_{i\alpha} v_{i\beta} = \mu \delta_{\alpha\beta} \qquad (3.225)$$

and

$$\sum_{i=1}^{8} m_i v_{i\alpha} v_{i\beta} v_{i\gamma} v_{i\delta} = \nu(\delta_{\alpha\beta}\delta_{\gamma\delta} + \delta_{\alpha\gamma}\delta_{\beta\delta} + \delta_{\alpha\delta}\delta_{\beta\gamma}) \qquad (3.226)$$

where μ and ν can be obtained with the help of some algebra. First, since $\vec{v}_1$ and $\vec{v}_3$ are orthogonal, any vector $\vec{x}$ can be written as $\vec{x} = (v^{-2})[(\vec{x} \cdot \vec{v}_1)\vec{v}_1 + (\vec{x} \cdot \vec{v}_3)\vec{v}_3]$ where $v = (\lambda/\tau)$ is the modulus of $\vec{v}_i$, for i odd. Consequently one must have

$$v_{1\alpha}v_{1\beta} + v_{3\alpha}v_{3\beta} = v^2\delta_{\alpha\beta} \qquad (3.227)$$

and similarly for $\vec{v}_5$ and $\vec{v}_7$. This is also true for $\vec{v}_i$ and v_{i+2} when i is even, except that the coefficient v^2 is replaced by $2v^2$, because of equation 3.223.

Therefore, with $m_i = 4$ and $m_i = 1$ for i odd and even, respectively, we obtain

$$\sum_{i=1}^{8} m_i v_{i\alpha} v_{i\beta} = 12v^2 \delta_{\alpha\beta} \tag{3.228}$$

The fourth-order tensor $\sum_i m_i v_{i\alpha} v_{i\beta} v_{i\gamma} v_{i\delta}$ can be computed explicitly by the method described by equations 3.155 through 3.164. Since the summation can be split into the odd and even values of i, we first compute

$$T_{\alpha\beta\gamma\delta} = \sum_{i=1}^{4} c_{i\alpha} c_{i\beta} c_{i\gamma} c_{i\delta} \tag{3.229}$$

where, for this calculation, $\check{c}_1$, $\check{c}_2$, $\check{c}_3$ and $\check{c}_4$ are orthogonal unit vectors defined as

$$\check{c}_k = (\cos(k\theta + \phi), \sin(k\theta + \phi)) \tag{3.230}$$

with $\theta = \pi/2$ and $\phi \in [0, 2\pi]$. Thus, we have

$$\sum_{k=1}^{4} (c_{k1} + ic_{k2})^4 = 4\exp(4i\phi) \tag{3.231}$$

$$\sum_{k=1}^{4} (c_{k1} + ic_{k2})^3 (c_{k1} - ic_{k2}) = 0 \tag{3.232}$$

and

$$\sum_{k=1}^{4} (c_{k1} + ic_{k2})^2 (c_{k1} - ic_{k2})^2 = 4 \tag{3.233}$$

Therefore, we have

$$\sum_{k=1}^{4} \left[(c_{k1}^4 + c_{k2}^4 - 6c_{k1}^2 c_{k2}^2) \right] = 4\cos 4\phi \tag{3.234}$$

$$\sum_{k=1}^{4} \left[4(c_{k1}^3 c_{k2} - c_{k1} c_{k2}^3) \right] = 4\sin 4\phi \tag{3.235}$$

$$\sum_{k=1}^{4} \left[(c_{k1}^4 - c_{k2}^4) \right] = \sum_{k=1}^{4} \left[(c_{k1}^3 c_{k2} + c_{k1} c_{k2}^3) \right] = 0 \tag{3.236}$$

and

$$\sum_{k=1}^{4} \left[c_{k1}^4 + c_{k2}^4 + 2c_{k1}^2 c_{k2}^2 \right] = 4 \tag{3.237}$$

Solving this system yields

$$\sum_{k=1}^{4} c_{k1}^2 c_{k2}^2 = \frac{1}{2}(1 - \cos 4\phi) \tag{3.238}$$

$$\sum_{k=1}^{4} c_{k1}^3 c_{k2} = -\sum_{k=1}^{4} c_{k1} c_{k2}^3 = \frac{1}{2}\sin 4\phi \tag{3.239}$$

and, finally

$$\sum_{k=1}^{4} c_{k1}^4 = \sum_{k=1}^{4} c_{k2}^4 = \frac{3}{2} + \frac{1}{2}\cos 4\phi \tag{3.240}$$

This result shows that, as expected, the fourth-order tensor $T_{\alpha\beta\gamma\delta}$ is not isotropic on a two-dimensional lattice. However, here, we also want to consider the contribution of the other vectors, $\vec{v}_2$, $\vec{v}_4$, $\vec{v}_6$ and $\vec{v}_8$, which are rotated by $\pi/4$ with respect to the vectors, $\vec{c}_i$.

With $\phi' = \phi + \pi/4$ in the previous expression, we obtain $\cos 4\phi' = -\cos 4\phi$ and $\sin 4\phi' = -\sin 4\phi$. Therefore

$$\begin{aligned}
\sum_{k=1}^{8} m_k v_{k1}^2 v_{k2}^2 &= 4 \sum_{k \text{ odd}} v_{k1}^2 v_{k2}^2 + \sum_{k \text{ even}} v_{k1}^2 v_{k2}^2 \\
&= 4v^4 \frac{1}{2}(1 - \cos 4\phi) + 4v^4 \frac{1}{2}(1 + \cos 4\phi) \\
&= 4v^4
\end{aligned} \tag{3.241}$$

The two contributions come with a factor of $4v^4$ because, in the first term $m_k = 4$ and, in the second $\|\vec{v}_k\|^4 = (\sqrt{2})^4 v^4 = 4v^4$. In a similar way, we obtain

$$\sum_{k=1}^{8} m_k v_{k1}^3 v_{k2} = \sum_{k=1}^{8} m_k v_{k1} v_{k2}^3 = 0 \tag{3.242}$$

and

$$\sum_{k=1}^{8} m_k v_{k1}^4 = \sum_{k=1}^{8} m_k v_{k2}^4 = 12v^4 \tag{3.243}$$

All these results can be combined to give

$$\sum_{i=1}^{8} m_i v_{i\alpha} v_{i\beta} v_{i\gamma} v_{i\delta} = 4v^4(\delta_{\alpha\beta}\delta_{\gamma\delta} + \delta_{\alpha\gamma}\delta_{\beta\delta} + \delta_{\alpha\delta}\delta_{\beta\gamma}) \tag{3.244}$$

The local equilibrium. The lattice Boltzmann dynamics is given by equation 3.221

$$f_i(\vec{r} + \tau\vec{v}_i, t + \tau) - f_i(\vec{r}, t) = \Omega_i = \frac{1}{\xi}\left(f_i^{(0)}(\vec{r}, t) - f_i(\vec{r}, t)\right) \qquad (3.245)$$

We will find a solution to this equation using, as before, a Chapman–Enskog expansion

$$f_i = f_i^{(0)} + \epsilon f_i^{(1)} + \dots$$

The main difference is that, here, $f_i^{(0)}$ is given, or more precisely, chosen, in an appropriate way.

A natural choice for specifying $f_i^{(0)}$ as a function of the physical quantities ρ and $\rho\vec{u}$ is to adopt a similar expression to that obtained for the FHP model, namely equation 3.152. Accordingly, we define

$$f_i^{(0)} = a\rho + \frac{b}{v^2}\rho\vec{v}_i \cdot \vec{u} + \rho e\frac{u^2}{v^2} + \rho\frac{h}{v^4}v_{i\alpha}v_{i\beta}u_\alpha u_\beta \qquad (3.246)$$

where a, b, e and h are coefficients which will now be determined. There are several constraints to fulfill. First, mass and momentum conservation impose

$$\sum_{i=1}^{8} m_i\Omega_i = 0 \quad \text{and} \quad \sum_{i=1}^{8} m_i\vec{v}_i\Omega_i = 0 \qquad (3.247)$$

This implies that

$$\sum_{i=1}^{8} m_i f_i^{(0)} = \rho \quad \text{and} \quad \sum_{i=1}^{8} m_i\vec{v}_i f_i^{(0)} = \rho\vec{u} \qquad (3.248)$$

because ρ and $\rho\vec{u}$ are defined through relation 3.224. This choice is also compatible with the Chapman–Enskog expansion which requires that the local equilibrium solution completely defines the physical variables (see equation 3.62). A second condition we will use to determine a, b, e and h is the Galilean invariance. This is a freedom which is possible in the framework of the lattice Boltzmann method.

Using the fact that $\sum m_i = 20$, $\sum\vec{v}_i = 0$ and $\sum m_i v_{i\alpha}v_{i\beta} = 12v^2\delta_{\alpha\beta}$, we obtain

$$\sum_{i=1}^{8} m_i f_i^{(0)} = 20a\rho + (12h + 20e)\rho\frac{u^2}{v^2}$$

$$\sum_{i=1}^{8} m_i v_{i\alpha} f_i^{(0)} = 12b\rho u_\alpha \qquad (3.249)$$

Therefore, mass and momentum conservation requires that

$$a = 1/20, \qquad 12h + 20e = 0 \qquad \text{and} \qquad b = \frac{1}{12} \qquad (3.250)$$

We can similarly compute the momentum tensor at zeroth order. We have, by definition

$$\Pi_{\alpha\beta}^{(0)} = \sum_{i=1}^{8} m_i v_{i\alpha} v_{i\beta} f_i^{(0)}$$

and, after substitution of $f_i^{(0)}$, we find

$$\begin{aligned}
\Pi_{\alpha\beta}^{(0)} &= 12av^2 \rho \delta_{\alpha\beta} + \frac{\rho h}{v^4} u_\gamma u_\delta \sum_{i=1}^{8} m_i v_{i\alpha} v_{i\beta} v_{i\gamma} v_{i\delta} + 12e\rho u^2 \\
&= 12v^2 \left(a + (e + \frac{h}{3}) \frac{u^2}{v^2} \right) \rho \delta_{\alpha\beta} + 8h\rho u_\alpha u_\beta \qquad (3.251)
\end{aligned}$$

In order to satisfy Galilean invariance, the coefficient in front of the quadratic term $\rho u_\alpha u_\beta$ must be 1. Accordingly, we must have

$$h = \frac{1}{8}$$

Since $12h + 20e = 0$, we obtain

$$e = -\frac{3}{40}$$

and the expression for $\Pi_{\alpha\beta}^{(0)}$ reads

$$\Pi_{\alpha\beta}^{(0)} = p \delta_{\alpha\beta} + \rho u_\alpha u_\beta \qquad (3.252)$$

where the pressure is defined as

$$p = 12v^2 \left(a + (e + \frac{h}{4}) \frac{u^2}{v^2} \right) \rho = \frac{3}{5} v^2 \left(1 - \frac{2}{3} \frac{u^2}{v^2} \right) \rho \qquad (3.253)$$

Using the fact that $\Pi_{\alpha\beta}^{(0)}$ can be written as (see 3.176)

$$\Pi_{\alpha\beta}^{(0)} = c_s^2 \rho \delta_{\alpha\beta} + O(u^2)$$

we conclude that

$$c_s^2 = 12av^2 = \frac{3}{5} v^2$$

We notice that a contribution proportional to u^2 is present in the expression of p. This is an artifact of the model which has no physical meaning. This contribution can be removed when rest particles are introduced in the system [101].

The Navier–Stokes equation. In section 3.2.3, we obtained the following result (see equation 3.87)

$$\partial_t \rho u_\alpha + \partial_\beta \left[\Pi_{\alpha\beta} + \frac{\tau}{2} \left(\epsilon \partial_{t_1} \Pi^{(0)}_{\alpha\beta} + \partial_\gamma S^{(0)}_{\alpha\beta\gamma} \right) \right] = 0 \qquad (3.254)$$

where $t = \frac{t_1}{\epsilon} + \frac{t_2}{\epsilon^2}$ and $\vec{r} = \vec{r}_1/\epsilon$ takes into account the different time scales of the problem (see section 3.2.3).

The derivation of 3.254 relies only on the fact that $\sum m_i \Omega_i = 0$ and $\sum m_i \Omega_i \vec{v}_i = 0$ (in section 3.2.3, $m_i = 1$, for all i). Therefore, this equation is still valid here. As we previously mentioned, the generality of this result is due to the conservation laws which define the form of the equation. The precise expression of the collision term, on the other hand, determines the transport coefficient.

We obtained $\Pi^{(0)}$ in the previous section. We still need to compute $S^{(0)}_{\alpha\beta\gamma}$ and $\Pi^{(1)}$. The quantity $f^{(1)}$ is defined by a similar equation to that obtained in relation 3.94, namely

$$\frac{1}{\tau} \sum_{j=1}^{8} \left(\frac{\partial \Omega_i(f^{(0)})}{\partial f_j} \right) f_j^{(1)} = \partial_{t_1} f_i^{(0)} + \partial_{1\alpha} v_{i\alpha} f_i^{(0)}$$

where $\partial_{1\alpha}$ denotes the derivative with respect to the α-component of $\vec{r}_1$. Since $\Omega_i = \frac{1}{\xi} \left(f_i^{(0)}(\vec{r}, t) - f_i(\vec{r}, t) \right)$, the above equation simply reads

$$-\frac{1}{\tau\xi} f_i^{(1)} = \partial_{t_1} f_i^{(0)} + \partial_{1\alpha} v_{i\alpha} f_i^{(0)}$$

Following the derivation of equation 3.186, we have

$$-\frac{1}{\tau\xi} f_i^{(1)} = \partial_{t_1} f_i^{(0)} + \partial_{1\alpha} v_{i\alpha} f_i^{(0)}$$

$$= -\frac{\partial f_i^{(0)}}{\partial \rho} \operatorname{div}_1 \rho \vec{u} - \frac{\partial f_i^{(0)}}{\partial \rho u_\alpha} \partial_{1\beta} \Pi^{(0)}_{\alpha\beta} + \partial_{1\alpha} v_{i\alpha} f_i^{(0)} \qquad (3.255)$$

with $\operatorname{div}_1 = \sum_\alpha \partial_{1\alpha}$.

As earlier, we shall compute $f_i^{(1)}$ to the first order in $\vec{u}$. We have

$$f_i^{(0)} = a\rho + \frac{b}{v^2} \rho v_{i\alpha} u_\alpha + O(u^2)$$

and

$$\Pi^{(0)}_{\alpha\beta} = 12 a v^2 \rho \delta_{\alpha\beta}$$

Thus,

$$\frac{\partial f_i^{(0)}}{\partial \rho} = a \qquad \frac{\partial f_i^{(0)}}{\partial \rho u_\alpha} = \frac{b}{v^2} v_{i\alpha}$$

and

$$-\frac{1}{\tau\xi}f_i^{(1)} = -a\mathrm{div}_1\rho\vec{u} - 12ab\partial_{1\alpha}v_{i\alpha}\rho + \partial_{1\alpha}v_{i\alpha}\left(a\rho + \frac{b}{v^2}\rho v_{i\beta}u_\beta\right)$$

$$= -a\mathrm{div}_1\rho\vec{u} + \frac{b}{v^2}v_{i\alpha}v_{i\beta}\partial_{1\alpha}\rho u_\beta$$

$$= b\left(\frac{v_{i\alpha}v_{i\beta}}{v^2} - \frac{a}{b}\delta_{\alpha\beta}\right)\partial_{1\beta}\rho u_\alpha \tag{3.256}$$

Thus

$$f_i^{(1)} = -\tau\xi\frac{b}{v^2}\left(v_{i\gamma}v_{i\delta} - \frac{av^2}{b}\delta_{\gamma\delta}\right)\partial_{1\gamma}\rho u_\delta \tag{3.257}$$

Using $\epsilon\partial_{1\gamma} = \partial_\gamma$, the order $O(\epsilon)$ contribution to Π reads

$$\epsilon\Pi_{\alpha\beta}^{(1)} = \epsilon\sum_{i=1}^{8}m_i f_i^{(1)}v_{i\alpha}v_{i\beta}$$

$$= \tau\xi\left[a\mathrm{div}\rho\vec{u}\sum m_i v_{i\alpha}v_{i\beta} - \frac{b}{v^2}\sum m_i v_{i\alpha}v_{i\beta}v_{i\gamma}v_{i\delta}\partial_\gamma\rho u_\delta\right]$$

$$= \tau v^2\xi\left[(12a - 4b)\delta_{\alpha\beta}\mathrm{div}\rho\vec{u} - 4b(\partial_\beta\rho u_\alpha + \partial_\alpha\rho u_\beta)\right] \tag{3.258}$$

where we have used expressions 3.228 and 3.244. We then obtain

$$\partial_\beta\epsilon\Pi_{\alpha\beta}^{(1)} = -\tau v^2\xi\left[(8b - 12a)\partial_\alpha\mathrm{div}\rho\vec{u} + 4b\partial_\beta^2\rho u_\alpha\right] \tag{3.259}$$

From this expression, we obtain two viscosity coefficients (shear and bulk viscosity), as is usual in compressible fluids.

Our final step is the calculation of the lattice viscosity. The first term in 3.254 giving a contribution to the lattice viscosity is $\partial_\beta(\tau\epsilon/2)\partial_{t_1}\Pi_{\alpha\beta}^{(0)}$. With $\Pi_{\alpha\beta}^{(0)} = 12av^2\rho\delta_{\alpha\beta} + O(u^2)$, we have

$$\frac{\tau\epsilon}{2}\partial_{t_1}\Pi_{\alpha\beta}^{(0)} = 6\tau v^2 a\epsilon\partial_{t_1}\rho\delta_{\alpha\beta} = -6\tau v^2 a\delta_{\alpha\beta}\mathrm{div}\rho\vec{u}$$

where we have used $\partial_{t_1}\rho + \mathrm{div}_1\rho\vec{u} = 0$ (see equation 3.77) and the definition of the length scale $\epsilon\mathrm{div}_1 = \mathrm{div}$.

Therefore, we have

$$\frac{\tau\epsilon}{2}\partial_\beta\partial_{t_1}\Pi_{\alpha\beta}^{(0)} = -6\tau v^2 a\partial_\alpha\mathrm{div}\rho\vec{u} \tag{3.260}$$

Similarly, we must compute the contribution due to $S_{\alpha\beta\gamma}^{(0)}$ in equation 3.254

$$S_{\alpha\beta\gamma}^{(0)} = \sum_{i=1}^{8}m_i v_{i\alpha}v_{i\beta}v_{i\gamma}f_i^{(0)}$$

$$= \frac{b}{v^2}\rho u_\delta\sum_{i=1}^{8}m_i v_{i\alpha}v_{i\beta}v_{i\gamma}v_{i\delta}$$

$$= \frac{v^2}{3} \rho u_\delta (\delta_{\alpha\beta} \delta_{\gamma\delta} + \delta_{\alpha\gamma} \delta_{\beta\delta} + \delta_{\alpha\delta} \delta_{\beta\gamma})$$

$$= \frac{v^2}{3} \rho (u_\gamma \delta_{\alpha\beta} + u_\beta \delta_{\alpha\gamma} + u_\alpha \delta_{\beta\gamma}) \qquad (3.261)$$

Consequently, we obtain

$$\frac{\tau}{2} \partial_\beta \partial_\gamma \left(S^{(0)}_{\alpha\beta\gamma} \right) = \frac{\tau v^2}{3} \left(\partial_\alpha \mathrm{div} \rho \tilde{u} + \frac{1}{2} \nabla^2 \rho u_\alpha \right)$$

The dissipative lattice contributions to the lattice Boltzmann dynamics is obtained by combining the two terms we have obtained

$$\frac{\tau}{2} \partial_\beta \left(\epsilon \partial_{t_1} \Pi^{(0)}_{\alpha\beta} + \partial_\gamma S^{(0)}_{\alpha\beta\gamma} \right) = \tau v^2 \left[\frac{1}{6} \nabla^2 (\rho u_\alpha) + \left(\frac{1}{3} - 6a \right) \partial_\alpha \mathrm{div} \rho \tilde{u} \right] \qquad (3.262)$$

Finally, after substituting 3.262, 3.259 and 3.252 into the balance equation for momentum 3.254, we obtain

$$\partial_t \rho u_\alpha + \partial_\beta \left[p \delta_{\alpha\beta} + \rho u_\alpha u_\beta \right] = \tau v^2 \xi \left[(8b - 12a) \partial_\alpha \mathrm{div} \rho \tilde{u} + 4b \nabla^2 \rho u_\alpha \right]$$

$$-\tau v^2 \left[\frac{1}{6} \nabla^2 (\rho u_\alpha) + \left(\frac{1}{3} - 6a \right) \partial_\alpha \mathrm{div} \rho \tilde{u} \right]$$

where p is the scalar pressure given by 3.253. After rearranging the terms, we have

$$\partial_t \rho u_\alpha + \rho u_\beta \partial_\beta u_\alpha + u_\alpha \mathrm{div} \rho \tilde{u} = -\partial_\alpha p + \tau v^2 \left(\frac{\xi}{3} - \frac{1}{6} \right) \nabla^2 \rho u_\alpha$$

$$+\tau v^2 \left[\xi \left(\frac{2}{3} - 12a \right) - \left(\frac{1}{3} - 6a \right) \right] \partial_\alpha \mathrm{div} \rho \tilde{u} \qquad (3.263)$$

In the case of an incompressible fluid (at low Mach number, for instance) one has $\mathrm{div} \rho \tilde{u} = 0$ and we recover the usual Navier–Stokes equation

$$\partial_t \tilde{u} + (\tilde{u} \cdot \nabla) \tilde{u} = -\frac{1}{\rho} \nabla p + v_{\mathrm{lb}} \nabla^2 \tilde{u} \qquad (3.264)$$

where v_{lb} is the kinematic viscosity of our lattice Boltzmann fluid

$$v_{\mathrm{lb}} = \frac{\tau v^2}{3} \left(\xi - \frac{1}{2} \right) \qquad (3.265)$$

From this result, we observe that the lattice Boltzmann viscosity is independent of the particle density ρ. In addition, ξ is a free parameter which can be tuned to adjust the viscosity within some range. We can see that when ξ is small, relaxation to $f^{(0)}$ is fast and viscosity small. However, ξ cannot be made arbitrarily small since $\xi < 1/2$ would imply a negative viscosity. Practically, more restrictions are expected, because the dissipation length scale should be much larger than the lattice spacing.

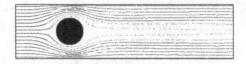

Fig. 3.16.   Laminar flow past a disk obtained from the 2-D lattice Boltzmann model. System size is 512×128, $\xi = 10$ and $u_\infty = 0.1$.

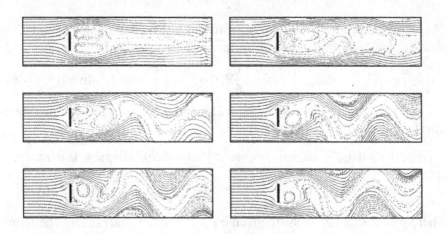

Fig. 3.17.   Non-stationary flow past a plate obtained from the 2-D lattice Boltzmann model. System size is 512×128, $\xi = 1$. and $u_\infty = 0.025$. From left to right and top to bottom, the figure shows the different stages of evolution.

A more detailed discussion of the lattice Boltzmann approach to fluid dynamics is given by Benzi, Succi and Vergassola [96].

3.5.3   *Lattice Boltzmann flows*

Finally, this section presents the results of simulations with the LBGK model discussed in the previous section. Laminar flows, such as the one presented in figure 3.16, are easy to obtain. On the other hand, a Karman vortex street requires some care to set up the boundary conditions. First, the initial condition $f_i(\vec{r}, t = 0)$ should be random, in order to break the symmetry of the numerical scheme. If not, the pattern shown in figure 3.17 cånnot develop (everything being symmetric with respect to horizontal axis). Second, boundary conditions at the upper and lower limits of the system are periodic. Other solutions can be considered, though, and a condition of zero vertical velocity flow gave good oscillation patterns, too. On the other hand, the flow is more sensitive to the boundary conditions imposed at the left and right extremities.

A simple, yet effective, solution is the following: periodic boundary conditions are first imposed along the horizontal direction. The fluid exiting on the right is re-injected on the left. Then, the driven force is applied on the left to ensure the fluid motion and simulate an unperturbed flow at velocity u_∞ coming from the left. This is obtained by re-distributing equally all the particles at the system input and systematically removing some of them from direction $\vec{c}_5$ to put them in direction $\vec{c}_1$.

3.6 Problems

3.1. Consider an HPP gas at equilibrium with M particles on V lattice sites. Assume that all possible configurations of M particles on V sites have the same probability p (equilibrium). Show that $(1/p) = (4V!)/(4V - M)!M!$. Deduce that the average occupation number $\langle n_i(\vec{r}) \rangle$ for any site $\vec{r}$ and direction i is $\rho/4$. Show that $\langle n_i(\vec{r})n_j(\vec{r}\,') \rangle$ is zero for an infinite system, unless $i = j$ and $\vec{r} = \vec{r}\,'$. This shows that the Boltzmann factorization hypothesis is obeyed, at least at equilibrum. Compute $\langle n_i(\vec{r})n_j(\vec{r}\,') \rangle$ for $i = j$ and $\vec{r} = \vec{r}\,'$.

3.2. Consider a three-speed HPP model at equilibrium on a two-dimensional square lattice. There are three populations of particles: speed 0, speed 1 and speed $\sqrt{2}$ particles. Suppose each of these populations has a total of N_{rest}, N_{slow} and N_{fast} particles, respectively. From the principle of equipartition of energy we may assume that each particle contributes of amount $(dk_B T)/2$ to the total kinetic energy E, where d is the spatial dimension, T the temperature and k_B the Boltzmann constant. Show how this temperature scale depends on N_{rest}, N_{slow} and N_{fast}. What is a natural choice for k_B in terms of the kinetics parameter of the model?

3.3. Consider another temperature scale for the previous problem, based on the microcanonical definition: $(k_B T)^{-1} = (\partial \ln W / \partial E)_V$ where W is the number of possible configurations with N_{rest}, N_{slow} and N_{fast} particles, E the kinetic energy and V the lattice size which is kept constant. Compare this temperature scale with that of the previous problem. Which one seems the most appropriate?

3.4. Write down the microdynamics for the HPP gas.

3.5. Derive the differential equation governing the FHP dynamics with some simplified assumptions. From the lattice Boltzmann equation

$$N_i(\vec{r} + \lambda\vec{c}_i, t + \tau) - N_i(\vec{r}, t) = \Omega_i(N)$$

where Ω is the FHP collision term, take the Taylor expansion of the left-hand side only to *first* order in λ and τ. Look for a solution of the form $N_i = N_i^{(0)} + \tau N_i^{(1)} + ...$ using the time step τ as the small expansion parameter. Follow the Chapman–Enskog formalism, neglecting the multiscale aspect of the dynamics. Show that the consequence of dropping higher orders in τ and λ, as well as the multiscale part of the development, only modifies the viscosity by a constant (the lattice viscosity).

3.6. Apply the same approximations as in the previous problem to the HPP microdynamics. Show that anisotropy shows up in two places: in front of the nonlinear velocity term $(\vec{u}\nabla)\vec{u}$ and in the viscosity term.

3.7. Write the collision term of the HPP rule in the form of a 16×16 collision matrix. Write the most general form of this matrix which is compatible with mass and momentum conservation. Check whether semi-detailed balance is obeyed.

3.8. Add a gravity field on a HPP gas and show that the *equilibrium* density $\rho(y)$ of particles at a height y follows the barometric law in the limit of low density. Hint: gravity can be added by writing the lattice Boltzmann dynamics as

$$N_i(\vec{r} + \lambda \vec{c}_i, t + \tau) - N_i(\vec{r}, t) = (1 - g)\Omega_i(N) + gF_i(N)$$

where Ω is the HPP collision term, g the probability that gravity operates at site $\vec{r}$ and F_i is defined as

$$F_1 = F_3 = 0 \qquad F_2 = -F_4 = -N_2(1 - N_4)$$

which means that gravity acts vertically and may transform a particle moving up (N_2) into a particle going down (N_4). Assume that $N_i = N_i(y) = \rho(y)/4$ and show that this is a solution of the lattice Boltzmann equation provided that

$$\rho(y) = (1 - g)^{y/\lambda}\rho(0)$$

(Note: in equilibrium the time dependence can be omitted and in the limit of low density, one can neglect the ρ^2 terms.) How should the probability g vary with the lattice spacing λ when taking the continuous limit $\lambda \to 0$?

3.9. Propose an LBGK model for a one-dimensional flow, with or without rest particles. Derive the associated hydrodynamic equation. Express the local equilibrium function directly in terms of the density of particle f_i and discuss the numerical stability of the equation. Can the population densities become negative? What happens if the gradient are large?

3.10. For the previous problem of a 1D flow, can you choose the local equilibrium distribution so as to model the Burger equation: $\partial_t u + u \partial_x u = v \partial_x^2 u$?

3.11. In the LBGK framework, discuss how a non-Newtonian fluid could be modeled.

4

Diffusion phenomena

4.1 Introduction

Diffusive phenomena play an important role in many areas of physics, chemistry and biology and still constitute an active field of research. There are many applications involving diffusion for which a particle based model, such as a lattice gas dynamics, could provide a useful approach and efficient numerical simulations.

For instance, processes such as aggregation, formation of a diffusion front, trapping of particles performing a random walk in some specific region of space [106,107], or the adsorption of diffusing particles on a substrate are important problems that are difficult to solve with the standard diffusion equation. A microscopic model, based on a cellular automata dynamics, is therefore of clear interest.

Diffusion is also a fundamental ingredient in reaction-diffusion phenomena that will be discussed in detail in the next chapter. Reaction processes such as $A + B \rightarrow C$, as well as growth mechanisms are naturally implemented in the framework of a point particle description, often with the help of threshold rules. Consequently, microscopic fluctuations are often relevant at a macroscopic level of observation because they make symmetry breaking possible and are responsible for triggering complex patterns.

Cellular automata particles can be equipped with diffusive and reactive properties, in order to mimic real experiments and model several complex reaction-diffusion-growth processes in the same spirit as a cellular automata fluid simulates fluid flow: these systems are expected to retain the relevant aspects of the microscopic world they are modeling. However, diffusion is much simpler than the motion of a fluid and this will be reflected in the cellular automata rule.

At the beginning of chapter 3, we proposed a one-dimensional cellular automata model of diffusion. Here, we shall generalize it to two- and three-dimensional systems and consider the presence of various types of complex processes. The purpose of this chapter is thus to investigate the cellular automata approach to modeling diffusion and growth phenomena. We shall discuss their reliability (the discrete artifacts of the models must not show up) and their applicability to various problems. We shall first spend some time on the pure diffusion rule and, in the next chapter, add reaction mechanisms.

4.2 The diffusion model

Phenomenologically, diffusion follows from the mass conservation law, i.e. the continuity equation

$$\partial_t \rho + \mathrm{div}\vec{J} = 0 \tag{4.1}$$

where $\vec{J}$, the particle current is given in terms of the particle density ρ as

$$\vec{J} = -D\mathrm{grad}\rho \tag{4.2}$$

The quantity D is the diffusion constant. The last relation expresses that the particle current is proportional to the spatial inhomogeneities of the density.

When D is independent of ρ and independent of the spatial coordinates, we have a linear and homogeneous diffusion. By substitution of (4.2) into (4.1), one obtains the famous Fick's law

$$\partial_t \rho = D\nabla^2 \rho \tag{4.3}$$

At a microscopic level, on the other hand, a diffusive phenomenon corresponds to the random walk of many particles. Particle number is the only quantity conserved by this dynamics.

This random motion is typically due to the properties of the environment the particles are moving in. When one is not interested in an explicit description of this environment, it can be considered as a source of thermal noise and its effective action on the particles can be assumed to be stochastic.

It is easy to imagine a simple mathematical model of a random walk on a lattice. The first idea is to assume that each particle will pick at random one neighbor lattice site and jump to it. This is the approach considered in many numerical models in which only one particle moves at a time. However, in the framework of a cellular automata dynamics this may cause some problems. All the particles move synchronously and several of them may choose to hop to the same lattice site. For instance, on a two-dimensional square lattice, suppose that site $\vec{r}$ is empty. Then, it may happen that all neighboring sites (north, west, south and west) are occupied by a particle and that all of them select precisely site $\vec{r}$ as their new destination.

In this example, if one allows four particles to sit at the same lattice site then, at the next random motion, they will all be able to jump to a neighboring site. There is some probability that two of them move to the same neighbor. But this neighbor could very well receive other particles from elsewhere and, finally, receive more than four particles.

Therefore, simultaneous motion may lead to an arbitrarily large number of particles in some lattice sites. This is in conflict with the exclusion principle which requires that the state of each site be coded with a small and finite number of bits.

There is a simple solution to this problem, as already sketched in section 3.1. We shall consider the same approach as used in the HPP or FHP dynamics: at any site, there may be as many particles as there are lattice links. Each particle has a velocity, defined by its direction of motion. At each time step, the particles move to a nearest neighbor site, according to its velocity. By reaching the site, the particles experience a "collision" during which they change randomly their direction of motion.

In other terms, each particle selects, at random, a new velocity among the possible values permitted by the lattice. But, as several particles may enter the same site (up to four, on a two-dimensional square lattice), the random change of directions should be such that there are never two or more particles exiting a site with the same travel direction. This would otherwise violate again the exclusion principle. In order to avoid such a situation, it is sufficient to shuffle the directions of motion or, more precisely, to perform a random permutation of the velocity vectors, independently at each lattice site and each time step.

This evolution rule requires random numbers and corresponds to a *probabilistic* cellular automaton.

4.2.1 *Microdynamics of the diffusion process*

In this section, we shall write down the microdynamics related to the diffusion rule we have just presented. Following the same formalism as developed in chapter 3, we shall derive its macroscopic behavior and show that Fick's law of diffusion is obtained in the limit of an infinite lattice and a time step going to zero. Corrections due to the finite size of the lattice spacing and the time step will be discussed in section 4.3.

Our model consists of particles moving along the main directions of a hypercubic lattice (a square lattice in two dimensions or a cubic lattice in three dimensions). As opposed to cellular automata fluids, we do not have to consider here more complicated lattices. The reason is that diffusion processes do not require a fourth-order tensor for their description.

The random motion is obtained by permuting the direction of the incoming particles. If d is the space dimension, there are $2d$ lattice directions. These $2d$ directions of motion can be shuffled in $2d!$ ways, which is the number of permutations of $2d$ objects. However, it is not necessary to consider all permutations. A subset of them is enough to produce the desired random motion. For the sake of simplicity, we restrict ourselves to cyclic permutations. Thus, at each time step, the directions of the lattice are "rotated" by an angle α_i chosen at random, with probability p_i, independently for each site of the lattice. With this mechanism, the direction a particle will take and exit a given site depends on the direction it had when entering the site. The modification of its velocity determines its next location on the lattice. The particle behaves as if it has some inertia, as in a model proposed by G.I. Taylor [108] in 1921.

By labeling the lattice directions with the unit vectors $\vec{c}_i$ we can introduce the occupation numbers $n_i(\vec{r})$ defined as the number of particles entering the site $\vec{r}$, at time t with a velocity pointing in direction $\vec{c}_i$.

With this notation, the CA rule governing the dynamics of our model reads

$$n_i(\vec{r} + \lambda \vec{c}_i, t + \tau) = \sum_{\ell=0}^{2d-1} \mu_\ell(\vec{r}, t) n_{i+\ell}(\vec{r}, t) \qquad (4.4)$$

where i is wrapped onto $\{1, 2, ..., 2d\}$. The $\mu_\ell \in \{0, 1\}$ are Boolean variables which select only one of the $2d$ terms in the right-hand side. Therefore they must obey the condition

$$\sum_{\ell=0}^{2d-1} \mu_\ell = 1 \qquad (4.5)$$

Practically, this condition can be enforced in a simulation by dividing the interval [0,1] into $2d$ bins of length p_ℓ, each assigned to one of the μ_ℓ. Then, at each lattice site and each time step, a real random number between 0 and 1 is computed (with a random number generator). The bin it falls in will determine which μ_ℓ is the one that will be non-zero.

In order to be more specific, we shall now consider the two-dimensional case. This will correspond to many of the situations studied later in this chapter.

On a two-dimensional square lattice, there are four possible random rotations, as shown in figure 4.1 and equation 4.4 reads

$$n_i(\vec{r} + \lambda \vec{c}_i, t + \tau) = \mu_0 n_i + \mu_1 n_{i+1} + \mu_2 n_{i+2} + \mu_3 n_{i+3} \qquad (4.6)$$

where the values of i are wrapped into $\{1,2,3,4\}$ and the vectors $\vec{c}_i$ are defined in figure 4.2. Only one of the μ_ℓ is equal to 1, at a given site and time step. When $\mu_0 = 1$, the particles are not deflected and, locally,

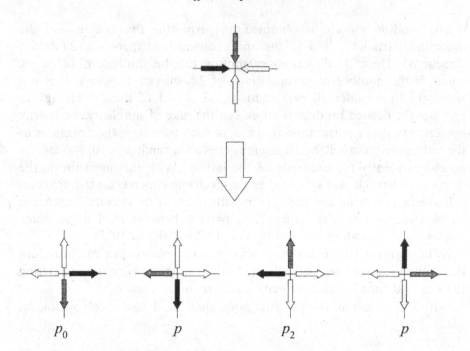

Fig. 4.1. How the entering particles are deflected at a typical site, as a result of the diffusion rule. The four possible outcomes occur with respective probabilities p_0, p_1, p_2 and p_3. The figure shows four particles, but the mechanism is data-blind and any one of the arrows can be removed when fewer entering particles are present.

there is a ballistic motion. On the other hand, when $\mu_2 = 1$ the particles bounce back to where they came from (rotation by 180° degrees). The other two cases ($\mu_1 = 1$ or $\mu_3 = 1$) give clockwise and counterclockwise 90° rotations. We also observed that if $\mu_1 = \mu_3 = 0$ everywhere on the lattice, equation 4.6 reduces to two one-dimensional random walks along the horizontal and vertical directions.

The μ_ℓ are 1 with probability p_ℓ, otherwise they are zero. Consequently their average value is

$$\langle \mu_\ell \rangle = p_\ell \tag{4.7}$$

For symmetry reasons, it is natural to impose the condition that

$$p_1 = p_3 = p \tag{4.8}$$

because the particle is equally likely to be deflected clockwise as counterclockwise. Thus, the probabilities p_ℓ satisfy

$$p_0 + 2p + p_2 = 1 \tag{4.9}$$

It is interesting to notice that a model similar to this one has been

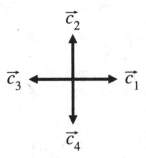

Fig. 4.2. The four lattice directions.

considered by Hauge [109], for the case of continuous space and time.

We shall now derive the macroscopic limit of equation 4.6 following the same procedure as developed in section 3.1.

The operation of random rotation is spatially uniform and data-blind, i.e. it is produced by an external mechanism which is the same for all sites and which is statistically independent of the actual occupation numbers n_i of particles at a given site. Consequently,

$$\langle \mu_\ell n_{i+\ell} \rangle = p_\ell \langle n_{i+\ell} \rangle = p_\ell N_{i+\ell} \tag{4.10}$$

and equation 4.6 simply becomes

$$N_i(\vec{r} + \lambda \vec{c}_i, t + \tau) - N_i(\vec{r}, t) = (p_0 - 1)N_i + p_1 N_{i+1} + p_2 N_{i+2} + p_3 N_{i+3} \tag{4.11}$$

This equation is the lattice Boltzmann equation associated with the diffusion rule. As opposed to the hydrodynamical situation discussed in the previous chapter, it is exact and has not required any factorization assumption. This is due to its linearity, of course.

We now write the left-hand side as a Taylor expansion

$$N_i(\vec{r} + \lambda \vec{c}_i, t + \tau) - N_i(\vec{r}, t) =$$
$$\left[\tau \partial_t + \frac{\tau^2}{2} \partial_t^2 + \lambda(\vec{c}_i \cdot \partial_{\vec{r}}) + \frac{\lambda^2}{2}(c_i \cdot \partial_{\vec{r}})^2 + \tau\lambda\partial_t(\vec{c}_i \cdot \partial_{\vec{r}}) \right] N_i \tag{4.12}$$

and we write, as usual,

$$N_i = N_i^{(0)} + \epsilon N_i^{(1)} + \dots \tag{4.13}$$

With the condition (see equation 3.26) that

$$\lambda \sim \epsilon\lambda \qquad \text{and} \qquad \tau \sim \epsilon^2 \tau$$

the order $O(\epsilon^0)$ of equation 4.11 reads

$$O(\epsilon^0): \qquad \Omega N^{(0)} = 0$$

where

$$\Omega = \begin{pmatrix} p_0 - 1 & p_1 & p_2 & p_3 \\ p_3 & p_0 - 1 & p_1 & p_2 \\ p_2 & p_3 & p_0 - 1 & p_1 \\ p_1 & p_2 & p_3 & p_0 - 1 \end{pmatrix} \tag{4.14}$$

and

$$N^{(0)} = \begin{pmatrix} N_1^{(0)} \\ N_2^{(0)} \\ N_3^{(0)} \\ N_4^{(0)} \end{pmatrix} \tag{4.15}$$

Since $\sum p_\ell = 1$, $N^{(0)} \propto E_0$ is clearly a solution, where $E_0 = (1,1,1,1)$ is the collisional invariant which expresses particle conservation. Except for pathological values of the p_ℓ this is the only solution. Using the Chapman–Enskog requirement

$$\rho = \sum_{i=1}^{4} N_i^{(0)} \tag{4.16}$$

we obtain that

$$N_i^{(0)} = \frac{\rho}{4} \tag{4.17}$$

The next order $O(\epsilon)$ in equation 4.11 is

$$O(\epsilon^1): \qquad \lambda(\vec{c}_i \cdot \partial_{\vec{r}}) N_i^{(0)} = \sum_j \Omega_{ij} N_j^{(1)} \tag{4.18}$$

In order to solve this equation, it is interesting to notice that

$$\sum_{\ell=0}^{3} p_\ell \vec{c}_{i+\ell} = (1 - 2p)\vec{c}_i \tag{4.19}$$

Indeed, we have $\vec{c}_i = -\vec{c}_{i+2}$ and

$$p_0 \vec{c}_i + p \vec{c}_{i+1} + p_2 \vec{c}_{i+2} + p \vec{c}_{i+3} = (p_0 - p_2)\vec{c}_i$$

Therefore, the vectors E_1 and E_2 defined as

$$E_\alpha = \begin{pmatrix} c_{1\alpha} \\ c_{2\alpha} \\ c_{3\alpha} \\ c_{4\alpha} \end{pmatrix} \tag{4.20}$$

are eigenvectors of the matrix Ω, with eigenvalue $p_0 - p_2 - 1 = -2(p + p_2)$.

In other terms,

$$\sum_{j=1}^{4} \Omega_{ij} c_{j\alpha} = -2(p + p_2)c_{i\alpha} \qquad (4.21)$$

With this observation it is easy to solve equation 4.18. We write

$$N_i^{(1)} = \phi c_{i\alpha} \partial_\alpha \rho$$

where we have used the summation convention over spatial coordinate index $\alpha = 1, 2$ and ϕ is a constant to be determined. By substitution into equation 4.18, we obtain

$$\begin{aligned}
\frac{\lambda}{4}(\vec{c}_i \cdot \partial_{\vec{r}})\rho &= \sum_j \Omega_{ij} N_j^{(1)} \\
&= -2(p + p_2)\phi(\vec{c}_i \cdot \partial_{\vec{r}})\rho
\end{aligned}$$

and

$$\phi = -\frac{\lambda}{8(p + p_2)} \qquad (4.22)$$

This yields the solution

$$N_i^{(1)} = -\frac{\lambda}{8(p + p_2)} c_{i\alpha} \partial_\alpha \rho \qquad (4.23)$$

Note that when $p + p_2 = 0$, this derivation is not possible. But, then, we are in the situation where $p_0 = 1$ which corresponds to a ballistic motion without any direction change.

The equation for ρ is then obtained in the usual way, by summing over i equation 4.11. Of course, $\sum \Omega_i = 0$ because particle number is conserved. Using the Taylor expansion 4.12, one has, at order $O(\epsilon)$

$$O(\epsilon): \quad \sum_{i=1}^{4} \left[\lambda(\vec{c}_i \cdot \partial_{\vec{r}})N_i^{(0)}\right] = 0 \qquad (4.24)$$

which is obviously satisfied since $N_i^{(0)} = \rho/4$ and $\sum \vec{c}_i = 0$. More interesting is the second order

$$O(\epsilon^2): \quad \sum_{i=1}^{4} \left[\tau\partial_t N_i^{(0)} + \lambda(\vec{c}_i \cdot \partial_{\vec{r}})N_i^{(1)} + \frac{\lambda^2}{2}(c_i \cdot \partial_{\vec{r}})^2 N_i^{(0)}\right] = 0 \qquad (4.25)$$

With 4.23, the equation reduces to

$$\partial_t \rho + \partial_\alpha \left[-\frac{\lambda^2}{\tau}\left(\frac{1}{8(p + p_2)} - \frac{1}{8}\right)\sum_{i=1}^{4} c_{i\alpha} c_{i\beta} \partial_\beta \rho\right] = 0 \qquad (4.26)$$

Since $\sum c_{i\alpha}c_{i\beta} = 2\delta_{\alpha\beta}$, as obtained in section 3.5, we have

$$\partial_t\rho + \operatorname{div}\left[-D\operatorname{grad}\rho\right] = 0 \tag{4.27}$$

where D is the diffusion constant

$$D = \frac{\lambda^2}{\tau}\left(\frac{1}{4(p+p_2)} - \frac{1}{4}\right) = \frac{\lambda^2}{\tau}\left(\frac{p+p_0}{4[1-(p+p_0)]}\right) \tag{4.28}$$

and $\vec{J} = -D\operatorname{grad}\rho$ is the particle current. It is interesting to observe that $\vec{J}$ does not coincide with the expression $\sum N_i\vec{v}_i$ which is the standard definition of the particle current. Indeed, we have

$$\sum_{i=1}^{4} N_i v_{i\beta} = -\frac{\lambda}{\tau}\sum_{i=1}^{4}\frac{\lambda}{8(p+p_2)}c_{i\beta}c_{i\alpha}\partial_\alpha\rho$$

$$= -\frac{\lambda^2}{\tau}\frac{1}{4(p+p_2)}\partial_\beta\rho$$

and thus we have

$$\sum_{i=1}^{4} N_i\vec{v}_i = -\left(D + \frac{\lambda^2}{4\tau}\right)\operatorname{grad}\rho \tag{4.29}$$

This discrepancy by a factor of $(\lambda^2/4\tau)$ can be interpreted as a renormalization of the diffusion constant due to the discreteness of the lattice (because it comes from the second-order terms in the Taylor expansion 4.12). It is analogous to the lattice viscosity obtained in section 3.2.3.

Relation 4.28 shows that the diffusion coefficient is adjustable in this model. It can be made very small when p_2 is close to 1 and very large when p_0 is close to 1. Practically speaking, however, not all choices of p_ℓ are as good as others in a finite system. We will return to this question in section 4.3.

In the case $p = 0$, the particles follow one-dimensional trajectories. It is important to notice that setting $p = 0$ into 4.28 does not give the correct diffusion coefficient derived in equation 3.38, namely

$$D_{1d} = \frac{\lambda^2}{\tau}\frac{p_0}{2(1-p_0)}$$

Instead, there is a factor of 4 in the denominator. The reason is that, when $p = 0$, there is another conserved quantity in addition to the particle number. Due to the one-dimensional motion, the number N_x of particles along the x-axis and the number N_y of particles along the y-axis are both constant. As a result $N_x - N_y$ is a conserved quantity. It is associated to the extra collisional invariant $E_4 = (1,-1,1,-1)$ which is the fourth eigenvector of Ω, with eigenvalue $-4p$. Therefore, if $p = 0$, the local equilibrium solution $N_i = \rho/4$ is not the most general solution (which is then $N_1 = N_3 = \rho_x/2$ and $N_2 = N_4 = \rho_y/2$).

4.2.2 The mean square displacement and the Green–Kubo formula

In the previous section, we obtained the value of the diffusion coefficient D by considering the continuous limit of the dynamics of a system of particles performing a synchronous random walk, with an exclusion principle. The same coefficient D (as found in 4.28) can also be obtained by considering only one particle and calculating its mean square displacement

$$\sigma^2(t) = \langle x^2(t) + y^2(t) \rangle$$

where

$$(x(t), y(t)) = \tau \sum_{n=0}^{t-1} \vec{v}(n)$$

and $\vec{v}(n)$ is the velocity the particle has at time step n (which is one of the four possible velocities $\vec{v}_i = (\lambda/\tau)\vec{c}_i$). Thus,

$$\sigma^2(t) = \langle \tau^2 \sum_{n=0}^{t-1} \sum_{m=0}^{t-1} \vec{v}(n) \cdot \vec{v}(m) \rangle$$

There is a well-known relation between the mean square displacement and the diffusion constant. When the probability of having a particle at a given site obeys a diffusion equation, it is easy to show the following identity, provided the probability and its gradient vanish on the boundaries

$$D = \frac{1}{2d} \lim_{t \to \infty} \frac{\sigma^2(t)}{t} \tag{4.30}$$

where d is the dimension of the space. This is Einstein's relation [110].

The value of σ^2 can be calculated in our model. According to the cellular automata rule, each particle performs a random walk, with probability p_0 of going straight, p_2 of bouncing back and p of being deflected by 90° degrees right or left.

For the purpose of the calculation, let us assume that the particles diffuse in the complex plane. Let us write

$$v(n) = \frac{\lambda}{\tau} e^{i\theta_n}$$

where θ_n is the angle formed by the velocity vector at iteration n. One has

$$\theta_n = \sum_{j=0}^{n} \Delta_j$$

where $\Delta_j \in \{0, \pm\pi/2, \pi\}$ is the random variable giving the rotation at time j. With these definitions and the notation $\bar{v}(m) = (\lambda/\tau)e^{-i\theta_m}$, we can write

$$\sigma^2 = \tau^2 \sum_{n,m=0}^{t-1} \langle v(n)\bar{v}(m)\rangle$$

$$= \lambda^2 \sum_{n<m}^{t-1} \langle e^{i(\theta_n-\theta_m)}\rangle + \lambda^2 \sum_{n=m}^{t-1} \langle e^{i(\theta_n-\theta_n)}\rangle + \lambda^2 \sum_{n>m}^{t-1} \langle e^{i(\theta_n-\theta_m)}\rangle$$

$$= \lambda^2 t + \lambda^2 \sum_{n>m}^{t-1} \left(\langle e^{i(\theta_n-\theta_m)}\rangle + \langle e^{-i(\theta_n-\theta_m)}\rangle \right)$$

Now, we have

$$\langle e^{i(\theta_n-\theta_m)}\rangle = \langle e^{i\sum_{j=m+1}^{n}\Delta_j}\rangle = \prod_{j=m+1}^{n} \langle e^{i\Delta_j}\rangle$$

because the Δ_j are independent random variables. Since Δ_j is 0, $\pi/2$, π or $-\pi/2$ with probability p_0, p, p_2 and p, respectively, we have

$$\langle e^{i\Delta_j}\rangle = p_0 + ip - p_2 - ip = p_0 - p_2 = \langle e^{-i\Delta_j}\rangle$$

Therefore,

$$\langle v(n)\bar{v}(m)\rangle = \frac{\lambda^2}{\tau^2}\langle e^{i(\theta_n-\theta_m)}\rangle = (p_0 - p_2)^{n-m}$$

and we obtain

$$\sigma^2(t) = \lambda^2 t + 2\lambda^2 \sum_{n>m}^{t-1}(p_0 - p_2)^{n-m}$$

$$= \lambda^2 t + 2\lambda^2 \sum_{k=1}^{t-1}\sum_{l=1}^{k}(p_0 - p_2)^l$$

$$= \lambda^2 t + 2\lambda^2(p_0 - p_2) \sum_{k=1}^{t-1}\left[\frac{(p_0 - p_2)^k - 1}{(p_0 - p_2) - 1}\right] \quad (4.31)$$

After summation over k, it is found that the mean square displacement is

$$\sigma^2(t) = \lambda^2 t + 2\lambda^2(t-1)\frac{p_0 - p_2}{1 + p_2 - p_0}$$

$$+ 2\lambda^2 \left(\frac{p_0 - p_2}{1 + p_2 - p_0}\right)^2 \left[(p_0 - p_2)^{t-1} - 1\right] \quad (4.32)$$

provided that the lattice is infinite or the time short enough so that the particle has not reached the limit of the system. This relation leads to a value of

$$D = \frac{1}{4} \lim_{t \to \infty} (\sigma^2/t\tau)$$

in perfect agreement with equation 4.28.

In statistical mechanics, it is common in the framework of the Green–Kubo formalism to express transport coefficients in terms of the integral of autocorrelation functions [87,109]. We are now going to explicitly derived the Green–Kubo relation for diffusion. We shall see that here, due to the lattice discreteness, a correction to the usual formulation is present [111].

Using equation 4.30 and the expression for σ^2 in terms of $v(n)$ and $v(m)$, we can write D as (note that $\langle v(n)\bar{v}(m)\rangle = \langle v(n-m)\bar{v}(0)\rangle$ and $\langle v(0)\bar{v}(0)\rangle = \langle 1 \rangle = 1$)

$$D = \frac{\tau}{2d} \lim_{N \to \infty} \frac{1}{N} \sum_{k=0}^{N-1} \langle v(0)\bar{v}(0)\rangle + \frac{\tau}{d} \lim_{N \to \infty} \frac{1}{N} \sum_{n>m}^{N-1} \langle v(n-m)\bar{v}(0)\rangle \qquad (4.33)$$

Since $\langle v(n-m)\bar{v}(0)\rangle \propto (p_0 - p_2)^{n-m}$, the summation over n and m can be easily performed and, after the limit $N \to \infty$ is taken, we get

$$\lim_{N \to \infty} \frac{1}{N} \sum_{n>m}^{N-1} \langle v(n-m)\bar{v}(0)\rangle = \lim_{N \to \infty} \frac{1}{N} \sum_{k=1}^{N-1} \sum_{l=1}^{k} \langle v(l)\bar{v}(0)\rangle$$

$$= \sum_{j=1}^{\infty} \langle v(j)\bar{v}(0)\rangle$$

$$= \sum_{j=0}^{\infty} \langle v(j)\bar{v}(0)\rangle - \langle v(0)\bar{v}(0)\rangle \qquad (4.34)$$

Therefore, the diffusion constant is given as a sum over the velocity autocorrelation

$$D = \frac{\tau}{d} \sum_{l=0}^{\infty} \langle v(l)\bar{v}(0)\rangle - \frac{\tau}{2d} \langle v(0)\bar{v}(0)\rangle \qquad (4.35)$$

Equation 4.35 is equivalent to a Green–Kubo relation. In our case, however, there is a correction $-(\tau/2d)\langle v(0)\bar{v}(0)\rangle$ due to the discreteness of the lattice. This term has the same origin as the difference between the particle flux $\vec{J}$ and $\rho\vec{u}$ we mentioned in section 4.2.1. Mathematically, it comes from the fact that the contributions $k = l$ in the double sum over k and l is not negligible in a discrete system.

4.2.3 The three-dimensional case

In a three-dimensional space, our cellular automata model of diffusion can be easily extended if the "rotations" that take place at each node of the lattice are properly chosen. On a cubic lattice, we have six possible directions of motion which can be mixed in several different ways, in order to produce a random walk. Among the set of all the possibilities, one of them gives equations very close to the two-dimensional case. For this reason, we shall not repeat the details of the calculation. We shall simply describe how to produce the random walk and write down the corresponding macroscopic equation.

Let us first label the three main directions of the lattice $\vec{c}_1$, $\vec{c}_2$ and $\vec{c}_3$, with the property that $\vec{c}_3 = \vec{c}_1 \times \vec{c}_2$. The three other directions are defined according to the relation

$$\vec{c}_i = -\vec{c}_{i+3}, \qquad i \text{ being wrapped onto } \{1, 2, 3, 4, 5, 6\}$$

Now, we consider six operations R_k, $k = 0, .., 5$, that occur with probability p_k. The result of the operation R_k is to deflect the particle entering a site with velocity $\vec{v}_i = (\lambda/\tau)\vec{c}_i$ into direction $\vec{c}_{i+k}$. The lattice Boltzmann equation then reads

$$N_i(\vec{r} + \lambda\vec{c}_i, t + \tau) = \sum_{k=0}^{5} p_k N_{i+k}(\vec{r}, t)$$

Symmetry requires that we choose

$$p_1 = p_2 = p_4 = p_5 = p$$

Since the probabilities sum up to 1, we have

$$p_0 + 4p + p_3 = 1$$

Following the same calculations as we performed in section 4.2.1, we obtain that the particle density $\rho = \sum N_i$ obeys a Fick's equation with a diffusion coefficient

$$D_{3D} = \frac{\lambda^2}{\tau} \frac{(1 + p_0 - p_3)}{6(1 - p_0 + p_3)} = \frac{\lambda^2}{\tau} \frac{p_0 + 2p}{6(1 - (p_0 + 2p))}$$

4.3   Finite systems

The results obtained in the previous section are exact in the limit of an infinite lattice with $\lambda \to 0$, $\tau \to 0$ and $\lambda^2/\tau \to$ const. In practice, however, one never performs numerical simulation in this limit and some corrections to Fick's equation are expected. In this section, our purpose is to solve the lattice Boltzmann equation for finite systems, with various boundary conditions. As a result, we will be able to determine the length and time scale for which a fair modeling of diffusion is obtained.

Since the cellular automata dynamics is linear, the lattice Boltzmann equation is exact and also linear. For this reason, exact solutions can be obtained and discrepancies with the continuous limit discussed rigorously.

4.3.1 The stationary source–sink problem

There is a simple situation in which our lattice Boltzmann diffusion equation can be solved exactly: this is the stationary source–sink problem which consists of two plates separated by a distance L. One plate continuously injects particles in the system while the other absorbs all particles reaching it.

If one waits long enough, a density profile establishes in the system and the average occupation number N_i no longer depends on time. For the steady state, the lattice Boltzmann equation 4.11 associated with the diffusion rule reads

$$N_1(x + \lambda, y) = p_0 N_1(x, y) + p N_2(x, y) + p_2 N_3(x, y) + p N_4(x, y)$$
$$N_2(x, y + \lambda) = p_0 N_2(x, y) + p N_3(x, y) + p_2 N_4(x, y) + p N_1(x, y) \cdot$$
$$N_3(x - \lambda, y) = p_0 N_3(x, y) + p N_4(x, y) + p_2 N_1(x, y) + p N_2(x, y)$$
$$N_4(x, y - \lambda) = p_0 N_4(x, y) + p N_3(x, y) + p_2 N_4(x, y) + p N_1(x, y) \qquad (4.36)$$

This system of equations has to be supplemented by boundary conditions. We shall assume that the source and the sink are the vertical planes $x = -1$ and $x = L + 1$, respectively. Consequently, we have

$$N_1(0, y) = 1 \qquad N_3(L, y) = 0$$

For symmetry reasons, we shall look for a solution which does not depend on the y variable and such that $N_2 = N_4$.

In the case of a continuous diffusion equation, it is straightforward to check that the solution of $\nabla^2 \rho = 0$ is a linear expression in x. Following this indication we try the solution

$$N_1(x, y) = ax + 1 \qquad N_3 = a(x - L) \qquad N_2 = N_4 = ax + b$$

for $x \in 0, \lambda, 2\lambda, \ldots, L$. After substitution, we are left with the conditions

$$a(p_2 L + \lambda) - 2pb = p_0 - 1$$
$$apL + 2pb = p$$
$$a(p_0 L - \lambda - L) - 2pb = p_2 \qquad (4.37)$$

Among these equations, only two are independent since adding the first to the third gives the second. We then obtain

$$a = -\frac{1}{L} \frac{p + p_2}{(p + p_2 + \frac{\lambda}{L})} = -\frac{1}{L + \lambda(4D_* + 1)}$$

and

$$b = \frac{1 - aL}{2} = \frac{1}{2}\left(1 + \frac{1}{1 + \frac{\lambda}{L}(4D_* + 1)}\right)$$

where D_* is the dimensionless diffusion coefficient

$$D_* = \frac{\tau}{\lambda^2}D = \frac{p + p_0}{4(p + p_2)} = \frac{p + p_0}{4(1 - p - p_0)}$$

Therefore, we obtain

$$N_1(x, y) = 1 - \frac{1}{1 + \frac{\lambda}{L}(4D_* + 1)}\frac{x}{L}$$

$$N_3(x, y) = \frac{1}{1 + \frac{\lambda}{L}(4D_* + 1)}\frac{(L - x)}{L}$$

and finally, since $b = (1 - aL)/2$,

$$N_2 = N_4 = \frac{1}{2}(N_1 + N_3)$$

We observe that, in the limit $\lambda \to 0$, $a \to -(1/L)$ and $b = 1$. Thus, the density

$$\rho = N_1 + N_2 + N_3 + N_4 = 4ax + 2(1 - aL)$$

tends to $4(L - x)/L$, as expected for the solution of $(d^2\rho/dx^2) = 0$ with the condition $\rho(x = 0) = 4$ and $\rho(x = L) = 0$.

In contrast, when λ is finite, some corrections (of the order $\frac{\lambda}{L}$) are present. However, the density profile is always a straight line. But the values at $x = 0$ and $x = L$ differ from 4 and 0, respectively. At $x = L/2$, the discrete and continuous solution intersect, for the value $\rho = 2$.

Finally, it is interesting to compute, in this simple case, the particle current $\vec{J} = v[N_1(x + \lambda) - N_3(x)]$, that is the net number of particles that have crossed an imaginary vertical plan located at $x + (\lambda/2)$ (note: $N_1(x + \lambda)$ is the probability of having a particle move from x to $x + \lambda$ during the time τ while $N_3(x)$ is the probability of having a particle move from $x + \lambda$ to x during the same time interval; this not equivalent to $N_3(x + \lambda) - N_1(x)$).

We have

$$\vec{J} = \frac{\lambda}{\tau}(1 + aL + a\lambda) = 4\frac{\lambda^2}{\tau}D_*\frac{1}{L + \lambda(4D_* - 1)} = -4Da$$

Since

$$4a = \frac{d^2\rho}{dx^2}$$

we find that the particle current is

$$\vec{J} = -D\mathrm{grad}\rho$$

This relation, which is the usual one in a diffusion process is true even if the lattice spacing does not tend to zero.

4.3.2 Telegraphist equation

When both λ and τ are finite, the propagation velocity $v = \lambda/\tau$ is also finite (note that, in contrast, v is infinite when $\lambda, \tau \to 0$ and $\tau \propto \lambda^{-2}$). Therefore, information propagates at a finite speed in the cellular automata and the diffusion equation $\partial_t \rho = D\nabla^2 \rho$ cannot be true. Indeed, with this equation any variation of ρ is instantaneously transmitted arbitrarily far away from where it occurred. There is, however, an equation which combines diffusive behavior with a finite propagation speed of information. This is the well-known telegraphist equation. We will see that our cellular automata rule naturally leads to this equation when a finite lattice is considered.

Using a matrix notation for the N_i, the lattice Boltzmann diffusion equation can be cast in the following form

$$TN(t + \tau) = (\Omega + 1)N(t)$$

where T is the spatial translation operator. In one dimension, this reads

$$T = \begin{pmatrix} T_x & 0 \\ 0 & T_{-x} \end{pmatrix}$$

where $T_{\pm x}$ is a translation along the x-axis of $\pm\lambda$

$$T_{\pm x}N_i(x, t) = N_i(x \pm \lambda, t)$$

The collision operator Ω is the usual matrix made of the p_i. In one dimension, this reads

$$\Omega = \begin{pmatrix} p_0 - 1 & 1 - p_0 \\ 1 - p_0 & p_0 - 1 \end{pmatrix}$$

In two dimensions, T is diagonal and T_{ii} is composed of either a translation $T_{\pm x}$ or a translation $T_{\pm y}$ along the y-axis. The collision matrix Ω is given in equation 4.14.

The Ω matrix can be diagonalized easily. For the sake of simplicity, we shall consider the one-dimensional case. We introduce the variable M

$$N = PM$$

where P is the transformation matrix

$$P = \frac{1}{\sqrt{2}} \begin{pmatrix} 1 & -1 \\ 1 & 1 \end{pmatrix} \tag{4.38}$$

Its inverse reads

$$P^{-1} = \frac{1}{\sqrt{2}} \begin{pmatrix} 1 & 1 \\ -1 & 1 \end{pmatrix} \tag{4.39}$$

and one has for the two components of the column vector M

$$M_1 = \frac{1}{\sqrt{2}}(N_1 + N_2) = \frac{1}{\sqrt{2}}\rho$$

and

$$M_2 = \frac{1}{\sqrt{2}}(-N_1 + N_2) = -\frac{1}{\sqrt{2v}}\rho u$$

One immediately checks that

$$P^{-1}(\Omega + 1)P = D \equiv \begin{pmatrix} 1 & 0 \\ 0 & 2p_0 - 1 \end{pmatrix}$$

The microdynamics of the diffusion rule reads

$$TN(t + \tau) = (\Omega + 1)N(t)$$

With the transformation $N = PM$, this is equivalent to

$$P^{-1}TPM(t + \tau) = DM(t)$$

From this equation we can write two relations

$$P^{-1}TPM(t) = DM(t - \tau) \tag{4.40}$$

and

$$M(t + \tau) = P^{-1}T^{-1}PDM(t) \tag{4.41}$$

where T^{-1} is the inverse translation

$$T^{-1} = \begin{pmatrix} T_{-x} & 0 \\ 0 & T_x \end{pmatrix}$$

Now, it is easy to check that

$$P^{-1}TP = \frac{1}{2}\begin{pmatrix} T_x + T_{-x} & -T_x + T_{-x} \\ -T_x + T_{-x} & T_x + T_{-x} \end{pmatrix}$$

whereas

$$P^{-1}T^{-1}P = \frac{1}{2}\begin{pmatrix} T_x + T_{-x} & T_x - T_{-x} \\ T_x - T_{-x} & T_x + T_{-x} \end{pmatrix}$$

and we see that the off-diagonal terms come with opposite signs. More generally, in higher dimensions, a similar result holds: some terms change sign and others do not. This is related to the properties of the M_i to change sign under the transformation $x \to -x$ (i.e $N_1 \to N_2$ and $N_2 \to N_1$). This fact can be used to easily eliminate some of the M_i from the equations. In one dimension, there is just M_2 to eliminate and we obtain an equation for M_1 only.

The first equation of 4.40 and 4.41 gives

$$M_1(t + \tau) = \frac{1}{2}(T_x + T_{-x})M_1(t) - \frac{2p_0 - 1}{2}(T_x + T_{-x})M_2(t)$$

$$M_1(t + \tau) = \frac{1}{2}(T_x + T_{-x})M_1(t) + \frac{1}{2}(T_x + T_{-x})M_2(t) \qquad (4.42)$$

and by simple substitution of $(T_x + T_{-x})M_2(t)$ from the second line into the first, we obtain

$$M_1(t + \tau) = \frac{p_0}{2}(T_x + T_{-x})M_1(t) - (2p_0 - 1)M_1(t - \tau) \qquad (4.43)$$

In order to put this equation in a more convenient form, we introduce the discrete differential operators

$$\hat{\partial}_t M_1(t) = \frac{1}{2\tau}(M_1(t + \tau) - M_1(t - \tau))$$

$$\hat{\partial}_t^2 M_1(t) = \frac{1}{\tau^2}(M_1(t + \tau) - 2M(t) + M_1(t - \tau))$$

and

$$\hat{\nabla}^2 M_1(t, x) = \frac{1}{\lambda^2}(M_1(t, x + \lambda) - 2M(x, t) + M_1(t, x - \lambda))$$

These finite difference operators tend to their usual counterpart in the limit $\lambda \to 0$ and $\tau \to 0$. They also correspond to the standard way of discretizing space and time derivatives.

Equation 4.43 can be written as

$$\frac{a}{2}(M_1(t + \tau) - M_1(t - \tau)) + b(M_1(t + \tau) - 2M_1$$
$$+ M_1(t - \tau)) = p_0(T_x + T_{-x} - 2)M_1$$

with

$$a = 2(1 - p_0) \qquad b = p_0$$

Therefore, we obtain

$$\hat{\partial}_t M_1 + \tau \frac{p_0}{2(1 - p_0)} \hat{\partial}_t^2 M_1 = \frac{\lambda^2}{\tau} \frac{p_0}{2(1 - p_0)} \hat{\nabla}^2 M_1$$

We have $\rho = \sqrt{2}M_1$, $D = (\lambda^2/\tau)[p_0/(2 - 2p_0)]$, the diffusion coefficient we defined previously and $c^2 = (\lambda^2/\tau^2)$, the speed at which information travels in the system. With these definitions, we obtain that the density of particles obeys a discrete telegraphist equation

$$\hat{\partial}_t \rho + \frac{D}{c^2} \hat{\partial}_t^2 \rho = D\hat{\nabla}^2 \rho \qquad (4.44)$$

Equation 4.44 is exact and involves no approximation. It is true for all x except at the system limits where boundary conditions prevail.

The telegraphist equation describes a situation in which a signal propagates at finite speed and is subject to diffusion. It is the presence of the second time "derivative" which guarantees a finite propagation speed of information in the system. When c tends to infinity, it can be neglected and one obtains a finite difference Fick's equation.

In the case of periodic boundary conditions, another way to obtain the telegraphist equation is to consider again the dispersion relation derived in section 3.1 (equation 3.9). It can also be written as (with the current notation, p_0 stands for p)

$$\cos k\lambda = \frac{\exp(i\omega\tau) + (2p_0 - 1)\exp(-i\omega\tau)}{2p_0}$$

$$= \frac{\exp(i\omega\tau) + \exp(-i\omega\tau)}{2} + (1 - p_0)\frac{\exp(i\omega\tau) + \exp(-i\omega\tau)}{2p_0}$$

consequently we have

$$i\sin\omega\tau + \frac{p_0}{1 - p_0}\cos\omega\tau = \frac{p_p}{1 - p_0}\cos k\lambda$$

Expanding this relation up to second order in $\omega\tau$ and $k\lambda$ and using the fact that in Fourier space one has the correspondence

$$\partial_t \leftrightarrow i\omega \qquad \nabla^2 \leftrightarrow -k^2$$

one gets back to the telegraphist equation.

The agreement between our dispersion relation and the telegraphist equation can be seen graphically. It is straightforward to show that $f(x,t) = \exp(i\omega\tau - kx)$ is solution of

$$\partial_t\rho + \frac{D}{c^2}\partial_t^2\rho = D\nabla^2\rho$$

provided that

$$i\omega = -\frac{c^2}{2D}\left(1 \mp \sqrt{1 - \frac{4D^2k^2}{c^2}}\right)$$

or equivalently

$$e^{i\omega\tau} = \exp\left[-\frac{c_*^2}{2D_*}\left(1 \mp \sqrt{1 - \frac{4D_*^2(k\lambda)^2}{c_*^2}}\right)\right]$$

where $D_* = (\tau/\lambda^2)D$ and $c_* = (\tau/\lambda)c$ are the dimensionless diffusion coefficient and speed of sound, respectively. The above expression is the dispersion relation for the telegraphist equation. It can be checked against

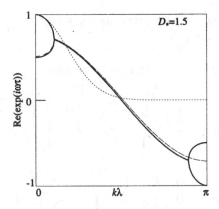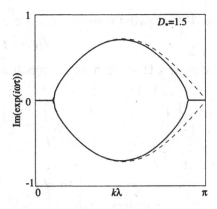

Fig. 4.3. Comparison of the real and imaginary parts of the dispersion relations of (i) the cellular automata model (solid line); (ii) the telegraphist equation (dashed line) and (iii) Fick's law of diffusion (dotted line). Notice that some curves may have two branches because of the $\pm$ sign in front of the square root.

equation 3.10. This is done in figure 4.3, where $e^{i\omega\tau}$ is given as a function of $k\lambda$. We also show the dispersion relation of the pure diffusion equation

$$e^{i\omega\tau} = e^{-D_\bullet(k\lambda)^2}$$

We can observe that our cellular automata diffusion model is much closer to the telegraphist equation than Fick's law. Of course, at large time, only small k contributions will survive (i.e those for which $e^{i\omega\tau}$ is close to 1) and in this limit all three evolutions correspond. Note however that $k \sim \pi$ also produces long-lived terms. As we previously mentioned, this is related to the checkerboard invariant that plagues cellular automata dynamics.

4.3.3   The discrete Boltzmann equation in 2D

The same approach as described in the previous section can be applied to a two-dimensional system. Of course, the algebra becomes more complicated. Here, we would like to mention some results, without giving a complete derivation. More details can be found in [112].

In the 2D case, the discrete Boltzmann equation reads

$$T N(\vec{r}, t + \tau) = (\Omega + 1)N(\vec{r}, t)$$

where N is the column vector composed of the N_i, $\Omega + 1$ the four by four matrix

$$\Omega + 1 = \begin{pmatrix} p_0 & p & p_2 & p \\ p & p_0 & p & p_2 \\ p_2 & p & p_0 & p \\ p & p_2 & p & p_0 \end{pmatrix}$$

and T the spatial translation matrix defined in the previous section

$$T_{ij}f(\vec{r}) \equiv f(\vec{r} + \lambda\vec{c}_i)\delta_{ij}$$

The matrix $\Omega + 1$ can be diagonalized with the linear transformation

$$P = \frac{1}{2}\begin{pmatrix} 1 & 1 & 1 & -1 \\ 1 & 1 & -1 & 1 \\ 1 & -1 & -1 & -1 \\ 1 & -1 & 1 & 1 \end{pmatrix} \quad \text{and} \quad P^{-1} = \frac{1}{2}\begin{pmatrix} 1 & 1 & 1 & 1 \\ 1 & 1 & -1 & -1 \\ 1 & -1 & -1 & 1 \\ -1 & 1 & -1 & 1 \end{pmatrix}$$

so that

$$P^{-1}(\Omega + 1)P = \begin{pmatrix} 1 & 0 & 0 & 0 \\ 0 & p_0 - p_2 & 0 & 0 \\ 0 & 0 & p_0 - p_2 & 0 \\ 0 & 0 & 0 & 1 - 4p \end{pmatrix}$$

In terms of the new quantities

$$M = P^{-1}N \tag{4.45}$$

and, following the procedure of section 4.3.2, one obtains

$$\tau\hat{\partial}_t M_1 + \frac{D_*}{c_*^2}\tau^2\hat{\partial}_t^2 M_1 = \lambda^2 D_*\hat{\nabla}M_1 + \lambda^2\frac{p_0 - p}{4(p + p_2)}(-\hat{\partial}_x^2 + \hat{\partial}_y^2)M_4 \tag{4.46}$$

$$a_4\tau\hat{\partial}_t M_4 + b_4\tau^2\hat{\partial}_t^2 M_4 + 8p(p + p_2)M_4 = \lambda^2\frac{p_0 + p}{2}(-\hat{\partial}_x^2 + \hat{\partial}_y^2)M_1$$
$$+ \lambda^2\frac{p_0 - p}{2}(\nabla^2)M_4 \tag{4.47}$$

where $\hat{\partial}_t$ and $\hat{\partial}_t^2$ are the discrete time derivative and $\hat{\partial}_x^2$, $\hat{\partial}_y^2$ and $\hat{\nabla}^2$ the second-order discrete space derivative.

The dimensionless diffusion constant D_* and speed of sound c_* are defined in agreement with section 4.2.1 and read

$$D_* = \frac{p_0 + p}{4(1 - (p + p_0))} \quad \text{and} \quad c_*^2 = \frac{1}{2}$$

and the coefficients a_4 and b_4 are

$$a_4 = 1 - (p_0 - p_2)(1 - 4p) \quad \text{and} \quad b_4 = p + p_0 - 2p(p_0 - p_2)$$

Similarly, it is found that M_2 and M_3 are related to M_1 and M_4. In the case $p_0 = p$, the equation for M_1 decouples from the others and we simply obtain that the density $\rho = 2M_1$ (see 4.45) obeys the discrete telegraphist equation

$$\hat{\partial}_t\rho + \frac{D}{c^2}\hat{\partial}_t^2\rho = D\hat{\nabla}^2\rho$$

From the equations for M_1 and M_2, one can derive the dispersion relation of our discrete diffusive process. In the case of a periodic system,

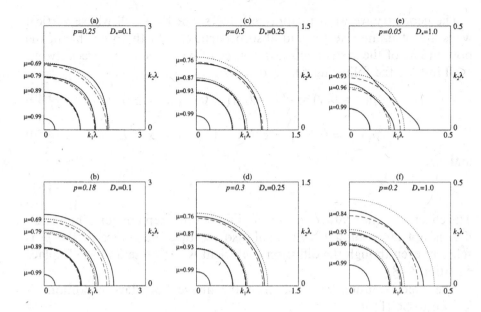

Fig. 4.4. Contour lines of the diffusive part of the dispersion relation ($\mu = \exp(i\omega\tau)$) for various values of D_* and p. The solid lines correspond to the CA model, the dashed ones to the telegraphist equation and the dotted ones to Fick's law of diffusion. For $\mu = 0.99$ the three dynamics are almost indistinguishable. Note that the scales are not the same in all figures.

equations 4.46 and 4.47 can be Fourier transformed and the solution takes the form

$$M_{1,4} = A_{1,4}e^{i(\omega\tau + \vec{k}\cdot\vec{r})}$$

provided that ω is a function of $\vec{k}$. The dependence $\omega = \omega(\vec{k})$ is shown in figure 4.4. We observe that the choice of the p_i affects the dynamics at short time and for small wave length. However, in the macroscopic limit, the dispersion relation of the automaton is in agreement with the behavior of a continuous diffusion process.

4.3.4 Semi-infinite strip

In this section, we consider the solution of the discrete Boltzmann equation in the case of a two-dimensional stationary problem. For a time-independent diffusion process, Fick's law reduces to the Laplace equation

$$\nabla^2\rho = 0 \tag{4.48}$$

with appropriate boundary conditions.

The geometry we want to study has the shape of a well whose vertical walls are of infinite height and absorb particles. On the other hand, the bottom line of the system acts as a source of particles. More precisely, we shall impose the following conditions

$$\rho(x = 0, y) = \rho(x = L, y) = 0 \quad \text{for} \quad y \geq 0 \tag{4.49}$$

$$\rho(x, y = 0) = \phi = \text{const} \quad \text{for} \quad 0 < x < L \tag{4.50}$$

and

$$\lim_{y \to \infty} \rho = 0 \tag{4.51}$$

The effect of a finite lattice spacing and finite time step are quite important if L is not large enough. We shall also see that the choice of the p_i is relevant even though the diffusion coefficient is not present in the Laplace equation 4.48.

The exact solution of 4.48 with the above boundary conditions is known to be [113]

$$\rho = \frac{4\phi}{\pi} \sum_{n \text{ odd}} \frac{1}{n} \exp(-\frac{n\pi y}{L}) \sin(\frac{n\pi x}{L}) = \frac{2\phi}{\pi} \tan^{-1} \left(\frac{\sin(\frac{\pi x}{L})}{\sinh(\frac{\pi y}{L})} \right) \tag{4.52}$$

For a stationary state, the lattice Boltzmann dynamics 4.46 and 4.47 read

$$\hat{\nabla}^2 M_1 + \alpha(-\hat{\partial}_x^2 + \hat{\partial}_y^2)M_4 = 0 \tag{4.53}$$

and

$$(-\hat{\partial}_x^2 + \hat{\partial}_y^2)M_1 + \alpha\hat{\nabla}^2 M_4 = \beta M_4 \tag{4.54}$$

where α and β are defined as

$$\alpha = \frac{p_0 - p}{p_0 + p} \qquad \beta = \frac{16p(p + p_2)}{p + p_0}$$

In order to solve 4.53 and 4.54 for the boundary conditions we defined at the beginning of this section, we first assume that $\vec{c}_1$ and $\vec{c}_2$ are, respectively, parallel to the x- and y-axis. Next, we assume a solution of the following form

$$M_{1,4} = \sum_n A_{1,4}(n) \exp(z_n y) \sin(\omega_n x) \tag{4.55}$$

where

$$\omega_n = \frac{\pi n}{L} \tag{4.56}$$

and z_n has to be determined from 4.53 and 4.54.

This calculation turns out to be straightforward because the discrete derivatives have simple eigenfunctions. Indeed, the functions $\exp(z_n y)$ and $\sin(\omega_n x)$ are eigenfunctions of the discrete operators $\hat{\partial}_x^2$ and $\hat{\partial}_y^2$

$$\hat{\partial}_x^2 \sin(\omega_n x) = \frac{b_\omega}{\lambda^2} \sin(\omega_n x)$$

$$\hat{\partial}_y^2 \exp(z_n y) = \frac{b_z}{\lambda^2} \exp(z_n y)$$

From the simple properties of the sin and exp functions, we find

$$b_\omega \equiv 2(\cos(\frac{\pi n}{L}\lambda) - 1) \tag{4.57}$$

and

$$b_z \equiv \exp(z_n \lambda) + \exp(-z_n \lambda) - 2 \tag{4.58}$$

After substitutions of 4.55 into 4.53 and 4.54, we obtain

$$(b_\omega + b_z)A_1 + \alpha(-b_\omega + b_z)A_4 = 0$$
$$(-b_\omega + b_z)A_1 + [\alpha(b_\omega + b_z) - \beta]A_4 = 0$$

In order to find a non-zero solution, we obtain a relation between ω_n and z_n, namely

$$b_z = \frac{b_\omega}{4\frac{\alpha}{\beta}b_\omega - 1} \tag{4.59}$$

Thus, the ratio

$$\frac{\alpha}{\beta} = \frac{p_0 - p}{16p(p + p_2)} \tag{4.60}$$

is a free parameter of the model. Its value plays a role when the quantity λ/L does not tend to zero. Clearly, α/β can be made arbitrarily large by choosing p small enough. On the other hand, its smallest value is

$$\frac{\alpha}{\beta} = -\frac{1}{8}$$

obtained for $p = 1/2$ and $p_0 = p_2 = 0$.

It is then straightforward to calculate the density $\rho = 2M_1$ from 4.55 because, for $y = 0$, our expansion is identical to 4.52. Therefore

$$\rho = \frac{4\phi}{\pi} \sum_{n \text{ odd}} \frac{1}{n} [\exp(z_n \lambda)]^{(y/\lambda)} \sin(\frac{\pi n x}{L}) \tag{4.61}$$

Strictly speaking, the density is only defined on the lattice sites but equation 4.61 provides an interpolation of ρ everywhere. Its similarity

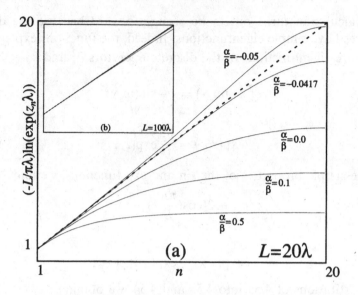

Fig. 4.5. Comparison between the solution of the Laplace equation and the solution of the CA dynamics for various values of α/β and for two system sizes (a) $L = 20\lambda$ and (b) $L = 100\lambda$. The dashed line corresponds to the exact solution $\exp -\pi n$ (in a log scale). This line turns out to coincide very well with $\alpha/\beta = -1/24 \approx -0.0417$.

with 4.52 makes it very appropriate for a comparison with the continuous solution of Laplace equation. Since

$$\exp(z_n y) = \left([\exp(z_n \lambda)]^{(L/\lambda)} \right)^{\frac{y}{L}}$$

the convergence of the discrete solution can be checked by comparing

$$\exp(-\pi n) \quad \text{with} \quad [\exp(z_n \lambda)]^{(L/\lambda)}$$

This is shown in figure 4.5, where the logarithm of these two quantities is shown for various values of λ/L and α/β. One sees that the significant deviations that occur when the system size is too small become negligible as the lattice spacing λ decreases. To draw this graph, we have solved equation 4.58 for $\exp(z_n \lambda)$,

$$\exp(z_n \lambda) = 1 + \frac{b_z}{2} \pm \frac{1}{2}\sqrt{b_z^2 + 4b_z}$$

and computed b_z as a function of b_ω, as given by 4.59. In figure 4.5, one observes that the dashed line, which represents the solution in the limit $\lambda \to 0$, is approximated with more or less accuracy, depending on the value of α/β. The curve $\alpha/\beta = -1/24$ turns out to be quite good, even for a small system $L = 20\lambda$. It can be shown [112] that this value yields a correct solution to the Laplace equation up to order $(\lambda/L)^6$. Therefore, in a finite system, the choice of the p_i may play a role in the quality of the solution.

4.4 Applications of the diffusion rule

In section 4.2, we presented a cellular automata model of synchronous random walks and discussed how well the discrete dynamics reproduces the behavior of a diffusive process. The examples we have studied are rather trivial from the view point of physics and there is no reason to use a cellular automata to solve Laplace equation in a one-dimensional source–sink problem, or in the semi-infinite strip.

However, there are several physical processes involving diffusing particles that are much more complex and difficult to describe in terms of a solution of the diffusion equation. On the other hand, the cellular automata approach is quite natural and well suited to perform numerical simulation.

This section describes three applications of the diffusion rule, namely the properties of a diffusion front, the so-called DLA aggregation process and, finally, the problem of random deposition on a surface.

4.4.1 Study of the diffusion front

The diffusion model we have introduced in this chapter can be used to create and study rough interfaces between two media in contact. Most solid interfaces observed in nature are irregular or rough on all length scales and the fractal geometrical properties of such interfaces are of great importance to materials science. The way to model an interface is to have particles diffuse in a two-dimensional system, from a source on the left, to a sink on the right. Before the density gradient reaches a steady state, two regions can be identified: that made up of particles and connected to the source and the other one, made up of holes (empty lattice sites), and connected to the sink.

As practical examples of physical systems in which the interface plays an important role, we can mention solid–solid chemical reactions, alloys formed out of the melt and many processes in solid state technology [114,115]. Catalysis, corrosion, crystal growth and the physics of superionic conduction [116] are other phenomena where the interfacial properties are directly relevant. Finally, fractal interfaces occur naturally in invasion percolation of a porous medium [117].

In addition to modeling fractal interfaces, the nonstationary diffusion process is related to the problem of site percolation [118], and diffusion of particles in a concentration gradient provide an ideal means to compute percolation properties, notably the percolation threshold p_c and the fractal dimension characterizing large clusters. The main advantage of this approach is that one does not have to know or guess the value of p_c in advance: it is determined self-consistently.

In the present example, we consider a system with a perfect line source at $y = 0$ which replaces immediately particles diffusing into the system and a line sink at $y = L_y$ which removes all arriving particles. In the perpendicular direction, the system has a size L_x with periodic boundary conditions. At time $t = 0$ the system is completely empty. Although the diffusion process does conserve the number of particles inside the system, the total number still changes, because of the source and the sink.

The diffusion front is defined through particle connectivity, as in site percolation: any two particles sitting on nearest neighbor sites belong to the same cluster. For percolation, a threshold p_c separates a region where only finite clusters are present from the region where an infinite cluster occupies a fraction of the system. In our model, p varies from $p = 0$ to $p = 1$ across the sample so that the finite cluster and the infinite cluster regimes are present simultaneously. Close to the source $p > p_c$ and one cluster spans the system. Close to the sink $p < p_c$ and only finite clusters are present. In a weak gradient, large finite clusters have the structure of percolation clusters and are located near p_c.

More precisely, the diffusion front, also called external perimeter or hull, is obtained according to the following procedure

(1) Determine the infinite cluster ("the land") as the set of all particles which are connected to the source by nearest-neighbor particle–particle bonds.

(2) The complementary infinite cluster ("the ocean") consists of the non-land sites which are connected by nearest-neighbor bonds to the sink. By this process, the islands (particles not connected to the source) are removed.

(3) The diffusion front (or the "shore") is built as the set of land sites which are nearest neighbor to at least one ocean site.

Different fronts can be defined by extending the connectivity from nearest neighbors to second nearest neighbors [118]. Different geometrical properties are then expected.

Note that the front construction, as given by the three above steps, is the result of a cellular automata rule. The land and the ocean are obtained by iterating a "coloring" rule: first, the source is painted, say, black. Then, each cell which is occupied by a diffusing particle and which contains a

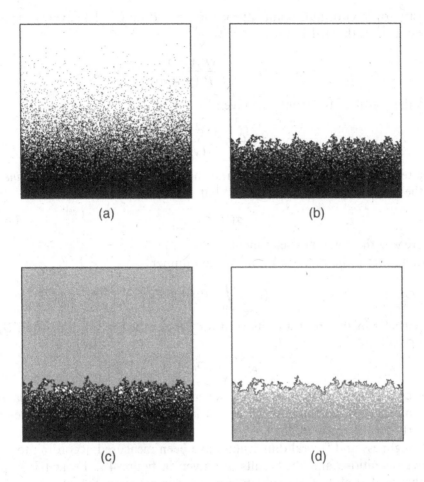

Fig. 4.6. The diffusion front is the boundary between the source connected cluster and the sink-connected cluster. The figures shows the four steps necessary to extract the front: (a) the location of diffusive particles; (b) the "land" cluster; (c) the "ocean" in gray; and (d) the fractal front in black.

black site in its von Neumann neighborhood turns black, too. A similar procedure is applied from the sink to determine the ocean.

Figure 4.6 illustrates these various steps, from the configuration of diffusing particles to the construction of the front.

As we can see from figure 4.6, the diffusion front is a complex object. Depending on the time at which it is obtained (i.e. the number of iterations of the diffusion rule before rules (1) to (3) are applied), its width varies. The longer the diffusion takes place, the wider the front is.

Practically, the width σ is defined as follows. By symmetry, the front properties are similar along the x-axis. Let us call $P_f(y, t)$ the average density (over x) of hull points at location y and time t. The average front position y_f is defined by:

$$y_f(t) = \frac{\int_0^\infty y P_f(y,t)\, dy}{\int_0^\infty P_f(y,t)\, dy}$$

and the width of the front σ is given by

$$\sigma^2 = \frac{\int_0^\infty (y - y_f)^2 P_f(y,t)\, dy}{\int_0^\infty P_f(y,t)\, dy}$$

The relation between the width σ and the time t determines the *roughness* of the front. Typically, one expects that

$$\sigma(t) \propto t^\alpha \tag{4.62}$$

where α is the roughness exponent.

In addition, the relation between the number

$$N_f(t) = \int_0^\infty P_f(y,t)\, dy$$

of particles in the front and its width σ defines the fractal dimension D_H of the hull as

$$N_f = A\sigma^{D_H - 1}$$

where A is some amplitude. Here, we have subtracted -1 to D_H because we consider the dependence of N_f on σ only. Clearly, N_f also depends linearly on the front length L_x.

Roughness and fractal dimension have been measured according to the above definitions and the results are given in figure 4.7. From the slope of the log–log plot of N_f againts σ and σ against t, we obtain

$$\alpha \approx 0.27 \qquad D_H \approx 1.41$$

respectively. These values are in good agreement with those predicted by independent theoretical arguments. The simulation which produced these results has been obtained with the values $p = 1/2$ and $p_0 = p_2 = 0$ in the diffusion rule. Other values of the p_i yield the same fractal dimension and roughness.

4.4.2   Diffusion-limited aggregation

Growth phenomena are frequent in many fields of science and technology. A generic aspect of several growth processes is a far-from-equilibrium formation situation and a fractal structure. Dendritic solidification in an undercooled medium, electrodeposition of ions on an electrode and viscous fingering are well-known examples of fractal growth [119].

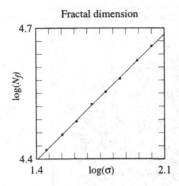
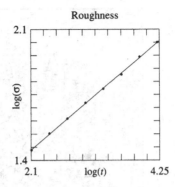

Fig. 4.7. Fractal dimension of the diffusion front and roughness. The slopes of the curves give $N_f \sim \sigma^{0.41}$ and $\sigma \sim t^{0.27}$.

Aggregation constitutes an important growth mechanism: like particles stick to each other as they meet and, as a result, form a complicated pattern with a branching structure.

In many cases, growth is governed by a spatial quantity such as an electric field, a local temperature, or a particle density field. Quite often, the behavior of these quantities can be approximated by solving the Laplace equation with moving boundary conditions.

A prototype model of aggregation is the so-called DLA model (diffusion-limited aggregation), introduced by Witten and Sander [120] in the early 1980s to simulate the following process: a negative electrode is placed in an electrolyte containing positive metallic ions. The ions diffuse in the electrolyte and, when they hit the negative electrode, stick to it. In addition, the ions can also stick to the already deposited aggregate and, as time goes on, a complex rigid tree-like structure emerges. Screening plays an important role in the growth process. As the aggregate gets larger, a new ion is more likely trapped around the perimeter than deep inside. Since its introduction, the DLA model has been investigated in great detail. However, diffusion-limited aggregation is a far-from-equilibrium process which is not described theoretically by first principles only. Spatial fluctuations that are typical of the DLA growth are difficult to take into account and a numerical approach is necessary to complete the analysis.

The DLA process can be readily modeled by our diffusion cellular automata, provided that an appropriate rule is added to take into account the particle–particle aggregation. The first step is to introduce rest particles to represent the particles of the aggregate. Therefore, in a two-dimensional system, a lattice site can be occupied by up to four diffusing particles, or by one "solid" particle. Figure 4.8 shows a DLA cluster grown by the cellular automata dynamics. The diffusion constant of the particles is $D_* = 1/4$, obtained with $p_0 = p = p_2 = 1/4$.

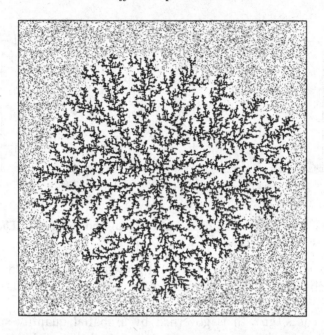

Fig. 4.8.   Two-dimensional cellular automata DLA cluster (black), obtained with $p_s = 1$, an aggregation threshold of one particle and a density of diffusing particle of 0.06 per lattice direction. The gray dots represent the diffusing particles not yet aggregated.

At the beginning of the simulation, one or more rest particles are introduced in the system to act as aggregation seeds. The rest of the system is filled with particles in average concentration ρ.

When a diffusing particle becomes nearest neighbor to a rest particle, it stops and sticks to it by transforming into a rest particle. Since several particle can enter the same site, we may choose to aggregate all of them at once (i.e. a rest particle is actually composed of several moving particles), or to accept the aggregation only when a single particle is present.

In addition to this question, the sticking condition is important. If any diffusing particle always sticks to the DLA cluster, the growth is very fast and can be influenced by the underlying lattice anisotropy. It is therefore more appropriate to stick with some probability p_s. Since up to four particles may be simultaneously candidate for the aggregation, we can also use this fact to modify the sticking condition. A simple way is to require that the local density of particles be larger than some threshold (say three particles) to yield aggregation. Similarly, the so-called technique of noise reduction can also be implemented in the cellular automata rule: the idea is to have a counter associated with each lattice site on which aggregation is possible. The counter is initialized to some given threshold

(a)

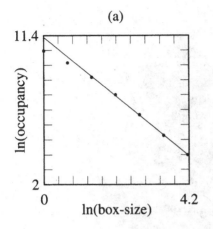

(b)

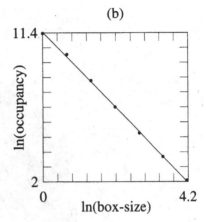

Fig. 4.9. Box-counting estimate of the fractal dimension of (a) a two-dimensional and (b) a three-dimensional cellular automata DLA, using the box-counting method. The slope of the lines give: (a) $d_f = 1.78$ and (b) $d_f = 2.3$.

(for instance four) and each time a diffusing particle reaches the site, the counter is decremented by one. When it reaches 0, aggregation takes place.

Our approach has also some other differences compared with the original Witten and Sanders model. All particles move simultaneously and can stick to different parts of the cluster and we do not launch them, one after the other, from a region far away from the DLA cluster. For this reason, we may expect some quantitative variation of the fractal dimension for the clusters generated with our method. Figure 4.9 shows the result of a box-counting algorithm in order to estimate the fractal dimension d_f of the DLA cluster obtained from our cellular automata rule. The principle of this technique is to tile the cluster with a regular array of square boxes of size $\ell \times \ell$. One then counts the number N_b of these boxes which contain a particle of the cluster. For a two-dimensional object, N_b is proportional to the number of boxes, namely ℓ^{-2}. For a fractal object, one expects that

$$N_b \sim \ell^{-d_f}$$

and, from figure 4.9 (a), one obtains

$$d_f \approx 1.78 \quad \text{(2-D)}$$

Similarly, a three-dimensional cellular automata DLA cluster can be generated by using the diffusion rule presented in section 4.2.3. In figure 4.10, such a cluster is shown, using a ray tracing rendering technique. Although it looks less dendritic than its two-dimensional counterpart, its fractal dimension is less than 3. The same box-counting method as before

Fig. 4.10. Three-dimensional cellular automata DLA cluster, on a 128^3 lattice, comprising around 80 000 particles (which is still not enough to obtain an accurate estimate of the fractal dimension).

(now with cubic boxes of size $\ell \times \ell \times \ell$) yields

$$d_f \approx 2.3 \qquad \text{(3-D)}$$

as shown in figure 4.9 (b).

These fractal dimensions (2-D and 3-D) correspond to measurements on a single cluster (that of figures 4.8 and 4.10, respectively). Due to this limited statistic, we can conclude to a reasonable agreement with the more classically established values [121,119] $d_f = 1.70$ (2-D) and $d_f = 2.53$ (3-D).[*]

The cellular automata approach is also well suited to study dynamical properties such as the DLA growth rate. The standard numerical experiment is to distribute uniformly the initial diffusing particles on the lattice with a single aggregation seed in the middle. As time t goes on, more and more particles solidify and the cluster mass $M(t)$ increases. Our simulations indicate (see figure 4.11) that this process has an intermediate regime governed by a power law

$$M(t) =\sim t^\alpha$$

---

[*] It is interesting to note that there is a mean-field prediction $d_f = (d^2+1)/(d+1)$ which corresponds rather well to the numerical measurements, where d is the space dimensionality.

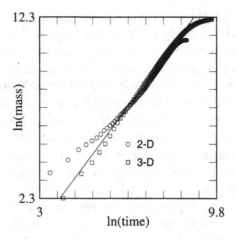

Fig. 4.11. Formation rate of cellular automata DLA clusters in two and three dimensions. The lattice has periodic boundary conditions.

where

$$\alpha \approx 2$$

in both two and three dimensions. Although these results are not sufficient to conclude definitely that the 2-D and 3-D exponents are the same, an explanation would be that in 3-D there is more surface to stick to than in 2-D, but also more space to explore before diffusing particles can aggregate. These two effects may just compensate.

A phenomena very similar to DLA is the so-called diffusion-limited deposition [119] which corresponds to a deposition-aggregation process on a surface. Our cellular automata model can be applied to model such a process by having a nucleation surface instead of a single seed.

In the next section we consider a different deposition situation, in which particles do not stick to each other but need an empty slot on the surface to become adsorbed.

4.4.3   Diffusion-limited surface adsorption

As a last example of the application of our cellular automata diffusion rule, we show here how it can be used to model irreversible particle adsorption on a surface. This problem is common in many areas of science. For instance, adsorption of proteins and other particles at the solid–liquid interface is a process of fundamental importance in nature.

The typical ingredients of irreversible adsorption are (i) the transport of particles from the bulk to the surface and (ii) their subsequent adsorption onto the surface. In this process, the interaction between an already adsorbed particle and a newly arriving one plays a crucial role.

For instance, two particles cannot overlap on the adsorption substrate. When one is already present, the second one can only be adsorbed if its center lands sufficiently far away from the first one. More generally, there is an *excluded surface* around each deposited particle, in which no more adsorption is possible. As the deposition process goes on, the surface coverage increases. But, due to this excluded surface, some "holes" on the surface will remain empty for ever because they are too small to accommodate a new particle. As a result, the coverage fraction $\theta(t)$ will tend to a maximum *jamming* limit $\theta_\infty < 1$ as the time t tends to ∞.

The excluded surface depends on the type of inter-particle interaction and their shape. Clearly, in the case of hard core interactions, one expects a different behavior when disks or needles are adsorbed on a surface.

The kinetics of this process (i.e. the time dependence of θ) and the limit value θ_∞ are, of course, quantities of particular importance when describing irreversible adsorption.

Another relevant aspect of surface adsorption is the transport process in the bulk. Particles which are brought in contact with the substrate have their own dynamics. At some point of their motion, they hit the adsorption surface and if they pick the right spot, are adsorbed. As a result, the surface coverage increases by a quantity s corresponding to the area of the particle. Otherwise, if the particle hits an already excluded part of the surface, it is rejected and continues its motion according to the bulk dynamics.

The most studied process which mimics these interactions, is the random sequential adsorption (RSA) model (see [122] for a review). The RSA model is characterized by sequential (i.e one after the other) deposition attempts of particles of unit area s at a constant rate c per unit time and area. Adsorption occurs only under the condition that there is no resulting overlap. After adsorption or rejection, the next attempt occurs at a location which is completely uncorrelated with respect to the previous one.

Even though RSA goes deeper into the description of adsorption processes than previous models such as the Langmuir isotherm [123], it cannot be entirely correct since coupling between deposition and transport is neglected.

A very natural transport mechanism in the bulk is diffusion. Several theoretical [124,125] and numerical[126,127] works have focused on this effect on the kinetics of the deposition and the structure of the jammed configuration and have established departures from the RSA predictions.

In this section we consider the problem of diffusion-driven surface adsorption, in the framework of a cellular automata dynamics. That is,

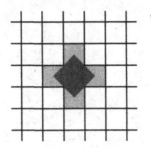

Fig. 4.12. Excluded surface on the lattice, due to the adsorption of a particle. We can imagine that the deposited particle is a diamond shaped object (dark gray) which overlaps the surrounding light gray cells, making them unavailable for further adsorption. In the diffusion plan, this picture is not the same, of course, and the particles are point-like.

we are going to assume that adsorption takes place on a two-dimensional lattice, which is of course different from the original case where a particle can deposit anywhere in a continuous space. However, if the substrate has some crystalline structure, this may correspond to a legitimate assumption. In addition, the same generic features as in the continuous case are present, but in a conceptually simpler form which may help developing an intuition.

The cellular automata we want to consider is the following: particles diffuse on the top a two-dimensional substrate, according to rule 4.4. These particles can be adsorbed on an empty substrate site. Adsorbed particles prevent further deposition on the entire von Neumann neighborhood surrounding them. Thus, the excluded surface amounts to five lattice sites, for each deposited particle. This is illustrated in figure 4.12.

More precisely, at each time step, adsorption obeys the following rules.

- When a site in the diffusion plan contains exactly one particle, this particle is candidate for adsorption.

- Adsorption takes place if the corresponding site on the substrate is empty and does not correspond to an excluded surface.

- Two candidate particles which are nearest neighbors are rejected. This rule is necessary to avoid two adjacent empty sites adsorbing simultaneously two particles.

- Particles that can be adsorbed are placed on the substrate and removed from the diffusion plan. The other ones diffuse according to the diffusion rule.

If the initial particle density in the diffusion plan is large enough, there are only a few sites containing exactly one particle. Only these sites will yield

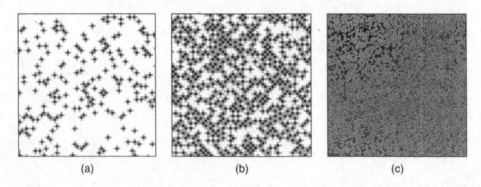

(a)   (b)   (c)

Fig. 4.13. Adsorption surface at different time steps. Black cells represent adsorbed particles and the surrounding excluded surface is shown in gray. White regions show the available space. Figure (a) and (b) are the upper left part of the system, four times magnified. They show the adsorption surface after one and five iterations, respectively. Figure (c) shows the entire surface, a 256×256 lattice, at the final stage. Light gray tortuous "paths" are the excluded surface which separate close-packed regions (dark gray) having a checkerboard structure

adsorption at the first iteration. Thus, it is not necessary to introduce an extra adsorption probability to avoid all particles becoming marked for adsorption at the first iteration.

Figure 4.13 shows the evolution of the adsorption surface.

Due to the shape of the excluded surface, the highest coverage fraction that can be theoretically achieved is $\theta_\infty = 1/2$. This close-packed configuration would be attained if the particles would be smart enough to organize themselves as the black squares of a checkerboard. But, diffusion has certainly not the global knowledge required to achieve this result. Random deposition is not better with this respect. One expects, on the other hand, to observe local checkerboard structures with defects, as shown in figure 4.13.

The global structure which emerges from the local adsorption process has a well defined jamming coverage, in the limit of an infinite system. From our cellular automata model, we have estimated $\theta_\infty = 0.366$. An interesting observation is that this jamming limit depends on the transport mechanism of the particle. For instance, for random sequential adsorption with nearest-neighbor exclusion on a square lattice, theoretical analysis yields the estimate $\theta_\infty = 0.364$ [122]. We found this result compatible with a *parallel* random adsorption, that is a cellular automata in which particle positions in the diffusion plan are decorrelated from one time step to the other. Note that these values correspond to a two-dimensional substrate and are not the same in one or three dimensions.

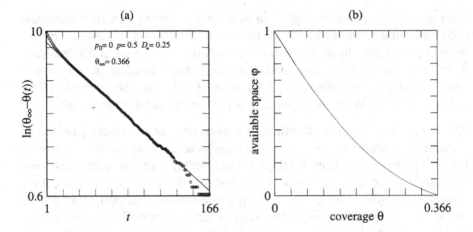

Fig. 4.14. (a) Kinetics of the cellular automata adsorption model in units of iterations. (b) Available space as a function of the surface coverage.

Our analysis also shows that the jamming limit depends slightly on the diffusion constant D and on the specific value of the p_i chosen for the diffusion rule. This may be an indication that, for this non-Markovian process, it is not sufficient to give D to uniquely specify the macroscopic behavior.

Finally, we show in figure 4.14 the kinetics of the diffusion-limited adsorption process and the relation between the available space (the white cells in figure 4.13) and the coverage (the black cells). From the first plot, we can conclude that the surface coverage obeys an exponential law

$$\theta_\infty - \theta(t) \sim \exp(-t/\tau_c)$$

where τ_c is some characteristic time. The second plot describes how the available space ϕ decreases with the coverage fraction θ. Clearly, for $\theta \approx 0$, we have $\phi(\theta) \approx -5\theta$ since each adsorption reduces the number of deposition sites by five cells. On the other hand, near the jamming limit $\theta \approx \theta_\infty$, only isolated empty sites remain and $\phi(\theta) \approx -\theta$. In between, ϕ is a non-trivial function of θ.

4.5 Problems

4.1. Simulate a CA system of particles which bounce into fixed obtacles that are initially distributed randomly across the lattice with probability q (Ehrenfest model). The obstacles act as mirrors and deflect the entering particles by $+\pi/2$, $-\pi/2$ or π. In the absence of obstacles the particles follow a straight line motion. How does this rule compare with the diffusion rule? Measure the relation between the mean square displacement and time (consider a simulation where all particles start moving off a vertical line and measure the positions x they have reached after t iterations). Does it obey the regular diffusion law $\langle x^2(t)\rangle \sim t$ or does it correspond to an anomalous random walk?

4.2. Propose a CA rule of diffusion where only one particle per site is permitted (instead of the usual four particles). Each particle is north-facing, east-facing, south-facing or west-facing, randomly at each time step and lattice site. It can move to the site it is pointing to only if no double occupation will occur. If such a conflict can happen, the particle remains in position. What is the size of the neighborhood required for this rule? Is the diffusion constant independent of the density? In the low density limit, what is the diffusion constant? What is, approximately, the critical density below which this limit is valid. Measure the diffusion constant using a simulation according to the method of the previous problem.

4.3. Show how the Margolus neighborhood can be used to express the diffusion rule with only one particle per site, with a much smaller neighborhood than in the previous problem.

4.4. Using the same approximations as in problem 3.5, derive the diffusion equation starting from the diffusion microdynamics presented at the beginning of this chapter. Compare the diffusion constant you obtain with the correct one.

4.5. Instead of a diffusion rule using random rotations, consider a rule based on random symmetries. Derive the diffusion coefficient.

4.6. Propose a nonlinear diffusion rule such that the rotation probabilities depend on the number of particles entering a site.

4.7. Chemotaxis describes the capacity of some organisms to adjust their movement in relation to the presence of chemical agents. Consider a population of diffusing ants. Each of them adds one unit of pheromone to the lattice site it visits. When a site is not visited at a given time step, the level of pheromone is reduced by one unit, until it reaches zero. When an ant is alone on a site, it jumps to the nearest (von Neumann) neighbor which contains the higher level of pheromone (except the site it just left at the previous iteration). If more than one ant are present (up to four are possible), they obey the usual diffusion rule. Simulate such a system and study its behavior. First, consider a homogeneous situation; then assume there are two sources of ants in the system. See if the ants can find a path from one source to the other.

4.8. Devise a CA model of driven diffusion to simulate for instance a situation where an electric field bias the diffusion towards a given direction. As in the previous problem, suppose that the field acts with probability E only on those sites which contain no more than one particle. When several particles enter the same site, the regular diffusion rule is applied.

4.9. Propose a LBGK equation for the diffusion rule with rest particles. What is the expression to be chosen for the local equilibrium distribution $f^{(0)}$? How is the diffusion constant related to the relaxation time?

4.10. In the diffusion model of the previous problem, how can we make the diffusion constant depend on the spatial coordinates?

4.11. Simulate a DLA process with a probability of re-dissolution of the aggregated particles. How is the fractal dimension affected?

4.12. One considers the diffusion of particles on a one-dimensional lattice. One assumes that there are traps (acting as sinks of particles) distributed on the lattice with a concentration c. Two situations have to be distinguished: the quenched case where the traps are at fixed positions during the diffusion process and the annealed case where the traps move on a time scale much faster than the diffusing species. Assuming that there is a source of particles at the origin $x = 0$ and a sink at $x = \infty$, find analytically the stationary diffusion profile $P(x)$ for both cases.

4.13. One considers the situation described in the previous exercise, but now on a two-dimensional lattice. Devise an algorithm (using one of the standard diffusion rule discussed in this chapter) which simulates the annealed case. Investigate the properties of $P(x)$ as a function of the concentration of traps c.

4.14. Find a CA model for the random deposition of dimers on a 2D lattice. A dimer is made of two adjacent cells and can be oriented horizontally or vertically. It can be adsorbed on the substrate only if two nearest-neighbor sites are empty (hint: see section 6.3 for the concept of matching nearest neighbor). What is the average jamming coverage of the lattice? What is the asymptotic time evolution (power law) of this process?

5

Reaction-diffusion processes

5.1 Introduction

Systems in which different chemical species diffuse (for instance in a solvent or a gel) and then react together are quite common in science. Lattice gases provide a very natural framework to model such phenomena. In the previous chapter, we discussed how diffusion can be implemented as a synchronous random walk of many particles, using a velocity shuffling technique.

The model can be extended so that several different "chemical" species coexist simultaneously on the same lattice. It just requires more bits of information to store the extra automaton state. Then, it is easy to supplement the diffusion rule with the annihilation or creation of particles of a different kind, depending on the species present at each lattice site and a given reaction rule.

In agreement with the cellular automata approach, chemical reactions are treated in an abstract way, as a particle transformation phenomena rather than a real chemical interaction. Only the processes relevant at the macroscopic level are taken into account.

Systems in which reactive particles are brought into contact by a diffusion process and transform, often give rise to very complex behaviors. Pattern formation [128–130], is a typical example of such a behavior in reaction-diffusion processes.

In addition to a clear academic interest, reaction-diffusion phenomena are also quite important in technical sciences and still constitute numerical challenges. As an example, we may mention the famous problem of carbonation in concrete [131,132].

Reaction processes are most of the time nonlinear and characterized by a threshold dynamics. Therefore, they may be very difficult to analyze theoretically and even numerically, with standard techniques, due to the important role that fluctuations may play.

For instance, even simple systems behave in an unexpected way: some simple reaction-diffusion processes exhibit an anomalous kinetics: they depart from the behavior predicted by the classical approach based on differential equations for the densities. The reason is that they are fluctuations-driven and that correlations cannot be neglected. In other words, one has to deal with a full N-body problem and the Boltzmann factorization assumption is not valid. For this kind of problem, a lattice gas automata approach turns out to be a very successful approach.

As already mentioned, lattice gas automata are a particular class of cellular automata. They are characterized by a two-phase dynamics: first, a completely local interaction on each lattice point, and then particle transport, or propagation, to nearest-neighbor sites. This is a way to partition the space, thus preventing the problem of having a particle simultaneously involved in several different interactions. Moreover, this approach yields a quite intuitive interpretation of the system as a kind of a fully discrete molecular dynamics and allows direct connection with statistical mechanics.

However, some reactive phenomena can also be described by simple cellular automata models which are not LGA. In the next chapter, we shall see an example of a surface reaction on a catalytic substrate. Here, we shall start our discussion with a model for an excitable media, in which chemical waves are observed. Then we shall concentrate on the lattice gas approach which is well suited to represent reaction-diffusion processes. The microdynamics will be given as well as its link to macroscopic rate equation. Various applications will be presented in this framework as well as in the framework of a lattice Boltzmann model and a multiparticle approach. The reader can find other examples and applications in [50].

A natural extension of what will be discussed in this chapter is to give the particles a hydrodynamical dynamics instead of a diffusive one. Thus, models for studying combustion phenomena can be devised.

5.2 A model for excitable media

In this section we present a simple cellular automata model for chemical waves in an excitable media. Strictly speaking, this is not an LGA since there is no explicit propagation phase. Rather, the rule is defined on the Moore neighborhood and the transport of information takes place because the neighborhood has some spatial extension.

An excitable medium is basically characterized by three states [50]: the resting state, the excited state and the refractory state. The resting state is a stable state of the system. But a resting state can respond to a local

Fig. 5.1. Excitable medium: evolution of a stable initial configuration with 10% of excited states $\phi = 1$, for $n = 10$ and $k = 3$. The color black indicates resting states. After a transient phase, the system sets up in a state where pairs of counter-rotating spiral waves propagate. When the two extremities come into contact, a new, similar pattern is produced.

perturbation and become excited. Then, the excited state evolves to a refractory state where it no longer influences its neighbors and, finally, returns to the resting state.

A generic behavior of excitable media is to produce chemical waves of various geometries, traveling in the system [133,134]. Ring and spiral waves are a typical pattern of excitations. Many chemical systems exhibits an excitable behavior. The Selkov model [135] and the Belousov–Zhabotinsky reaction are examples. Chemical waves play an important role in many biological processes (nervous systems, muscles) since they can mediate the transport of information from one place to another.

The Greenberg–Hasting model is an example of a cellular automata model of an excitable media. This rule, and its generalization, have been extensively studied [136,137]. It has some similarity with the "tube-worms" rule proposed by Toffoli and Margolus [26] (see problem 5.1).

The implementation we propose here for the Greenberg–Hasting model is the following: the state $\phi(\vec{r}, t)$ of site $\vec{r}$ at time t takes its value in the set $\{0, 1, 2, ..., n-1\}$. The state $\phi = 0$ is the resting state. The states $\phi = 1, ..., n/2$ (n is assumed to be even) correspond to excited states. The rest, $\phi = n/2 + 1, ..., n-1$ are the refractory states. The cellular automata evolution rule is the following:

1. If $\phi(\vec{r}, t)$ is excited or refractory, then $\phi(\vec{r}, t+1) = \phi(\vec{r}, t) + 1$ mod n.

2. If $\phi(\vec{r}, t) = 0$ (resting state) it remains so, unless there are at least k excited sites in the Moore neighborhood of site $\vec{r}$. In this case $\phi(\vec{r}, t) = 1$.

The n states play the role of a clock: an excited state evolves through the sequence of all possible states until it returns to 0, which corresponds to a stable situation.

The behavior of this rule is quite sensitive to the value of n and the excitation threshold k. Figures 5.1 and 5.2 show the evolution of this automaton for two different sets of parameters n and k. Both simulation are started with a uniform configuration of resting state, perturbed by some excited site randomly distributed over the system. If the concentration of perturbation is low enough, excitation dies out rapidly and the system returns to the rest state. Increasing the number of perturbed states leads to the formation of traveling waves and self-sustained oscillations may appear in the form of ring or spiral waves.

5.3 Lattice gas microdynamics

After the cellular automata model of an excitable media, we now return to the main purpose of this chapter, namely the discussion of a lattice gas automaton for describing a reaction-diffusion process.

We would like to model the following irreversible reaction

$$A + B \xrightarrow{K} C \tag{5.1}$$

where A, B and C are different chemical species, all diffusing in the same solvent, and K is the reaction constant. To account for this reaction,

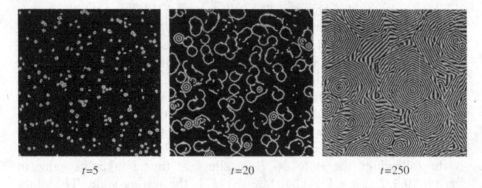

Fig. 5.2. Excitable medium: evolution of a configuration with 5% of excited states $\phi = 1$, and 95% of resting states (black), for $n = 8$ and $k = 3$.

one can consider the following mechanism: at the "microscopic" level of the discrete lattice dynamics, all the three species are first governed by a diffusion rule, as described in the previous chapter. When an A and a B particle enter the same site at the same time, they disappear and form a C particle.

Of course, there are several ways to select the events that will produce a C when more than one A or one B are simultaneously present at a given site. Also, when Cs already exist at this site, the exclusion principle may prevent the formation of new ones. A simple choice is to have A and B react only when they perform a head-on collision and when no Cs are present in the perpendicular directions. Other rules can be considered if we want to enhance the reaction (make it more likely) or to deal with more complex situations ($2A + B \rightarrow C$, for instance).

Finally, a parameter k can be introduced to tune the reaction rate K by controlling the probability of a reaction taking place.

In order to write down the microdynamic equation of this process, we shall denote by $a_i(\vec{r}, t)$, $b_i(\vec{r}, t)$ and $c_i(\vec{r}, t) \in \{0, 1\}$ the presence or the absence of a particle of type A, B or C, entering site $\vec{r}$ at time t pointing in lattice direction i.

We will assume that the reaction process first takes place. Then, the left-over particles, or the newly created ones, are randomly deflected according to the diffusion rule. Thus, using equation 4.4, we can write the reaction-diffusion microdynamics as (d is the dimensionality of the Cartesian lattice)

$$a_i(\vec{r} + \lambda \vec{e}_i, t + \tau) = \sum_{\ell=0}^{2d-1} \mu_\ell(\vec{r}, t) \left[a_{i+\ell}(\vec{r}, t) + R^a_{i+\ell}(a, b, c) \right] \tag{5.2}$$

and similarly for the two other species B and C.

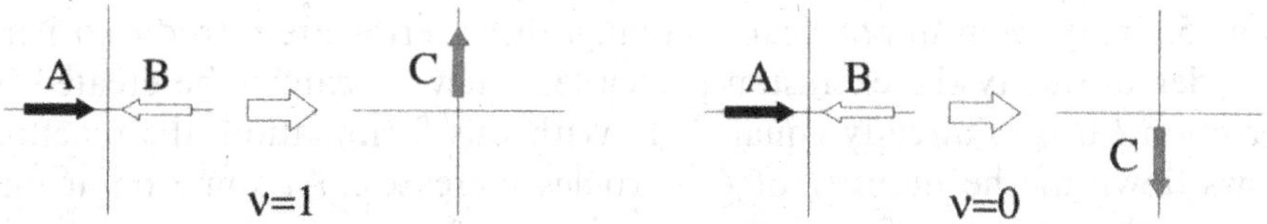

Fig. 5.3. Automata implementation of the $A + B \rightarrow C$ reaction process.

As before, the $\mu_\ell(\vec{r}, t)$ are independent random Boolean variables producing the direction shuffling. The lattice spacing λ and time steps τ are introduced as usual and the lattice directions $\vec{e}_i$ are defined in figure 4.2 (note that we have changed the notation in order to avoid confusion with the occupation number c_i of species C).

The quantity $R_j^a(a, b, c)$ is the reaction term: it describes the creation or the annihilation of an A particle in the direction j, due to the presence of the other species. In the case of an annihilation process, the reaction term takes the value $R_j^a = -1$ so that $a_j - R_j^a = 0$. On the other hand, when a creation process takes place, $a_j = 0$ and $R_j^a = 1$. When no interaction occurs, $R_j^a = 0$.

For instance, in the case of the reaction 5.1 (illustrated in figure 5.3), the reaction terms could be written as

$$
\begin{aligned}
R_i^a &= -\kappa a_i b_{i+2} \left[v(1 - c_{i+1}) + (1 - v)(1 - c_{i-1}) \right] \\
R_i^b &= R_{i+2}^a \\
R_i^c &= \kappa(1 - c_i) \left[v a_{i-1} b_{i+1} + (1 - v) a_{i+1} b_{i-1} \right]
\end{aligned}
\tag{5.3}
$$

R_i^a and R_i^b are annihilation operators, whereas R_i^c corresponds to particle creation. One can easily check that, for each A (or B) particle which disappears, a C particle is created. That is,

$$
\sum_{i=1}^{2d} R_i^a = \sum_{i=1}^{2d} R_i^b = -\sum_{i=1}^{2d} R_i^c
$$

The quantities $v(\vec{r}, t)$ and $\kappa(\vec{r}, t)$ in equations 5.3 are independent random bits, introduced in order to select among the various possible events: $v(\vec{r}, t)$ is 1 with probability $1/2$ and decides in which direction the reaction product C is created. When $v = 1$, the new C particle forms a $+90^o$ angle with respect to the old A particle. This angle is -90° when $v = 0$.

The occurrence of the reaction is subject to the value of the Boolean variable κ. With probability k, $\kappa = 1$. Changing the value of k is a way to adjust the reaction constant K. We shall see that k and K are proportional.

The presence of the terms involving c_i in the right-hand side of equations 5.3 may seem unphysical. Actually, these terms are introduced here in order to satisfy the exclusion principle: a new C cannot be created in direction i if c_i is already equal to 1. With this formulation, the reaction slows down as the number of C particles increases. At some point one may reach saturation if no more room is available to create new particles.

In practice, however, this should not be too much of a problem if one works at low concentration. Also, quite often, the C species also undergoes a transformation: the reaction can be reversible or C particles can precipitate if the concentration reaches some threshold. Or, sometimes, one is only interested in the production rate $\sum_j R_j^a = \sum_j R_j^b$ of the species C and one can forget about them once they are created. In this case, one simply puts $c_i = 0$ in the first two equations of 5.3.

Clearly, the exclusion principle may introduce some renormalization of the reaction rate. If for some reason, this is undesirable, multiparticle models offer an alternative to the LGA approach. This will be discussed in section 5.7.

Due to the simple microscopic interpretation, equation 5.3 is easily generalized to other reaction processes. A common situation is when one species is kept at a fixed concentration. This means that the system is fed a chemical by an external mechanism. In this case, the corresponding occupation numbers (for instance the b_is) can be replaced be random Boolean variables which are 1 with a probability given by the selected concentration of the species.

5.3.1 *From microdynamics to rate equations*

Several times, throughout this book, we have established a link between the discrete cellular automata dynamics and the corresponding macroscopic level of description. Here, again, we shall perform this calculation for the case of the $A + B \to C$ process. Generalization to other reaction schemes is straightforward. As in the previous chapters, we use the Boltzmann molecular chaos assumption, in which correlations are neglected. Within this approximation, we show that the microdynamics of reaction-diffusion processes yields the usual rate equation, which, in our case, reads

$$\partial_t \rho_A = D\nabla^2 \rho_A - K\rho_A\rho_B \tag{5.4}$$

To derive the macroscopic behavior of our automata rule, we first average equation 5.2

$$A_i(\vec{r} + \lambda\vec{e}_i, t + \tau) - A_i(\vec{r}, t) = \sum_{j=1}^{2d} \Omega_{ij} A_j(\vec{r}, t) + \sum_{j=1}^{2d} (\delta_{ij} + \Omega_{ij}) R_j^a(A, B, C) \tag{5.5}$$

where $A_i = \langle a_i \rangle$ is the average value of the occupation numbers a_i. The matrix Ω is the diffusion matrix defined in equation 4.14 for a two-dimensional system. Similar equations hold for B_i and C_i.

Using the Boltzmann hypothesis, the average value of the reaction term is written as

$$\langle R_i^a(a, b, c, \kappa, v) \rangle \approx R_i^a(A, B, C, \langle \kappa \rangle, \langle v \rangle) \tag{5.6}$$

that is the same function, but taken for the average values A, B and C instead of the occupation numbers a, b and c, and for the average values of the random Boolean fields. We shall see, in section 5.4 that this factorization may be wrong for simple annihilation reaction-diffusion. But, for the moment, we will assume that it is valid.

The second step is to replace the finite difference in the left-hand side of 5.5 by its Taylor expansion

$$A_i(\vec{r} + \lambda \vec{e}_i, t + \tau) - A_i(\vec{r}, t) =$$
$$\left[\tau \partial_t + \frac{\tau^2}{2} \partial_t^2 + \lambda(\vec{c}_i \cdot \partial_{\vec{r}}) + \frac{\lambda^2}{2}(c_i \cdot \partial_{\vec{r}})^2 + \tau \lambda \partial_t(\vec{c}_i \cdot \partial_{\vec{r}}) \right] A_i \tag{5.7}$$

and similarly for the other species B and C. As usual, we will consider a Chapman–Enskog-like expansion and look for a solution of the following form

$$A_i = A_i^{(0)} + \epsilon A_i^{(1)} + \epsilon^2 A_i^{(2)} + \dots \tag{5.8}$$

Since particle motion is governed by the diffusion process, we will use the result derived in section 3.1.1, namely that, when taking the continuous limit, the time and length scale are of the following order of magnitude

$$\lambda = \epsilon \lambda_1 \quad \text{and} \quad \tau = \epsilon^2 \tau_2 \tag{5.9}$$

In reactive systems, as opposed to hydrodynamics or pure diffusion, neither momentum nor particle number are conserved in general. For instance, in the annihilation process $A + A \rightarrow \emptyset$, no conservation law holds.

On the other hand, the reaction term can be considered as a perturbation to the diffusion process, which makes derivation of the macroscopic limit rather simple. In equation 5.4, the reaction constant K is proportional to the inverse of a time. This quantity defines at what speed the reaction occurs. At the level of the automaton, this reaction rate is controlled by the reaction probability $k = \langle \kappa \rangle$ introduced in the previous section.

Now, when the continuous limit is taken, the automaton time step τ goes to zero. Thus, the number of reactions per second will increase as τ decreases, unless the reaction probability k also diminishes in the right ratio. In other words, to obtain a finite reaction constant K in the macroscopic limit, it is necessary to consider that $k \propto \tau$. Since τ is of the order ϵ^2 in our Chapman–Enskog expansion, the reaction term R_i^a is also to be considered as an $O(\epsilon^2)$ contribution

$$R_i^a = \epsilon^2 R_{2i}^a \tag{5.10}$$

At the macroscopic level the physical quantities of interest are the particle densities of each species. Following the usual method, we define the density ρ_A of the A species as

$$\rho_A = \sum_{i=1}^{2d} A_i^{(0)}$$

with the condition

$$\sum_{i=1}^{2d} A_i^{(\ell)} = 0 \qquad \text{if } \ell \geq 1$$

Now we have to identify the different orders in ϵ which appear in equation 5.5, using the expressions 5.7, 5.8, 5.9 and 5.10. We obtain

$$O(\epsilon^0): \quad \sum_j \Omega_{ij} A_i^{(0)} = 0 \tag{5.11}$$

$$O(\epsilon^1): \quad \lambda_1 (\vec{e}_i \cdot \nabla) A_i^{(0)} = \sum_j \Omega_{ij} A_j^{(1)} \tag{5.12}$$

These equations are exactly similar to those derived in the case of pure diffusion (see equations 4.11 and following). Thus we know that

$$A_i^{(0)} = \frac{\rho_A}{2d}$$

and

$$A_i^{(1)} = \frac{\lambda_1}{2d} \frac{1}{V} e_{i\alpha} \partial_\alpha \rho_A$$

where V is the eigenvalue of the diffusion matrix Ω for the eigenvector

$$E_\alpha = (e_{1,\alpha}; e_{2,\alpha}; ...; e_{2d,\alpha})$$

The equation for the density ρ_A is now obtained by summing over i equation 5.5, remembering that

$$\sum_i \Omega_{ij} = 0$$

Collecting all the terms up to $O(\epsilon^2)$, we see that the orders $O(\epsilon^0)$ and $O(\epsilon)$

vanish and we are left with

$$\epsilon^2 \tau_2 \partial_t \sum_i A_i^{(0)} + \epsilon^2 \lambda_1 \sum_i (\vec{e}_i \cdot \nabla) A_i^{(1)} + \epsilon^2 \frac{\lambda_1}{2} \sum_i (\vec{e}_i \cdot \nabla)^2 A_i^{(0)}$$
$$= \epsilon^2 \sum_j R_{2j}^a (A^{(0)}, B^{(0)}, C^{(0)})$$

Using the definition of τ_2, λ_1, R_{2j}^a and performing the summations yield

$$\partial_t \rho_A = D\nabla^2 \rho_A + \frac{1}{\tau} \sum_j R_j^a \left(\frac{\rho_A}{2d}, \frac{\rho_B}{2d}, \frac{\rho_C}{2d}\right) \tag{5.13}$$

where D is the same diffusion constant as would be obtained without the chemical reactions (see equation 4.28).

It is interesting to note that the expression 5.13 has been obtained without knowing the explicit expression for the reaction terms R and independently of the number of species. Actually, from this derivation, we see that the reaction term enters in a very natural way in the macroscopic limit: we just have to replace the occupation numbers by $\rho/2d$, the random Boolean fields by their average values and sum up this result for all lattice directions.

For the case of the $A + B \rightarrow C$ process in two dimensions, with the reaction term given by 5.3, equation 5.13 shows that the macroscopic behavior is described by the rate equations

$$\partial_t \rho_A = D_A \nabla^2 \rho_A - \frac{k}{4\tau}\left(1 - \frac{\rho_C}{4}\right)\rho_A \rho_B$$
$$\partial_t \rho_B = D_B \nabla^2 \rho_B - \frac{k}{4\tau}\left(1 - \frac{\rho_C}{4}\right)\rho_A \rho_B$$
$$\partial_t \rho_C = D_C \nabla^2 \rho_C + \frac{k}{4\tau}\left(1 - \frac{\rho_C}{4}\right)\rho_A \rho_B \tag{5.14}$$

where, in principle a different diffusion constant can be chosen for each species. We also observe that the reaction constant K is related to the reaction probability k by

$$K = \frac{k}{4\tau}$$

But, as explained previously, the exclusion principle introduces a correction $(1 - \rho_c/4)$ which remains small as long as C is kept in low concentration.

5.4 Anomalous kinetics

In the previous section, we derived the partial differential rate equation governing the behavior of our cellular automata reagents with the assumption that many-body correlations can be neglected. When the reaction process is rare enough so that diffusion is efficient enough to mix the particles thoroughly, this derivation makes sense. However, correlations or intrinsic fluctuations often play a crucial role in reaction-diffusion processes. Whereas the cellular automata approach naturally takes into account this phenomenon, the rate equation may not always capture correctly the effect of these fluctuations, especially if the space dimensionality is low.

In this section we present simple reaction-diffusion systems in which fluctuations cannot be neglected or many-body correlations are non-zero. This will be the annihilation processes $A + B \rightarrow \emptyset$ or $A + A \rightarrow \emptyset$ where $\emptyset$ indicates that the reaction products of A and B do not play any role in the system and can be forgotten.

5.4.1   The homogeneous $A + B \rightarrow \emptyset$ process

The simple irreversible reaction-diffusion process $A + B \rightarrow \emptyset$ has received a great deal of attention among chemists and physicists (see for instance [138–143]).

Traditionally, this process is described in terms of rate equations for the local concentrations a and b of the various species

$$\partial_t a = D\nabla^2 a - R \qquad \partial_t b = D\nabla^2 b - R \qquad (5.15)$$

where D is the diffusion constant and R the reaction term. Usually, R is assumed to be proportional to the product of the a and b concentrations, multiplied by K, a reaction constant:

$$R = Kab \qquad (5.16)$$

The approximation is often referred to as a mean-field approximation since it assumes that the probability of a reaction taking place simply factorizes into the product of the probabilities of the presence of an A and a B particle.

Equations 5.15 with $R = kab$ are easy to solve when the system is homogeneous at time $t = 0$ (i.e a and b do not depend on space variables). The Laplacian terms vanish and one is left with ordinary differential equations. If the initial concentrations a_0 and b_0 are equal, $a(t)$ will be equal to $b(t)$ at any time and one has to solve

$$\frac{da}{dt} = -Ka^2$$

whose solution is simply

$$a(t) = \frac{a_0}{1 + a_0 kt} \qquad (5.17)$$

Note that if $a_0 \neq b_0$, say $b_0 > a_0$, the solution is different and is found to be

$$a(t) = a_0 \frac{(b_0 - a_0)\exp(-(b_0 - a_0)kt)}{b_0 - a_0 \exp(-(b_0 - a_0)kt)} \qquad (5.18)$$

Equation 5.17 predicts a t^{-1} behavior in the long time regime, whereas equation 5.18 gives an exponential decay. Unfortunately, these results are wrong if $a_0 = b_0$. It is now widely recognized that when the number of A and B particles are initially equal, the correct dynamics is (see [139], for instance)

$$a(t) \sim t^{-d/4} \qquad (5.19)$$

where $d \leq 4$ is the dimensionality of the system. Therefore, in dimension one, two and three, the predictions of the mean-field equations are wrong and an anomalous kinetics is observed. This is due to fluctuations and correlations which always exist in a many-particle system. In low dimensions, diffusion is not a good enough mechanism to mix correctly the particles.

This $t^{-d/4}$ law can be understood as follows. Let us consider a time interval τ. Due to diffusion, a particle explores a typical distance $l = \sqrt{D\tau}$. Suppose we divide our system in volumes of size l. If the size of the system is L, there are $(L/l)^d$ such volumes. Assuming that the average initial density of particles is $a_0 = b_0$, the average number N of A or B particles in our smaller volumes is $N = a_0 l^d = a_0(D\tau)^{d/2}$. However, this quantity is subject to statistical fluctuations of the order $\sqrt{N}$. Therefore, if the distribution of A and B particles is *not correlated* the number of A and B particles in our volumes typically differs from $\sqrt{N} = \sqrt{a_0}(D\tau)^{d/4}$. If, during the time interval τ, all the A and B particles located in the same volume l^d react together (which is reasonable since τ is the time needed to explore the volume), the total number of particles left after that time will be the excess of particles of one of the two species. Thus

$$a(\tau) = \sqrt{N}\left(\frac{L}{l}\right)^d = L^d\sqrt{a_0}(D\tau)^{-d/4} \qquad (5.20)$$

In addition to the time exponent, equations 5.20 and 5.17 differ by their amplitudes: equation 5.20 depends on $\sqrt{a_0}$ and D, whereas 5.17 depends on k but no longer on the density a_0, in the long time regime.

This discussion suggests the following physical picture: when the species A and B are randomly distributed in the solvent and the position of A and B particles are uncorrelated, then local density fluctuations make some region richer in A than in B particles. The reverse is true in some other regions. If the reaction is fast enough compared to the diffusion, the species which is more abundant will eliminate the other. A segregation occurs, producing clusters of A surrounded by B clusters. The reaction is then no longer a bulk reaction but a surface reaction since it is only at the domain boundaries that annihilation takes place. This slows down the dynamics and yields the $t^{-d/4}$ decay law instead of the expected mean-field t^{-1} behavior.

We have insisted on the fact that the initial A and B particle distributions should be noncorrelated in order for the $t^{-d/4}$ to be true. When this does not hold, another law is observed [142,143]. The system looks more like the $A + A \rightarrow \emptyset$ reaction, for which a $t^{-d/2}$ decay law holds [139,144]. Thus, in one dimension, a non-mean-field behavior (i.e a deviation from the predictions of equation 5.17) occurs.

At this point we may notice that there are two reasons to depart from a mean-field behavior: (i) the initial condition or, more precisely, the nature of the fluctuations in the initial distribution of the the A and B particles; and (ii) the factorization of the joint probability that an A particle meets a B particle into the product of the local density of A and B.

The fact that the initial condition plays such a crucial role makes the $A + B \rightarrow \emptyset$ system quite intricate. It is sometime difficult to differentiate between the two effects. Actually, both initial condition and many-body correlations may simultaneously affect the system. The slowest decay law will be longer lived and mask the other behavior. A $t^{-d/4}$ will certainly screen out a $t^{-d/2}$ decay.

5.4.2   Cellular automata or lattice Boltzmann modeling

The $A + B \rightarrow \emptyset$ reaction-diffusion process is therefore a good candidate for cellular automata modeling where no mean-field assumptions are made for the initial condition, or for the joint probability that an A and a B annihilate. Using the reaction term given in relation 5.3, with c_i always zero provides a natural implementation of this process.

However, some care is needed to obtain reliable simulations. Earlier in this book, we mentioned the checkerboard spurious invariant that plagues many lattice gas automata. In our case, the checkerboard invariant means that an A particle entering a "black" site of the lattice at time t will never interact with a B particle entering a "white" site, and conversely. Thus the total number of A particles left on the lattice as a function of time is composed of two contributions, one from the black sublattice and the other from the white sublattice. Since the two sublattices may not have the same number of particles, the measurement of particle densities may be erroneous.

The simplest solution is to consider an initial condition where all A and B particles occupy only one sublattice, the other being left empty.

The second technical difficulty is to build an initial condition with an equal number of *A* and *B* particles. This is a crucial condition to observe a power law instead of an exponential decay. The problem is to make both *A* and *B* distributions statistically independent. One possibility to achieve this result is to first take the *B* distribution equal to that of *A*. Then, the *B* particles should be sent somewhere in the considered sublattice, so that the original distribution is forgotten. The diffusion cellular automata rule can be used for this purpose but for a number of iterations at least as large as the square of the system size. On a general purpose computer, faster ways can be devised: a random permutation of the lattice sites can be performed to mix the position of the *B* particles [145].

Now, as already mentioned, we would like to identify more precisely the role of the fluctuations in the initial configuration against the role of the Boltzmann factorization assumption. For this purpose, a lattice Boltzmann model (evolution rule 5.5) can be simulated and compared with the cellular automata dynamics. The lattice Boltzmann approach makes it possible (and natural) to prepare a random initial condition, in which $a_i(\vec{r}, t)$ and $b_i(\vec{r}, t)$ are either 1 or 0, in exactly the same way as in the cellular automata case.

But the lattice Boltzmann method neglects many-body correlations, and when both the cellular automata rule and the lattice Boltzmann dynamics agree (and still depart from the mean-field predictions), we may conclude that we are observing an effect of the fluctuations of the initial configuration. Otherwise, we conclude that correlation cannot be factorized as a product of the one-point densities.

5.4.3 Simulation results

The behavior of the CA model and the LBM simulations are illustrated in figure 5.4. The number of *A* particle is exactly the same as the number of *B* particles and there is no correlation between their respective initial distributions. The power law 5.19 is well obeyed in both approaches.

On the other hand, when the initial distributions of *A* and *B* are correlated, a different behavior is observed. Figure 5.5 shows the behavior of the total number of *A* particles as a function of time, in a log–log plot. The two curves correspond to the lattice Boltzmann and the cellular automata dynamics, respectively. In both cases, the initial condition is prepared identically: (i) the *A* particles are placed randomly on the lattice and (ii) the *B* particles are obtained from the *A* particles by two diffusion steps. In this way, a correlation between the positions of the *A* and *B* particles is created.

Thus, the lattice Boltzmann dynamics predicts a long time regime which is not the same as the one observed in the cellular automata model. The

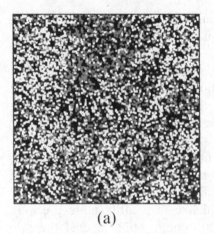
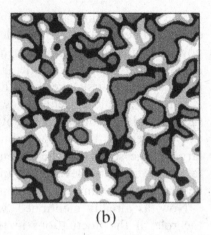

(a)   (b)

Fig. 5.4. Segregation in the $A + B \to \emptyset$ reaction-diffusion process, as obtained from (a) the cellular automata model and (b) the lattice Boltzmann dynamics. The situation corresponds to the evolution of a noncorrelated random, homogeneous initial condition. In (a) the white dots represent lattice sites with a majority of A particles and the gray ones a majority of B particles. The black dots correspond to empty sites. In (b) the gray levels indicate the regions where A (or B) is dominant: black when $a(\vec{r}, t)/b(\vec{r}, t) > 1$, dark gray when $a(\vec{r}, t)/b(\vec{r}, t) > 6$, light gray when $b(\vec{r}, t)/a(\vec{r}, t) > 1$, and white when $b(\vec{r}, t)/a(\vec{r}, t) > 6$.

$-1/2$ time exponent found in the cellular automata is not so surprising since the correlated $A + B \to \emptyset$ reaction presents many similarities with the $A + A \to \emptyset$ problem, for which $t^{-1/2}$ is the known decay law. It turns out that if one simulates the $A + A \to \emptyset$ process along the same lines, the cellular automata dynamics indeed reproduces the expected $t^{-1/2}$ behavior in one dimension, whereas the corresponding lattice Boltzmann simulation gives the mean-field result t^{-1}, in spite of a randomly distributed initial state.

Clearly, the factorization done in the lattice Boltzmann approach neglects an important aspect of the problem, which is the non-mean-field behavior of the dynamics itself, in one dimension. A cellular automata model captures this behavior because it makes no assumption on the joint probability of having an A particle meeting a B particle. We will return to this question in section 5.8, where it is explained how an exact solution of the $A + A \to \emptyset$ can be obtained analytically, in one dimension.

At this point it is interesting to note that the $A + B \to \emptyset$ reaction is even more intricate than just a correlated or noncorrelated initial configuration of the As and Bs. Depending on the way the initial correlation is built, different time exponents may be observed in both approaches.

When the initial correlation is created by a translation of the one species

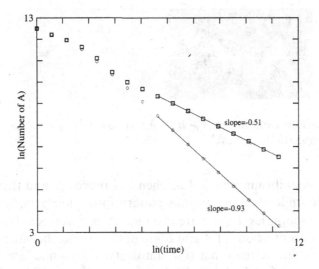

Fig. 5.5. Time behavior of the total number of A particles in the $A + B \rightarrow \emptyset$ process, in one dimension. The two plots correspond to the lattice Boltzmann (dots) and the cellular automata (squares) dynamics, both for a correlated initial condition. The lattice Boltzmann takes into account the fluctuations of the initial state but not the fact that the dynamics itself is not mean-field. The observed time behavior is t^α, with $\alpha \approx 1/2$ for the cellular automata and $\alpha \approx 0.9$ for the lattice Boltzmann dynamics.

with respect to the other, the predictions found by Lindenberg *et al.* [142] (namely a $t^{-(d+2)/4}$ long time regime and a $t^{-d/4}$ short time regime) are verified in the lattice Boltzmann dynamics. The cellular automata dynamics, on the other hand, does not agree well with these theoretical predictions which do not take the fluctuations fully into account.

5.5 Reaction front in the $A + B \rightarrow \emptyset$ process

The problem we discussed in the previous section corresponds to a homogeneous initial situation where all the reagents are uniformly (up to fluctuations) distributed in space. Another situation is obtained when the chemical species A and B are initially separated (inhomogeneous systems). For instance we may have the situation described in figure 5.6: the A particles occupy the left part of the system while the B particles are on the right. The A species try to diffuse into the B region, and conversely. A reaction take place in which an A–B pair is annihilated (or transformed into an inert substance).

The region where the reaction occurs is called the reaction front. In figure 5.6, this corresponds to the white pixels. Reaction fronts are present

Fig. 5.6.   Reaction front in the $A + B \rightarrow \emptyset$ reaction diffusion process (according to cellular automata dynamics).

in many nonequilibrium physical or chemical processes and their study is a first step to understanding various pattern formation phenomena.

On a macroscopic level, there are two important features to be described. First, if the concentrations of A and B are not the same, the more abundant species will penetrate the other (for simplicity, we assume here that both diffusion coefficients are identical). As a results, the position of the front will move and be a function of time. Second, as times goes on, the reaction front becomes larger and larger because, due to the depletion of A and B particles around the reaction zone, particles of one type have a greater chance to enter deeper into the other region before reacting.

Therefore the question is: what are the front width $W(t)$ and the front location $\vec{r}_0(t)$?

Pioneering work on this subject has been done by Gálfi and Rácz [146]. Using the rate equations 5.4, they considered two species A and B diffusing on a two-dimensional substrate and reacting to form an inert species C. For an appropriate choice of the initial conditions (at time $t = 0$, the system is uniformly filled with A for $x \leq 0$, and uniformly filled with B for $x > 0$), the problem reduces to one dimension. Gálfi and Rácz developed a scaling theory from which they found that the front position obeys $\vec{r}_0 \sim \sqrt{t}$ (with a coefficient depending on the initial concentration difference $A_0 - B_0$). Second, their scaling theory predicts that front width $W(t)$ follows, in the long time limit, the power law $W(t) \sim t^{1/6}$.

Whereas the $\sqrt{t}$ law giving the front position holds in all generality, the question of the width is more involved. It was shown [147], using cellular automata models, that the upper critical dimension of the problem is 2. That is, logarithmic corrections to the rate equation predictions (i.e. the $W \sim t^{1/6}$ behavior) are found for the space dimension $d = 2$ and an important correction is measured in $d = 1$. This reaction-diffusion process has now been intensively studied due to its departure from the mean-field description. The reason for this departure is of course that the average reaction term $\langle R \rangle$ is not the product of the A density times the B density. Instead, it is the joint probability of having one A and one B particle simultaneously present at the same position.

Unfortunately, this joint probability is not known in general. An equation for it could be written down but it would involve other higher-order correlation functions.

Thus the problem of finding the right behavior for $W(t)$ in low-dimensional systems is clearly a very difficult analytical task. Of course the reader may argue that, since the upper critical dimension is $d = 2$, everything will be fine in the 3d physical space and the mean-field result is enough. However, the problem is very interesting from a theoretical point of view and sheds some light on what may happen even in very simple nonequilibrium systems. In addition, some experiment have been realized in very thin tubes [148] but they are not thin enough to probe the one-dimensional regime.

5.5.1   The scaling solution

The cellular automata approach is a very successful tool in the analysis of reaction diffusion fronts. Moreover, a scaling theory can be considered in order to interpret, in a theoretical framework, the simulation results.

The idea is to write macroscopic rate equations (with dependence only on one spatial coordinate for symmetry reasons) as:

$$\partial_t a = D\nabla^2 a - KR, \qquad \partial_t b = D\nabla^2 b - KR \tag{5.21}$$

but without specifying an explicit form for the reaction term R. Here, as opposed to the homogeneous $A + B \to \emptyset$ case, the spatial fluctuations of the initial configuration does not play a crucial role since we assume that, at time $t = 0$, $a_i = 1$, $b_i = 0$ on the left part of the system and conversely on the right part. Thus, we may argue that the departure from mean-field behavior is due to the fact that

$$R \neq R_{\mathrm{mf}} = ab$$

where R_{mf} denotes the standard mean-field approximation. In general, R is the joint probability of having one particles of type A and one particles of type B simultaneously present at the same point. Of course, since R is not known, we cannot solve equation 5.21. However, in the spirit of a scaling approach, we may say something about the asymptotic time dependency of $a(x, t)$ and $b(x, t)$. Thus, we shall assume that

$$
\begin{aligned}
a(x,t) &= t^{-\gamma}\hat{a}(xt^{-\alpha}) \\
b(x,t) &= t^{-\gamma}\hat{a}(xt^{-\alpha}) \\
R(x,t) &= t^{-\beta}\hat{R}(xt^{-\alpha})
\end{aligned} \tag{5.22}
$$

where α, β and γ are the exponents we are looking for. From these exponents, we can obtained the behavior of the width front $W(t)$ defined as

$$W^2(t) = \frac{\sum_{x=-L}^{L-1}(x - \langle x \rangle)^2 R(x,t)}{\sum_{x=-L}^{L-1} R(x,t)} \tag{5.23}$$

Using the scaling form for R we have

$$W(t) \sim t^\alpha$$

Similarly, we also obtain that the total number of particle annihilations is related to α and β

$$\int_{volume} R \sim t^{-\beta+\alpha} \tag{5.24}$$

Now, the nice thing is that some general relations between these exponents can be derived from two assumptions [146,149] based on the fact that A and B react only in a localized region:

(i) The reaction zone increases more slowly than the characteristic length of the diffusion process. This means that, in the long time limit, a and b vary over a typical length governed by diffusion, i.e. grad$a \sim$ grad$b \sim t^{-1/2}$ and thus

$$\gamma + \alpha = 1/2 \tag{5.25}$$

(ii) Due to the gradient concentration of A and B, a flux of particles towards the reaction region is observed. Assuming that the reaction is fed by these particle currents, the reaction-diffusion equations take on a quasi-stationary form $D\nabla^2 a = KR$ and $D\nabla^2 b = KR$. For the exponents, this amounts to the relation

$$\gamma + 2\alpha = \beta \tag{5.26}$$

For consistency with assumptions (i) and (ii), we should also have

$$\alpha < 1/2 \tag{5.27}$$

so that the width of the front grows more slowly than the depletion zone produced by the diffusion. To ensure the quasi-stationarity of the solution, we need furthermore that

$$D\nabla^2 a \sim t^{-\gamma-2\alpha}$$
$$\gg \partial_t a = -t^{-\gamma-1}(\gamma\hat{a} + \alpha x t^{-\alpha}\hat{a}') \tag{5.28}$$

which is again consistent with the condition 5.27.

Relations 5.25, 5.26 and 5.29 are still quite general since no explicit form of the reaction term R has been used. But, they are not enough to find the value of α, β and γ. One of these exponent must be determined numerically, using our cellular automata model.

Note, however, that we can already deduce that

$$\beta - \alpha = 1/2 \tag{5.29}$$

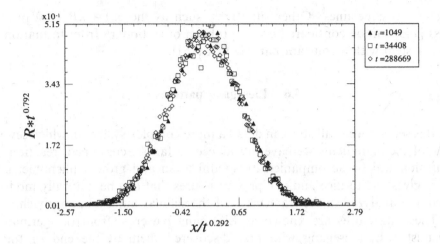

Fig. 5.7. Scaling profile of the reaction term, as obtained with a data collapse from the cellular automata simulations.

Therefore, according to equation 5.24, the total annihilation rate decreases with time as $t^{-1/2}$.

If we consider the mean-field approximation $R = ab$, this introduces another relation among our exponents, namely

$$\beta = 2\gamma \tag{5.30}$$

and we obtain the following the values for the mean-field exponents

$$\alpha_{mf} = \frac{1}{2} \qquad \gamma_{mf} = \frac{1}{3} \tag{5.31}$$

From our one-dimensional cellular automata simulations, the values of the non-mean-field exponents are found to be [150]

$$\alpha = 0.292 \pm 0.005 \tag{5.32}$$
$$\beta = 0.792 \pm 0.006 \tag{5.33}$$
$$\gamma = 0.216 \pm 0.003 \tag{5.34}$$

which are clearly different from their mean-field counterparts. On the other hand, in two- and three-dimensional systems the mean-field exponents are well obeyed (up to a logarithmic correction in 2D).

The shape of the reaction function $\hat{R}$ is not given by the scaling analysis. However, it can be measured in the simulation. Figure 5.7 shows the observed profile. Superposition of the data obtained for different time steps confirms our scaling hypotheses.

The analysis we presented about the $A + B \to \emptyset$ reaction-diffusion process, in both homogeneous and inhogeneous cases, can be generalized along the same lines. Other situations, such as the $mA + nB \rightarrow \emptyset$ process [143] can be considered or the problem of stationary front formation in a system with a constant particle flux [151].

5.6   Liesegang patterns

In this section we shall study in detail a more complex system in which several of the ingredients we have introduced so far are combined: reaction-diffusion will be accompanied by solidification and growth phenomena. This gives rise to nice and complex structures that can be naturally modeled and analyzed in the framework of the cellular automata approach.

These structures are known as Liesegang patterns, from the german chemist R.E. Liesegang who first discovered them at the end of the nineteenth century [152].

5.6.1   *What are Liesegang patterns*

Liesegang patterns are produced by precipitation and aggregation in the wake of a moving reaction front. Typically, they are observed in a test tube containing a gel in which a chemical species B (for example $AgNO_3$) reacts with another species A (for example HCl). At the beginning of the experiment, B is uniformly distributed in the gel with concentration b_0. The other species A, with concentration a_0 is allowed to diffuse into the tube from its open extremity. Provided that the concentration a_0 is larger than b_0, a reaction front propagates in the tube. As this $A + B$ reaction goes on, formation of consecutive bands of precipitate (AgCl in our example) is observed in the tube, as shown in figure 5.8. Although this figure is from a computer simulation, it is very close to the picture of a real experiment.

The presence of bands is clearly related to the geometry of the system. Other geometries lead to the formation of rings or spirals. The formation mechanism producing Liesegang patterns seems quite general. It is assumed to be responsible for the rings present in agate rocks [153, 154] to yield the formation of rings in biological systems such as a colony of bacterium [152] or to play a role the the process of cement drying.

Although the Liesegang patterns have been known for almost one hundred years, they are still subject to many investigations. The reason is that they are not yet fully understood and can present unexpected structures depending on the experimental situation (gravity, shape of the container, etc.).

On the other hand, it has been observed that, for many different substances, generic formation laws can be identified. For instance, after a

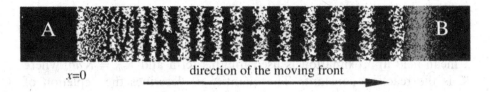

$x=0$ direction of the moving front

Fig. 5.8. Example of the formation of Liesegang bands in a cellular automata simulation. The white bands correspond to the precipitate which results from the $A + B$ reaction.

transient time, these bands appear at some positions x_i and times t_i and have a width w_i. It is first observed that the center position x_n of the nth band is related to the time t_n of its formation through the so-called *time law* $x_n \sim \sqrt{t_n}$.

Second, the ratio $p_n \equiv x_n/x_{n-1}$ of the positions of two consecutive bands approaches a constant value p for large enough n. This last property is known as the Jablczynski law [153] or the *spacing law*. Finally, the *width law* states that the width w_n of the the nth band is an increasing function of n.

In section 5.5, we explained that, in the $A + B \rightarrow \emptyset$ reaction process, the front moves according to a square root t law with a coefficient factor which depends on the difference of the concentrations a_0 and b_0.

As Liesegang patterns are formed in the wake of a moving reaction front produced by an $A + B \rightarrow C$ process, the time law appears to be a simple consequence of the diffusive dynamics. On the other hand, spacing and width laws cannot be derived with reaction-diffusion hypotheses alone. Extra nucleation–aggregation mechanisms have to be introduced, which makes any analytical derivation quite intricate [155–157].

From an abstract point of view, the most successful mechanism that can be proposed to explain the formation of Liesegang patterns is certainly the supersaturation assumption based on Ostwald's ideas [158]. This mechanism can be understood using the formation scenario proposed by Dee [159]: the two species A and B react to produce a new species C which also diffuses in the gel (this C species represents a colloidal state which is observed in several experiments). When the local concentration of C reaches some threshold value, nucleation occurs: that is, spontaneously, the C particles precipitate and become solid D particles at rest. This process is described by the following equations

$$
\begin{aligned}
\partial_t a &= D_a \nabla^2 a - R_{ab} \\
\partial_t b &= D_b \nabla^2 b - R_{ab} \\
\partial_t c &= D_c \nabla^2 a + R_{ab} - n_c \\
\partial_t d &= n_c
\end{aligned}
\tag{5.35}
$$

where, as usual, a, b, c, d stand for the concentration at time t and position $\vec{r}$ of the A, B, C and D species, respectively. The term R_{ab} expresses the production of the C species due to the $A + B$ reaction. Classically, a mean-field approximation is used for this term and $R_{ab} = Kab$, where K is the reaction constant. The quantity n_c describes the depletion of the C species resulting from nucleation and aggregation on existing D clusters. An analytical expression for this quantity is rather complicated. However, at the level of a cellular automata model, this depletion term can be included quite naturally. We will return to this point later in the discussion.

Now, within this framework, the supersaturation hypothesis can be stated as follows: due to aggregation, the clusters of nucleated D particles formed at the reaction front deplete their surroundings of the reaction product C. As a result, the level of supersaturation drops dramatically and the nucleation and solidification processes stop. To reach again suitable conditions to form new D nuclei, the $A - B$ reaction has to produce sufficient new C particles. But, the reaction front moves and this happens at some location further away. As a result, separate bands appear.

A similar scenario can be considered even if the intermediate C species does not exist as an individual diffusing molecule. In this case, the A and B species coexist in the gel until the solubility product ab reaches a critical value above which nucleation occurs according to the reaction $A + B \rightarrow AB(solid)$. As nucleation has started, depletion of A and B occurs in the surrounding volume and the same mechanism as described above takes place to produce distinct bands [128].

Not all Liesegang structures obey the spacing and width laws. The situation of inverse banding [160] can be observed in some cases, as well as some exotic behavior [161]. Although the fundamental mechanism described so far is probably still relevant, other effects should be taken into account, such as local properties of the gel, spatial variation of the diffusion coefficients and so on.

An important aspect of the mechanism of Liesegang patterns formation is the role of spontaneous fluctuations. In section 5.5 we mentioned that, in a one-dimensional system, the reaction front has an anomalous, non-mean-field behavior. Therefore, since the motion of the front is responsible for the creation of the bands of precipitate, it is likely that, in a true 1D system, Liesegang patterns no longer obey the spacing law. But, in two- and three-dimensional systems, we may safely ignore this problem. On the other hand, the precipitation process and the aggregation (such as a DLA process) are clearly dependent on local density fluctuations. For instance, even if the average particle concentration of C particles is less that the supersaturation threshold, it may be higher locally and give rise to spontaneous nucleation. Similarly, aggregation is a function of the particle density in the vicinity of an existing solid cluster, which is also a locally fluctuating quantity.

The cellular automata approach naturally accounts for these fluctuation phenomena and, in addition, captures the mesoscopic nature of the precipitate cluster, that can be fractal. As we shall see in the next section, this method turns out to be quite successful in modeling the formation of Liesegang patterns. After that, we show that a mixed lattice Boltzmann and lattice gas model can also be developed with the advantage of yielding much faster simulations.

5.6.2   *The lattice gas automata model*

Most of the ingredients we need for modeling the formation of Liesegang pattern have already been introduced in this chapter when describing the $A+B \rightarrow C$ reaction-diffusion process. In the case of Dee's scenario, we also need to provide a mechanism for spontaneous nucleation (or precipitation) in order to model the transformation of a diffusing C particle into a solid D particle. Finally, aggregation of C particles on an existing D cluster will be modeled in very much the same spirit as the DLA growth described in section 4.4.2. The key idea will be to introduce threshold values to control both of these processes.

More precisely, nucleation and aggregation phenomena are implemented according to general principles of supersaturation theory, but applied at a microscopic level. First, the C particles, once created, diffuse until their local density (computed as the number of particles in a small neighborhood divided by its total number of sites and lattice directions), reaches a threshold value k_{sp}. Then they spontaneously precipitate and become D particles at rest (nucleation). Here, we typically consider 3×3 Moore neighborhoods centered around each lattice site.

Second, C particles located in the vicinity of one or more precipitate D particles aggregate provided that their local density (computed as before) is larger than an aggregation threshold $k_p < k_{sp}$. Finally, a C particle sitting on top of a D particle always becomes a D (a third threshold k could also be used here).

The parameters k_p and k_{sp} are the two main control parameters of the model. The introduction of these critical values refers to the qualitative models of solidification theory, relating supersaturation and growth behavior [162].

As we said in section 4.4.2, it is common, from a microscopic point of view, to describe the aggregation process in terms of a noise reduction algorithm: C particles aggregate on a D cluster only after several encounters. The algorithm we use here is slightly different but allows us to produce a fast enough aggregation process, compare to the speed of the reaction front. This is important because a difference in these time scales is crucial for the formation of separate bands of precipitate.

From experiments, it is observed that Liesegang patterns are obtained only for a narrow range of parameters and fine tuning is necessary to produce them. In particular, it is important that the initial A concentration be significantly larger than the initial B concentration. In a cellular automata model with an exclusion principle, such a large difference implies having very few B particles. As a consequence, the production rate of C particles is quite low because very few reactions take place. For this reason, we have considered a pseudo-three-dimensional system composed of several two-dimensional layers. The reaction has been implemented so that particles of different layers can interact.

As a first case, we consider the situation of Liesegang bands (test tube with axial symmetry). The initial condition is built as follows: at time $t = 0$, the left part of the system ($x \leq 0$) is randomly occupied by A particles, with a density a_0 and the right part ($x > 0$) is filled with B particles with a density b_0, as described in figure 5.8.

5.6.3   Cellular automata bands and rings

Figure 5.8 shows a typical example of a cellular automata simulation with C particles, giving rise to bands. From the positions x_n and the formation time t_n of each band, we can verify the spacing and the time laws. For instance, the plot given in figure 5.9 shows very good agreement for the relation $x_n/x_{n-1} \to p$. It is found that the so-called Jablczynski coefficient p is 1.08, a value well within the range of experimental findings. Indeed, the values obtained in different experiments with axial symmetries show that $1.05 \leq p \leq 1.20$. The way the value of p depends on the parameters of the model has not been investigated yet.

As we said earlier, Liesegang patterns are found only if the parameters of the experiment are thoroughly adjusted. In our simulation, k_p and k_{sp} are among the natural quantities that control supersaturation and aggregation. In practice, however, one cannot directly modify these parameters. On the other hand, it is experimentally possible to change some properties of the gel (its pH for example) and thus influence the properties of the aggregation processes or the level of supersaturation.

Outside of the region where Liesegang patterns are formed, our simulations show that, when k_p and k_{sp} vary, other types of patterns are obtained. These various patterns can be classified in a qualitative phase diagram, as shown in figure 5.10. An example of some of these "phases" is illustrated in figure 5.11. Note that the limits between the different "phases" do not correspond to any drastic modification of the patterns. There is rather a smooth crossover between the different domains. The

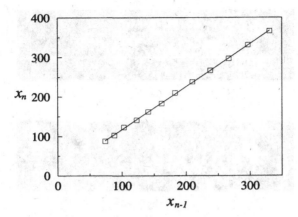

Fig. 5.9. Verification of the spacing law for the situation with C particles. The ratio x_n/x_{n-1} tends to $p = 1.08$.

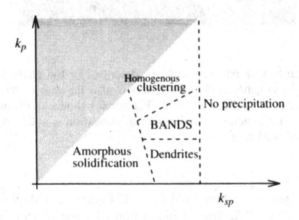

Fig. 5.10. Qualitative phase diagram showing the different possible patterns that can be obtained with our cellular automata model, as a function of the values of k_{sp} and k_p.

associated names are borrowed from the phenomenological theory of solidification [162].

The terminology of dendrite comes from the tree-like structures that are sometimes found on the surfaces of limestone rocks or plates and that can be confused with fossils. The plant-shaped deposit is made of iron or manganese oxides that appear when at some point in the geological past the limestone was penetrated by a supersaturated solution of manganese or iron ions. It turns out that the formation of these mineral dendrites can be simulated by the same scenario as Liesegang patterns,

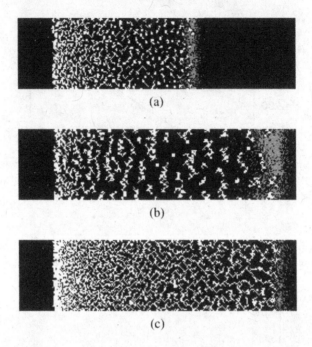

(a)

(b)

(c)

Fig. 5.11. Examples of patterns that are described in the phase diagram: (a) corresponds to homogeneous clustering; this is also the case of pattern (b) but closer to the region of band formation. Pattern (c) shows an example of what we called a dendrite structure. Amorphous solidification would correspond to a completely uniform picture.

but with an aggregation threshold $k_p = 0$. Figure 5.12 shows the results of such a modeling. The fractal dimension of these clusters is found to be around 1.77, a value which is very close to that measured in a real sample [163].

The patterns we have presented so far show axial symmetry, reflecting the properties of the experimental setup. But the same simulations can be repeated with different initial conditions. A case of interest is the situation of radial symmetry responsible for the formation of rings or spirals. The reactant A is injected in the central region of a two-dimensional gel initially filled with B particles. The result of the cellular automata simulation is shown in figure 5.13. In (a) concentric rings are formed, starting from the middle of the system and appearing as the reaction front radially moves away. In (b) a spiral-shaped structure is created. Although the two situations are similar as far as the simulation parameter are concerned, the appearance of a spiral stems from a spontaneous spatial fluctuation which breaks the radial symmetry.

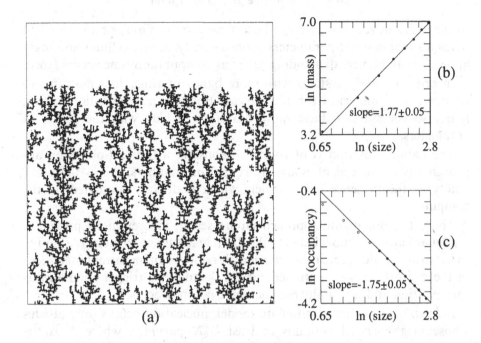

Fig. 5.12. (a) Examples of mineral dendrite obtained from a cellular automata simulation with $k_p = 0$; in this figure, the reaction front moves from upward. The two graphs on the right show the numerical measurement of the fractal dimension using: (b) a sand-box method and (c) a box-counting technique.

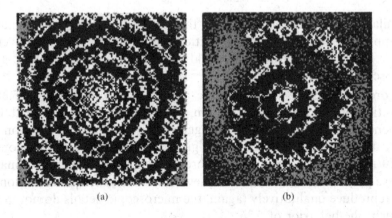

Fig. 5.13. Liesegang rings (a) and spiral (b), as obtained after 2000 iterations of the cellular automata model with C particles (indicated in gray).

5.6.4   *The lattice Boltzmann model*

In the previous section, we had to use a pseudo-3D simulation in order to reach a suitable set of parameters. Consequently, a pure cellular automata approach can be very demanding as far as computation time is concerned.

In order to study systems with more bands or rings, it is desirable to have a faster numerical scheme. In addition, for a systematic analysis, it is also necessary to have more flexibility to play with the values of parameters.

The natural solution is of course to consider a lattice Boltzmann approach where, instead of Boolean occupation numbers, this is the probability of the presence of a particle which is directly simulated on the computer.

The lattice Boltzmann approach suppresses the inherent noise present in a cellular automata model, due to the discrete nature of particles. Although this is usually a desirable property, here, it removes an important ingredient of the process: as we mentioned earlier, fluctuations trigger spontaneous precipitation and also influence the aggregation processes.

Indeed, in the cellular automata model, nucleation occurs only at sites whose neighborhood contains at least $k_{sp}N$ particles, where N is the product of the number of sites in the neighborhood times the number of directions of motion (typically, for a 3×3 neighborhood, $N = 3^2 \times 4 = 36$). If, at a given site, the average particle density per direction is ρ, the probability that nucleation occurs is given by:

$$p_{\text{nucl}}(\rho) = \sum_{\ell \geq k_{sp}N} \binom{N}{\ell} \rho^{\ell}(1-\rho)^{N-\ell} \qquad (5.36)$$

A similar relation can be derived for the aggregation probability. Thus, in this approach, nucleation or aggregation may happen even if the average density is below the critical thresholds, that is even if the system is not macroscopically supersturated.

In order to restore the stochastic aspects given by equation 5.36 in the lattice Boltzmann approximation, we may introduce a probabilistic component for the nucleation and aggregation processes: nucleation and aggregation will take place only with given probabilities, when the concentration reaches some renormalized threshold k_{sp} or k_p. The determination of these probabilities, as well as the new values of k_{sp} or k_p, can be done so as to reproduce qualitatively (again, the microscopic details do not matter so much) the behavior of 5.36.

The lattice Boltzmann method allows us to gain a factor of 100 in the speed of the simulation, as compared with the equivalent, pseudo-3D cellular automata model discussed in the previous sections.

Fig. 5.14. Example of the formation of Liesegang bands in a lattice Boltzmann simulation. The model considered here corresponds to a scenario with intermediate C particle.

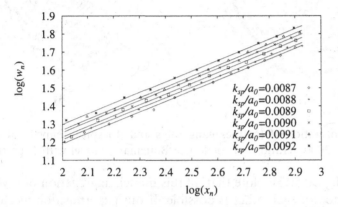

Fig. 5.15. Dependence of the width w_n of the Liesegang bands as a function of their position x_n, for various values of k_{sp}, for the concentration ratio $b_0/a_0 = 0.01$. We observe $w_n = x_n^{0.59}$, independently of the value of k_{sp}.

Figure 5.14 shows an example of a lattice Boltzmann simulation containing up to 30 consecutive bands, in a system of sizes 1024×64.

From the lattice Boltzmann simulations we may study how the width w_n of a band depends on its position x_n along the tube. Here, the width has been measured as the total spatial extension of a band.

Whereas the time and spacing laws have been investigated in great detail in many experiments, much less seems to be known about the width law. In most experiments, it is observed that w_n increases with x_n. A linear relation, derived from the width of only a few consecutive bands, has been proposed in the literature [159,164]. Our simulations predict a more general relation

$$w_n \sim x_n^{\alpha}$$

from which we conclude (using the spacing law) that

$$\frac{w_n}{w_{n-1}} \rightarrow \left(\frac{x_n}{x_{n-1}} \right)^{\alpha} \rightarrow p^{\alpha} \tag{5.37}$$

where α is an exponent which depends on the initial concentration a_0 and b_0. We find, for several simulations, that α is typically in the range

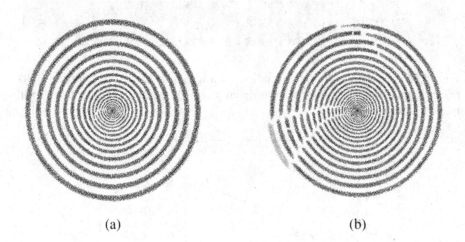

(a)   (b)

Fig. 5.16.   Formation of (a) Liesegang rings and (b) spiral-shaped pattern, as obtained after 2000 iterations of the lattice Boltzmann model with C particles.

0.5–0.6 (see figure 5.15). Note that, in this model, aggregation of several D particles on top of each other is possible. If this superposition mechanism is not permitted, it is found that the bands are wider and an exponent $\alpha \approx 1$ is then obtained, in agreement with experimental observations [164].

5.6.5   *Lattice Boltzmann rings and spirals*

We can consider again the case of Liesegang rings and spirals, but now in the framework of the lattice Boltzmann model.

For instance, figure 5.16(a) shows the situation where concentric rings of precipitate are formed. The numerical parameters are: $a_0 = 1$, $b_0/a_0 = 0.013$, $D_b/D_a = 0.1$, $k_{sp}/a_0 = 0.0087$, $k_p/a_0 = 0.0065$. The nucleation process takes place with a probability of 0.05 and aggregation with a probability close to 1. This pattern turns out to be quite similar to real Liesegang structures obtained in a similar experimental situation [152].

For the same set of parameters, but $b_0/a_0 = 0.016$, a different pattern is observed in figure 5.16(b). There, a local defect produced by a fluctuation develops and a spiral of precipitate appears instead of a set of rings. Such a spiral pattern will never be obtained from a deterministic model without a stochastic component.

From these data, we can check the validity of the spacing law for ring formation. The relation

$$r_n/r_{n-1} \rightarrow p$$

holds, where r_n is the radius of the nth ring. In figure 5.17, the Jablczynski coefficient p is plotted as a function of the concentration of B particles

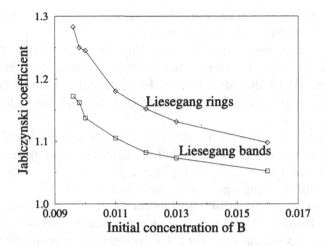

Fig. 5.17. Jablczynski coefficients p as a function of the concentration b_0 of B particles (for $a_0 = 1$), for bands (lower curve) and rings (upper curve).

b_0 (for $a_0 = 1$) both for axial (bands) and radial (rings) symmetries. We notice that p decreases when b_0 increases in agreement with experimental data. Moreover, for the same set of parameters, the value of p is found to be larger in the case of rings than it is for bands.

5.7 Multiparticle models

Several times in this book, we have presented a cellular automata (or lattice gas) model and then introduced a lattice Boltzmann version, in order to improve the numerical efficiency, for the price of losing the many-body correlations.

Clearly, lattice Boltzmann models are less noisy and provide more flexibility than their Boolean counterpart. But they may exhibit some bad numerical instabilities (that is the case of lattice BGK models of fluids) and they sometimes fail to account for relevant physical phenomena. An example is provided by the anomalous kinetics in simple reaction-diffusion processes.

The objective we would like to consider here is to find a dynamics in between lattice gas automata and lattice Boltzmann models. Multiparticle models, as we call them, provide such a possible alternative. The main idea is to conserve the point-like nature of particles, as in cellular automata, but to allow an arbitrary number of them to be present at a lattice site. In other words, we want to relax the exclusion principle that plagues the cellular automata approach and which appears as a numerical artifact rather than a desirable physical property.

Mathematically speaking, this means that the state of each lattice site cannot be described with a finite number of information bits. However, in practice, it is easy to allocate a 32- or 64-bit computer word to each lattice site, to safely assume that "any" number of particles can be described at that site.

If the number of particles per site is N, the intrinsic fluctuations due to the discrete nature of the particles will typically be of the order $\sqrt{N}$. This is small compared to N, if N is large enough. Therefore, we do not have to perform much averaging to get a meaningful result.

In addition, with an arbitrary number of particle per site, we have much more freedom to enforce a given boundary condition, or tune a parameter of the simulation. Thus, for modeling a reaction process, it is almost necessary to get rid of the exclusion principle. In equation 5.14 we showed how the exclusion principle slows down the reaction when the concentration of C is not small. To describe more complicated situations, such as $mA + nB \rightarrow C$, it is desirable to have more than four particles per site.

Unfortunately, the numerical implementation of multiparticle models is much more involved than lattice Boltzmann or cellular automata models and the computation time is also much higher. On the other hand, we restore in a natural way the fluctuations that are absent in lattice Boltzmann simulations and provide an intrinsically stable numerical scheme. Besides, when compared to cellular automata, the extra computational time may be well compensated by the fact that less averaging is required.

The purpose of this section is to introduce in more detail multiparticle models. Here, the discussion will focus on the case of a reaction-diffusion system, which is the topic of this chapter. In section 7.3, we shall also present a multiparticle fluid model.

5.7.1 *Multiparticle diffusion model*

Our algorithm is defined on a d-dimensional Cartesian lattice of spacing λ. Each lattice site $\vec{r}$ is occupied, at time t, by an arbitrary number of particles $n(\vec{r}, t)$. The discrete time diffusion process is defined as follows: during the time interval τ, each particle can jump to one of its $2d$ nearest-neighbor sites along lattice direction i with probability p_i, or stay at rest with a probability $p_0 = 1 - \sum_{i=1}^{2d} p_i$.

An advantage of dealing with multiparticle dynamics is that advection mechanisms can be added to the diffusion process. When the probabilities of jumping to a nearest-neighbor site are different in each direction, a drift is introduced. This adds a density gradient term to the diffusion equation which then reads

$$\partial_t \rho = \vec{V}\nabla\rho + D\nabla^2\rho$$

where $\vec{V}$ is the advection velocity. Such an advection effect is difficult to produce without an artifact when an exclusion principle holds.

For the sake of simplicity, we shall now consider a two-dimensional case. The generalization is straightforward and follows the same reasoning.

The idea is to loop over every particle at each site, decide where it goes and move it to its destination site. In terms of the particle numbers $n(\vec{r}, t)$, our multiparticle rule can be expressed as

$$
\begin{aligned}
n(\vec{r}, t + \tau) &= \sum_{\ell=1}^{n(\vec{r},t)} p_{0\ell}(\vec{r}, t) + \sum_{\ell=1}^{n(\vec{r}+\lambda\vec{e}_3,t)} p_{1\ell}(\vec{r} + \lambda\vec{e}_3, t) \\
&+ \sum_{\ell=1}^{n(\vec{r}+\lambda\vec{e}_1,t)} p_{3\ell}(\vec{r} + \lambda\vec{e}_1, t) + \sum_{\ell=1}^{n(\vec{r}+\lambda\vec{e}_4,t)} p_{2\ell}(\vec{r} + \lambda\vec{e}_4, t)) \\
&+ \sum_{i=1}^{n(\vec{r}+\lambda\vec{e}_2,t)} p_{4\ell}(\vec{r} + \lambda\vec{e}_2, t))
\end{aligned}
\tag{5.38}
$$

The vectors $\vec{e}_1 = -\vec{e}_3$, $\vec{e}_2 = -\vec{e}_4$ are the four unit vectors along the main directions of the lattice. The stochastic Boolean variable $p_{i\ell}(\vec{r}, t)$ is 1 with probability p_i and selects whether or not particle ℓ chooses to move to site $\vec{r} + \lambda\vec{e}_i$. Since each particle has only one choice, we must have

$$p_{0\ell} + p_{1\ell} + p_{2\ell} + p_{3\ell} + p_{4\ell} = 1$$

The macroscopic occupation number $N(\vec{r}, t) = \langle n(\vec{r}, t) \rangle$ is obtained by averaging the above evolution rule over an ensemble of equivalent systems. Clearly, one has

$$\langle \sum_{\ell=1}^{n(\vec{r},t)} p_{i\ell}(\vec{r}, t) \rangle = p_i N(\vec{r}, t)$$

Thus, we obtain the following equation of motion:

$$
\begin{aligned}
N(\vec{r}, t + \tau) &= p_0 N(\vec{r}, t) + p_1 N(\vec{r} + \lambda\vec{e}_3, t) \\
&+ p_3 N(\vec{r} + \lambda\vec{e}_1, t) + p_2 N(\vec{r} + \lambda\vec{e}_4, t) + p_4 N(\vec{r} + \lambda\vec{e}_2, t)
\end{aligned}
\tag{5.39}
$$

Assuming N varies slowly on the lattice, we can perform a Taylor expansion in both space and time to obtain the continuous limit. Using $\sum p_i = 1$

and $\vec{e}_i = -\vec{e}_{i=2}$, we obtain

$$\tau \partial_t N(\vec{r},t) + \frac{\tau^2}{2}\partial_t^2 N(\vec{r},t) + \mathcal{O}(\tau^3) = \lambda \left[(p_3 - p_1)\vec{e}_1 + (p_4 - p_2)\vec{e}_2 \right] \cdot \nabla N(\vec{r},t)$$
$$+ \frac{\lambda^2}{2}(p1 + p3)(\vec{e}_1 \cdot \nabla)^2 N(\vec{r},t) + \frac{\lambda^2}{2}(p2 + p4)(\vec{e}_1 \cdot \nabla)^2 N(\vec{r},t) + \mathcal{O}(\lambda^3)$$
$$(5.40)$$

Since, $\vec{e}_1$ and $\vec{e}_2$ are orthonormal, we have

$$(\vec{e}_1 \cdot \nabla)^2 + (\vec{e}_2 \cdot \nabla)^2 = \nabla^2$$

In order to use this property it is necessary that $p_1 + p_3 = p_2 + p_4$, otherwise the lattice directions will be "visible". Thus we impose the isotropy condition

$$p_1 + p_3 = p_2 + p_4 = \frac{1 - p_0}{2}$$

and we obtain

$$\partial_t N(\vec{r},t) + \frac{\tau}{2}\partial_t^2 N(\vec{r},t) + \mathcal{O}(\tau^2) = \vec{V} \cdot \nabla N(\vec{r},t)$$
$$+ D\nabla^2 N(\vec{r},t) + \mathcal{O}(\lambda^3) \qquad (5.41)$$

where $\vec{V}$ is the advection velocity

$$\vec{V} = \frac{\lambda}{\tau} \left[(p_3 - p_1)\vec{e}_1 + (p_4 - p_2)\vec{e}_2 \right]$$

and D the diffusion constant

$$D = \frac{\lambda^2}{4\tau}(1 - p_0)$$

We may now consider the limit $\lambda \to 0$ and $\tau \to 0$ with $\lambda^2/\tau \to$ constant, as usual in a diffusion process. However, here, some additional care is needed. If $p_3 \neq p_1$ or $p_4 \neq p_2$, the advective term will diverge in this limit. This means that $p_3 - p_1$ or $p_4 - p_2$ must decrease proportionally to λ when the limit is taken. Thus, with a halved lattice spacing, the difference between p_i and p_{i+2} must also be halved in order to produce the same advection. With these assumptions, we obtain, in the macroscopic limit

$$\partial_t N = \vec{V} \cdot \nabla N + D\nabla^2 N$$

5.7.2 Numerical implementation

The main problem when implementing our algorithm on a computer (for instance, for the two-dimensional case we described in the previous section) is to find an efficient way to select the particles at rest and distribute randomly the others among the four possible directions of motion. More precisely, we have to compute quantities such as

$$n_i = \sum_{\ell=1}^{n(\vec{r},t)} p_{i\ell}(\vec{r}, t)$$

In practice, we can loop over all ℓ particles at every site and, for each of them, choose a random number r, uniformly distributed in the interval $[0, 1]$. Then, we consider a division of this interval in subintervals $[r_j, r_{j+1}[$, $j = 0, ..., 5$, so that $p_i = r_{i+1} - r_i$. We say that $p_{i\ell} = 1$ if and only if $r_{i+1} \leq r < r_i$. The quantities n_i are thus distributed according to a multinomial distribution.

This procedure is acceptable for small values of n but, otherwise, very time consuming. However, when n is large (more precisely when $np_i(1 - p_i) \gg 1$, the statistical distributions of the n_i is expected to approach Gaussian distributions of mean np_i and variance $np_i(1 - p_i)$. This Gaussian approximation allows us to be much more efficient because we no longer have to generate a random number for each particle at each site.

For simplicity, take the case $p_0 = p$ and $p_1 = p_2 = p_3 = p_4 = (1 - p)/4$. The n_i's can be approximated as follows: we draw a random number n_0 from a Gaussian distribution of mean np and variance $np(1 - p)$ (for instance using the Box–Muller method [165]). This number is then rounded to the nearest integer.

Thus, in one operation, this procedure splits the population into two parts: n_0 particles that will stay motionless and $n - n_0$ that will move. In a second step, the $n - n_0$ moving particles are divided into two subsets according to a Gaussian distribution of mean $n_m/2$ and variance $n_m(1/2)(1/2)$. Splitting up each of these subsets one more time yields the number n_i of particles that will move in each of the four lattice directions.

If advection is present, we can also proceed similarly. First, we divide up the moving particle population into two parts: on the one hand, those going to north and east, for instance, and on the other hand, those going south and west. Second, each subpopulation is, in turn, split into two subsets according to to the values of the p_is. Of course, as in traditional lattice gas automata, these splitting operations can be performed simultaneously (in parallel) at each lattice site.

Empirical considerations, supported by theoretical arguments on binomial distributions, show that $n_i = 40$ is a good threshold value in two dimensions, above which the Gaussian procedure can be used. Below this critical value, it is safer to have the algorithm loop over all particles. Note that in a given simulation, important differences in the particle number can be found from site to site and the two different algorithms may have to be used at different places.

5.7.3   The reaction algorithm

We will now discuss how reaction processes can be implemented in the framework of multiparticle models (see also [166]). Reaction-diffusion phenomena can then be simulated by alternating the reaction process between the different species and then the diffusion of the resulting products, according to the multiparticle diffusion algorithm just described.

A reaction process couples locally the different species A_l, $l = 1, .., q$ to produce new species B_j, $j = 1, .., m$ according to the reaction:

$$\alpha_1 A_1 + \alpha_2 A_2 + \ldots + \alpha_q A_q \xrightarrow{k} \beta_1 B_1 + \beta_2 B_2 + \ldots + \beta_m B_m$$

The quantities α_l, β_j are the stoichiometric coefficients, and k is the reaction constant.

In order to model this reaction scheme with a multiparticle dynamics, one considers all the q-tuples that can be formed with α_1 particles of A_1, α_2 particles of A_2, etc. These q-tuples are transformed into m-tuples of B_j particles with probability k. At site $\vec{r}$ and time t, there are

$$\mathcal{N}(\vec{r}, t) \equiv \binom{n_{A_1}}{\alpha_1} \binom{n_{A_2}}{\alpha_2} \ldots \binom{n_{A_q}}{\alpha_q} (\vec{r}, t)$$

ways to form these q-tuples, where $n_X(\vec{r}, t)$ denotes the number of particles of species X present at $(\vec{r}, t)$. If one of the $n_{A_i} < \alpha_i$ then obviously $\mathcal{N} = 0$.

This techniques offers a natural way to consider all possible reaction scenarios. For instance, in the case of the annihilation reaction $2A \to \emptyset$, suppose we have three particles (labeled a_1, a_2, a_3) available at a given lattice site. Then, there are three possible ways to form a reacting pair: (a_1, a_2), (a_1, a_3) and (a_2, a_3). In principle, all these combinations have the same chance of forming and reacting. However, if (a_1, a_2) react, then only a_3 is left and there is no point in considering (a_1, a_3) or (a_2, a_3) as possible candidates for reaction. Thus $\mathcal{N}$ is the maximal number of possible events, but it is likely that the available particles are exhausted before reaching the end of this list of possible reactions.

The multiparticle reaction rule can therefore be summarized as follows:

- As long as there are enough particles left (i.e. at least α_l of species A_l, for each l), but no more than $\mathcal{N}$ times, choose a Boolean random κ which is 1 with probability k.

- If $\kappa = 1$, remove from each species A_l a number α_l of particles ($n_{A_l} \to n_{A_l} - \alpha_l$) and add a number β_j of particles to each species B_j, $j = 1, ..., m$ ($n_{B_j} \to n_{B_j} + \beta_j$).

This algorithm can easily be extended to a reversible reaction.

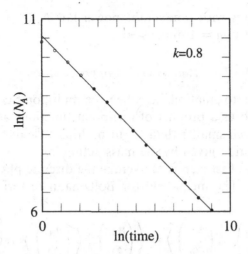

Fig. 5.18. Time decay of N_A, the total number of A particles in the $A + A \rightarrow \emptyset$ reaction-diffusion process, with the multiparticle method. A non-mean-field power law $t^{-d/2}$ is observed.

When k is very small, we may assume that all the $\mathcal{N}$ q-tuples need to be considered and the above reaction rule can be expressed as

$$n_{A_l}(\vec{r}, t + \tau) = n_{A_l}(\vec{r}, t) - \alpha_l \sum_{h=1}^{\mathcal{N}(\vec{r},t)} \kappa_h$$

$$n_{B_j}(\vec{r}, t + \tau) = n_{B_j}(\vec{r}, t) + \beta_j \sum_{h=1}^{\mathcal{N}(\vec{r},t)} \kappa_h \qquad (5.42)$$

where κ_h is 1 with probability k.

This algorithm may become quite slow in terms of computer time if the n_X are large and $k \ll 1$. In this case, the Gaussian approximation described in the previous section can be used to speed up the numerical simulations: the number of accepted reactions can be computed from a local Gaussian distribution of mean $k\mathcal{N}(\vec{r}, t)$ and variance $k(1-k)\mathcal{N}(\vec{r}, t)$.

In order to check that our multiparticle reaction rule captures the true nature of fluctuation and correlation, we simulated the $A + A \rightarrow \emptyset$ reaction-diffusion process. We explained in section 5.4 that this reaction exhibits a non-mean-field decay law in one-dimensional systems.

Figure 5.18 gives the behavior of a simulation performed on a line of 64 536 sites, with an initial number of about 100 particles per site. Diffusion and reaction processes are simulated with our multiparticle algorithms with a probability $1/2$ that each particle moves left or right and a reaction probability $k = 0.8$. We observe that the total number of

A particles decreases with time as the power law $N_A(t) \sim t^{-1/2}$, which is the expected result in $d = 1$ dimension.

5.7.4   Rate equation approximation

In a mean-field approximation, i.e. when the multipoint correlation functions are factorized as a product of one-point functions and the reaction probability k is much smaller than 1, our multiparticle dynamics gives the expected rate equation given by the mass action law.

To show this, the first step is to average the discrete physical quantities over an ensemble. The mean-field (or Boltzmann factorization approximation) yields

$$\left\langle \binom{n_{A_1}}{\alpha_1} \binom{n_{A_2}}{\alpha_2} \cdots \binom{n_{A_q}}{\alpha_q} \right\rangle = \left\langle \binom{n_{A_1}}{\alpha_1} \right\rangle \left\langle \binom{n_{A_2}}{\alpha_2} \right\rangle \cdots \left\langle \binom{n_{A_q}}{\alpha_q} \right\rangle$$

The first important result is that, if one assumes that each configuration with a total of M particles over L sites appears with the same probability, the quantities $\binom{n_{A_l}}{\alpha_l}$ can be computed

$$\left\langle \binom{n_{A_l}}{\alpha_l} \right\rangle = \frac{1}{L^{\alpha_l}} \binom{M_{A_l}}{\alpha_l} \sim \frac{1}{\alpha_l!} N_{A_l}^{\alpha_l} \tag{5.43}$$

where N_{A_l} is the average particle number per site of species A_l and the number of lattice sites L is supposed to be large enough.

This result can be derived from the following combinatory arguments: let us suppose that a number M of particles is homogeneously distributed among L sites of the lattice. The average number N of particles on these sites is then $N = M/L$.

The probability for a particle to pick a given site $\vec{r}$ to stand on is $1/L$ (and $1 - 1/L$ to choose any other site). Thus the probability that a given site contains n particles is

$$P(n) = \binom{M}{n} \frac{1}{L}^n \left(1 - \frac{1}{L}\right)^{M-n} = \binom{M}{n} \frac{(L-1)^{M-n}}{L^M}$$

The average we want to compute is

$$\left\langle \binom{n}{\alpha} \right\rangle = \sum_{n=0}^{M} \binom{n}{\alpha} P(n) = \sum_{n=0}^{M} \binom{n}{\alpha} \binom{M}{n} \frac{(L-1)^{M-n}}{L^N}$$

$$= \frac{1}{\alpha!} \sum_{n=\alpha}^{M} \frac{M!}{(M-n)!(n-\alpha)!} \frac{(L-1)^{M-n}}{L^M}$$

$$= \frac{1}{\alpha!} \sum_{n'=0}^{M'} \frac{(M'+\alpha)!}{(M'-n')!n'!} \frac{(L-1)^{M'-n'}}{L^{M'+\alpha}}$$

$$= \binom{M'+\alpha}{\alpha} L^{-\alpha} \sum_{n'=0}^{M'} \frac{M'!}{(M'-n')!n'!} \frac{(L-1)^{M'-n'}}{L^{M'}}$$

$$= \binom{M'+\alpha}{\alpha} L^{-\alpha} \frac{(1+L-1)^{M'}}{L^{M'}} \tag{5.44}$$

where $M' = M - \alpha$ and $n' = n - \alpha$. Thus

$$\left\langle \binom{n}{\alpha} \right\rangle = \binom{M}{\alpha} L^{-\alpha} = \frac{1}{\alpha!} N^{\alpha} \left(1 + \mathcal{O}\left(\frac{1}{NL}\right) \right)$$

where $N = M/L$ is the average number of particle per site. When the system size is large ($M \gg 1$) we can approximate

$$\left\langle \binom{n}{\alpha} \right\rangle = \frac{1}{\alpha!} N^{\alpha}$$

which proves equation 5.43.

With this result, we can now obtain the mean-field approximation of equation 5.42:

$$N_{A_i}(t+\tau) - N_{A_i}(t) = -k' N_{A_1}^{\alpha_1} N_{A_2}^{\alpha_2} \dots N_{A_q}^{\alpha_q}$$

$$N_{B_j}(t+\tau) - N_{B_j}(t) = k' N_{A_1}^{\alpha_1} N_{A_2}^{\alpha_2} \dots \bar{n}_{A_q}^{\alpha_q}$$

where the reaction constant k' is

$$k' = \frac{k}{\alpha_1! \alpha_2! \dots \alpha_q!}$$

In the limit $\tau \to 0$, we obtain the usual form of the rate equations for the reaction process considered, namely

$$\partial_t N_{A_i}(t) = -\frac{k'}{\tau} N_{A_1}^{\alpha_1} N_{A_2}^{\alpha_2} \dots N_{A_q}^{\alpha_q}$$

$$\partial_t N_{B_j}(t) = \frac{k'}{\tau} N_{A_1}^{\alpha_1} N_{A_2}^{\alpha_2} \dots N_{A_q}^{\alpha_q}$$

5.7.5 Turing patterns

In this section, we use our multiparticle reaction-diffusion model to simulate the formation of the so-called Turing structures. Turing [167] was the first to suggest that, under certain conditions, chemicals can react and diffuse so as to produce steady-state heterogeneous spatial patterns of chemical or concentrations [50]. Turing structures are believed to play an important role in biological pattern formation processes, such as the stripes observed on the zebra skin [130]. In contrast to most hydrodynamical instabilities, the structure of Turing patterns is not related to any imposed macroscopic length scales (like the size of the container). Turing patterns exhibit regular structure with an intrinsic wavelength depending on the diffusion constants and reaction rates. Typical examples of inhomogeneous stationary states observed in experiments have a hexagonal or a striped structure [168].

For the sake of simplicity, we consider here only one of the simplest models showing Turing patterns: the Schnackenberg reaction-diffusion model [169] in two dimensions. It describes the following autocatalytic reaction:

$$A \xrightarrow{k_1} X \qquad\qquad X \xrightarrow{k_2} \emptyset$$

$$2X + Y \xrightarrow{k_3} 3X \qquad\qquad B \xrightarrow{k_4} Y \tag{5.45}$$

where the densities of the species A and B are kept fixed (for instance by external feeding of the system). This situation of having a fixed concentration of some chemical is quite common in reaction-diffusion processes. As a result, there is no need to include all the dynamics of such reagents in cellular automata or multiparticle models. It is usually enough to create randomly a local population of these particles at each lattice sites.

Here we consider a two-dimensional multispecies multiparticle lattice gas model with alternating reaction and diffusion steps, as explained in sections 5.7.1 and 5.7.3, namely

- Reaction step: for the reaction $nS_1 + mS_2 \xrightarrow{k_j} qS_3$ all possible families composed of n S_1 and m S_2 particles have a probability k_j to create q particles S_3.

- Diffusion step: the particles jump to a nearest-neighbor site with an equal probability for each direction. This gives a bare diffusion coefficient of $\lambda^2/4\tau$. A simple way to tune it without having a probability for a particle to be at rest is to vary the number ℓ of consecutive diffusion steps for a given species: this amounts to introducing a different time step $\tau_m = \tau/\ell$ for this species and yields $D = \ell\lambda^2/4\tau$.

The instability of the homogeneous state leading to Turing structures can be understood using the corresponding macroscopic rate equations [130] for the local average densities x and y

$$\partial_t x = k_1 a - k_2 x + k_3 x^2 y + D_x \nabla^2 x$$
$$\partial_t y = k_4 b - k_3 x^2 y + D_y \nabla^2 y \tag{5.46}$$

where a and b represent the densities of particles A and B, respectively. A conventional analysis shows that for some values of the parameters a

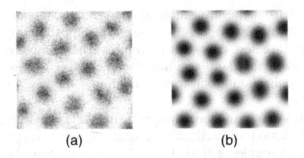

(a)   (b)

Fig. 5.19. Turing patterns obtained in the Schnackenberg reaction in the long time regime. (a) Multiparticle model and (b) mean-field rate equations.

homogeneous stationary state is unstable towards local density perturbations. Inhomogeneous patterns can evolve by diffusion-driven instabilities providing that the diffusion constants D_x and D_y are not the same. The region of the parameter space (a, b, D_y/D_x,...) for which homogeneous states of the system are unstable is called the deterministic Turing space.

Figure 5.19 shows the configuration obtained in the long time regime with our multiparticle model and the corresponding rate equations 5.46. In both cases, a hexagonal geometry is selected. The right panel corresponds to the solution of the rate equations, while the left panel corresponds to the multiparticle simulation. As we can see, the two pictures are quite similar. Thus, it is not clear that the multiparticle (which brings fluctuations into play) adds anything compared with the predictions of the mean-field rate equations (which use less computer time). However, there are some indications [170] that the Turing space may be enlarged when fluctuations are considered.

5.8 From cellular automata to field theory

As we have seen, reaction-diffusion problems have a rich structure which can lead to complex cooperative behaviors in the long time regime. These cooperative behaviors have some universal properties. This means that some aspects of the dynamics are "forgotten" or irrelevant in the long time regime, while more basic aspects related to symmetries and conservation laws govern the long time dynamics. Thus, different reaction-diffusion systems can have a similar long time behavior. In a language borrowed from the renormalization group [171] approach of critical phenomena, one says that these systems belong to the same universality class.

It is difficult from the inspection of the cellular automaton rules (or multiparticle dynamics) to predict what will happen in the long time regime. There are too many microscopic variables involved, and the collective behavior shows up at a macroscopic scale. It is also difficult to define a renormalization group transformation at the level of the cellular automaton rule.

A better strategy is to find, starting from the microscopic rules, the equations of motion of well chosen coarse-grained variables. These "macroscopic" variables $\phi_i(\vec{r}, t)$, $i = 1, 2, ..$ take continuous values and vary continuously in space and time. They describe the system at the macroscopic scale, while keeping track of all the fluctuations related to the disregarded microscopic degrees of freedom. Such a description is called a continuous field theory. These field theories can then be studied in their own right, using standard methods such as perturbation techniques or renormalization group.

The first step to perform this program is to notice that the probabilistic cellular automaton rules used for reaction-diffusion can be expressed as a master equation (in discrete time) for generalized birth–death processes.

The way to associate a continuous field to a master equation for a birth–death process (in continuous time) has been studied by many people [172–174].

For the sake of simplicity, and to avoid as much as possible technical difficulties, we shall start with a very simple one-cell problem (zero-dimensional system). Moreover, we shall assume that only one type of particle is present. The state of the cell is characterized by $\phi_n(t)$, the probability of having n particles at time t. We consider a multiparticle model, with no exclusion principle, so that any value of n is allowed.

One considers a dynamical process described by the following master equation

$$\partial_t \phi_n(t) = \sum_{n' \neq n} [w(n' \to n)\phi_{n'}(t) - w(n \to n')\phi_n(t)] \qquad (5.47)$$

where $w(n' \to n)$ is the transition rate from the state with n' particles to the state with n particles. One recognizes the degenerate form (only one site) of a probabilistic cellular automaton rule. This stochastic process is Markovian: there is no memory effect.

A natural way, in statistical physics, to deal with a problem in which the particle number is not conserved is to introduce a so-called Fock space structure. One works in a phase space which is the direct sum of all the phase spaces corresponding to an arbitrary number of particles. Creation and destruction of particles is described by creation and annihilation operators acting on this Fock space. To each state $\{\phi_n\}$ one can associate the state $|\Phi\rangle$ considered as an element of a real vector space or a Fock space $\mathscr{F}$

$$|\Phi\rangle = \sum_n \phi_n |n\rangle \qquad (5.48)$$

where $|n\rangle$ is the state with exactly n particles. A useful quantity is the generating function defined as

$$G(z,t) = \sum_n \phi_n(t) z^n \qquad (5.49)$$

where z is a complex number. The derivatives of $G(z,t)$ generate the factorial moments

$$n_k(\phi) = \langle n(n-1)...(n-k+1)\rangle = \frac{\partial^k}{\partial z^k} G(z,t)|_{z=1} \qquad (5.50)$$

The definition of a scalar product on this vector space gives it a Hilbert structure. Different scalar products are usually considered. We consider here the so-called exclusive scalar product defined as

$$\langle n|m\rangle = n! \delta_{nm} \qquad (5.51)$$

One then introduces the annihilation operator a and the creation operator π defined as

$$a|n\rangle = n|n-1\rangle \qquad (5.52)$$

and

$$\pi|n\rangle = |n+1\rangle \qquad (5.53)$$

These (bosonic like) operators obey commutation relations

$$[a, \pi]_- = 1 \qquad (5.54)$$

and are Hermitian conjugate with respect to the exclusive scalar product. It then follows that

$$\phi_n(t) = \frac{1}{n!}\langle n|\Phi\rangle = \frac{1}{n!}\langle 0|a^n|\Phi(t)\rangle \qquad (5.55)$$

Finally, the generating function can be written down in this formalism as

$$G(z,t) = \langle z|\Phi\rangle \qquad (5.56)$$

where the $|z\rangle$ are the coherent states of Bargmann and Fock [175]

$$|z\rangle = \exp(z^*\pi)|0\rangle \qquad (5.57)$$

One can then prove several useful properties

$$\langle z|\zeta\rangle = e^{z\zeta^*} \qquad (5.58)$$

$$\int \frac{dz dz^*}{2\pi i} e^{-zz^*} |z\rangle\langle z| = I \qquad (5.59)$$

$$|\Phi\rangle = \int \frac{dz\,dz^*}{2\pi i} e^{-zz^*} G(z,t)|z\rangle \tag{5.60}$$

$$a|z\rangle = z^*|z\rangle \tag{5.61}$$

In this approach the master equation takes the form

$$\partial_t|\Phi\rangle = \mathscr{L}^{\mathscr{F}}|\Phi\rangle \tag{5.62}$$

the Liouvillian $\mathscr{L}^{\mathscr{F}}$ is a polynomial in the creation and annihilation operators and can always be written in normal form (all the creation operators are on the left of the annihilation operators)

$$\mathscr{L}^{\mathscr{F}} = \sum_{i,j} C_{ij} \pi^i a^j \tag{5.63}$$

The exact form of $\mathscr{L}^{\mathscr{F}}$ i.e. the values that the constants C_{ij} take on, will depend on the form of the transition rates $w(n' \to n)$ for the model under consideration. Thus one can take either the w or the Liouvillian $\mathscr{L}^{\mathscr{F}}$ to define the model.

Before developing further this general formalism, let us consider a very simple example, namely the simple decay $A \to \emptyset$ with a reaction rate w. In this simple case, the different steps can be made explicitly. The master equation is

$$\partial_t \phi_n(t) = -wn\phi_n(t) + w(n+1)\phi_{n+1}(t) \tag{5.64}$$

From equation 5.64 it follows that the generating function $G(z,t)$ obeys the equation

$$\partial_t G(z,t) = w(1-z)\partial_z G(z,t) \tag{5.65}$$

whose solution is

$$G(z,t) = G((z-1)\exp(-wt)) \tag{5.66}$$

Thus the factorial moment of order one is

$$n_1(\phi) = \langle n\rangle(t) = \frac{\partial}{\partial z} G(z,t)|_{z=1} = \langle n\rangle(0)\exp(-wt) \tag{5.67}$$

as expected from physical grounds. We can now construct the Liouvillian $\mathscr{L}^{\mathscr{F}}$. We have

$$\partial_t|\Phi\rangle = -\sum_n wn\phi_n|n\rangle + \sum_n w(n+1)\phi_{n+1}|n\rangle = -w\pi a|\Phi\rangle + wa|\Phi\rangle \tag{5.68}$$

and the Liouvillan takes the simple form

$$\mathscr{L}^{\mathscr{F}} = w(a - \pi a) \tag{5.69}$$

Let us now return to the general case in zero dimension. To obtain a field continuous field theory, one constructs from 5.62 and 5.63 a path-integral representation for

$$P(z,t|z_0,0) = \langle z| \exp t\mathcal{L}^{\mathscr{F}} |z_0\rangle \tag{5.70}$$

by inserting 5.59 into 5.70 $(N-1)$ times, in the usual way [175]

$$P(z,t|z_0,0) = (\prod_{k=1}^{N-1} \int \frac{dz_k dz_k^*}{2\pi i} e^{-z_k z_k^*}) \langle z_N| \exp \epsilon \mathcal{L}^{\mathscr{F}} |z_{N-1}\rangle$$

$$\times \langle z_{N-1}| \exp \epsilon \mathcal{L}^{\mathscr{F}} |z_{N-2}\rangle ... \langle z_1| \exp \epsilon \mathcal{L}^{\mathscr{F}} |z_0\rangle \tag{5.71}$$

where $z_N = z$ and $\epsilon = t/N$. For small ϵ a typical term may easily be evaluated

$$\langle z_k| \exp \epsilon \mathcal{L}^{\mathscr{F}} |z_{k-1}\rangle \sim \exp(\epsilon \sum_{ij} C_{ij} z_k^i (z_{k-1}^*)^j + z_k z_{k-1}^*) \tag{5.72}$$

Substituting 5.72 into 5.71 and taking the limit $N \to \infty$ and $\epsilon \to 0$ with t fixed gives for 5.71, after some algebra and up to boundary terms

$$P(z,t|z_0,0) = \int d[z,z^*] \exp -\{\int_0^t dt[z\dot{z}^* - \sum_{ij} C_{ij} z^i (z^*)^j]\} \tag{5.73}$$

where the dot signifies differentiation with respect to time. One recovers an object familiar in field theory. The long time properties are extracted by studying the action

$$S(t,0) = \int_0^t dt' [z\dot{z}^* - \sum_{ij} C_{ij} z^i (z^*)^j] \tag{5.74}$$

The basic formalism being established, we can now consider general problems with many cells on a d-dimensional lattice. At each site r of the lattice, one can have n_r particles. One starts with a master equation for $\phi(n_1, n_2, ..., n_r, ...; t)$, the joint probability of finding n_j particles at site j $(j = 1, 2, ...)$.

In the Fock space representation, the basis states of the system are of the form $|\underline{n}\rangle \equiv |n_1, n_2, ...n_r, ...\rangle$. At each site r of the lattice, one associates a pair of creation π_r and annihilation a_r operators satisfying the commutation relations

$$[a_r, \pi_s]_- = \delta_{rs} \tag{5.75}$$

The macroscopic state is

$$|\Phi\rangle = \sum_{\underline{n}} \phi_{\underline{n}} |\underline{n}\rangle \tag{5.76}$$

and the generating function

$$G(z_1, z_2, ...z_r, ..; t) = \prod_{\underline{n}} \prod_r \phi_{\underline{n}} z_r^{n_r} \tag{5.77}$$

The Liouvillian $\mathscr{L}^{\mathscr{F}} = \mathscr{L}(\pi_1, a_1, ..., \pi_j, a_j, ...)$ contains terms which couple site r and its nearest neighbors. Such non-local terms are needed to describe diffusion of the particles on the lattice. Having the Liouvillian $\mathscr{L}^{\mathscr{F}}$, one can find its path-integral form.

As a simple illustration, let us consider the diffusion of particles in one dimension. Particles at a given site r can jump with a rate w to a nearest-neighbor site (labeled $r \pm e$). The masters equation reads

$$\partial_t \phi(\underline{n}, t) = \sum_r \sum_e w[(n_r + 1)\phi(..., n_{r+e} - 1, n_r + 1, .., t)$$
$$-n_r \phi(..., n_{r+e} - 1, n_r, .., t)] \tag{5.78}$$

The Liouvillan $\mathscr{L}^{\mathscr{F}}$, is

$$\mathscr{L}^{\mathscr{F}} = \sum_r \sum_e w(\pi_{r+e} a_r - \pi_r a_r) \tag{5.79}$$

The action of the continuous field theory becomes

$$S(t, 0) = \int_0^t dt' \{ \int d^d r [z(\dot{z}^* - D\Delta z^*)] \} \tag{5.80}$$

where $D = w\lambda^2$, λ being the lattice constant.

As a final example, we consider the case of the diffusion annihilation process $A + A \rightarrow \emptyset$. Let D be the diffusion constant of the A particles and k the reaction rate. The Liouvillian $\mathscr{L}^{\mathscr{F}}$ is easily found to be

$$\mathscr{L}^{\mathscr{F}} = D \sum_r \sum_e \{[\pi_{r+e} a_r - \pi_r a_r] - k(\pi_r^2 - 1)a_r^2\} \tag{5.81}$$

where e denotes the nearest-neighbor sites of r. At the coarse-grained level, the action is

$$S(t, 0) = \int_0^t dt' \{ \int d^d r [z(\dot{z}^* - D\Delta z^* + 2kz^{*2})] + kz^{*2} z^2 \} \tag{5.82}$$

The long time properties of this field theory have been analyzed using a dynamical renormalization group approach [176,177].

Until now, we have assumed that only one particle species was present. If two or more species participate or if the particles possess internal degrees of freedom, the formalism can be easily extended by introducing new fields [177].

There is another way to associate a field theory with the type of master equation we consider. This second approach that we shall call the Poisson approach has been introduced by Gardiner and Chaturvedi. Although equivalent to the Fock space formalism [178], the Poisson approach illustrates more explicitly the role played by the fluctuations. Let us consider again first the Poisson method for the one-cell problem. Specifically one assumes that a state of the system realized at time t

can be expanded as a superposition of multivariate uncorrelated Poisson distributions. One then writes

$$\phi_n(t) = \int d\alpha \frac{(\exp -\alpha)\alpha^n}{n!} f(\alpha, t) \tag{5.83}$$

In this formalism, the generating function takes the form

$$G(z, t) = \int d\alpha \exp[(z - 1)\alpha] f(\alpha, t) \tag{5.84}$$

If one is given the evolution equation for G in the form

$$\partial_t G(z, t) = \mathscr{L}^{\mathscr{G}} G(z, t) \tag{5.85}$$

then, by replacing G by its expression 5.84 and integrating by parts (assuming that $f(\alpha, t)$ and some of its derivatives vanish at the boundaries), one obtains a Fokker–Planck-like equation for $f(\alpha, t)$ [179,180]. This equation is not a standard Fokker–Planck equation because $f(\alpha, t)$ does not have to remain positive and thus is not a probability density. Moreover, α is not necessarily real, but can be complex. This Poisson representation leads to simple relations between the factorial moments of n and the moments of α. One has

$$n_k(f) = \int d\alpha \alpha^k f(\alpha, t) = \langle \alpha^k \rangle \tag{5.86}$$

The equivalence of the Fock and Poisson formalisms can be shown as follows. One has first to determine the form of the Louvillian, $\mathscr{L}^{\mathscr{G}}$, defined in 5.85, which describes the evolution of the generating function, $G(z, t)$, with time. Given the form 5.63 for $\mathscr{L}^{\mathscr{F}}$ this is straightforward. Indeed,

$$
\begin{aligned}
\partial_t G(z, t) &= \langle z | \partial_t | \Phi \rangle \\
&= \langle z | \mathscr{L}^{\mathscr{F}} | \Phi \rangle \\
&= \sum_{ij} C_{ij} \int \frac{d\zeta d\zeta^*}{2\pi i} e^{-\zeta\zeta^*} G(\zeta, t) \langle z | \pi^i a^j | \zeta \rangle
\end{aligned} \tag{5.87}
$$

where we have used 5.60. Now using 5.58 and 5.61 one has

$$
\begin{aligned}
\partial_t G(z, t) &= \sum_{ij} C_{ij} z^i (\frac{\partial}{\partial z})^j \langle z | \Phi \rangle \\
&= \mathscr{L}^{\mathscr{G}} G(z, t)
\end{aligned} \tag{5.88}
$$

where

$$\mathscr{L}^{\mathscr{G}} = \sum_{i,j} C_{ij} z^i (\frac{\partial}{\partial z})^j \tag{5.89}$$

One sees the close relationship between $\mathscr{L}^{\mathscr{F}}$ and $\mathscr{L}^{\mathscr{G}}$. To each creation operator π present in $\mathscr{L}^{\mathscr{F}}$ one associates a factor z in $\mathscr{L}^{\mathscr{G}}$ and to each annihilation operator a in $\mathscr{L}^{\mathscr{F}}$ one associates the operator $\partial/\partial z$ in $\mathscr{L}^{\mathscr{G}}$. It is now possible to proceed from 5.85 as described above to find the evolution equation for $f(\alpha, t)$. One finds

$$\partial_t f(\alpha, t) = \mathscr{L}^{\mathscr{P}} f(\alpha, t) \qquad (5.90)$$

where

$$\mathscr{L}^{\mathscr{P}} = \sum_{ij} C_{ij}(1 - \frac{\partial}{\partial \alpha})^i \alpha^j \qquad (5.91)$$

is the Liouvillian describing the evolution of the quasi-probability density $f(\alpha, t)$. This Fokker–Planck-like equation is exact, no approximation or truncation has been made in deriving it from the master equation 5.47. It can be used as the basis for the study of processes described by 5.47, or as the starting point for deriving a path-integral representation in terms of the α variables. To carry out this latter program, one introduces operators $\hat{\alpha}$ and $\hat{p}$ [181] satisfying

$$[\hat{\alpha}, \hat{p}]_- = i \qquad [\hat{\alpha}, \hat{\alpha}]_- = 0 \qquad [\hat{p}, \hat{p}]_- = 0 \qquad (5.92)$$

with

$$\hat{p} = -i\frac{\partial}{\partial \alpha} \qquad (5.93)$$

in the α-representation. Then just as the master equation 5.62 is the Fock space analogue of 5.85, one can write an operator analogue of 5.91

$$\partial_t |f\rangle = \mathscr{L}^0 |f\rangle \qquad (5.94)$$

where

$$\mathscr{L}^0 = \sum_{ij} C_{ij}(1 + i\hat{p})^i (\hat{\alpha})^j \qquad (5.95)$$

Introducing $|\alpha\rangle$ and $|p\rangle$, which are eigenkets of $\hat{\alpha}$ and $\hat{p}$ with eigenvalues α and p, respectively, one can derive a path-integral representation for

$$f(\alpha, t|\alpha_0, 0) = \langle \alpha| \exp t\mathscr{L}^0 |\alpha_0\rangle \qquad (5.96)$$

in exactly the same way as described in 5.70. One finds the analogous expression to 5.73 to be

$$\int d[p, \alpha] \exp -\{\int_0^t dt[ip\dot{\alpha} - \sum_{ij} C_{ij}(1 + ip)^i \alpha^j]\} \qquad (5.97)$$

The "action" in the path-integral 5.73 and that in 5.97 are identical as long as the identifications

$$\alpha \leftrightarrow z^*, \quad ip \leftrightarrow (z - 1) \qquad (5.98)$$

are made. Although it was not obvious *a priori* that the two actions would be identical, if 5.98 is expressed in the form

$$\alpha \leftrightarrow \frac{\partial}{\partial z}, \quad \frac{\partial}{\partial \alpha} \leftrightarrow (z-1) \tag{5.99}$$

it becomes very much more plausible, given the conjugate nature of the variables α and $(z-1)$ indicated in 5.84.

To show the equivalence of the two formalisms in the general d-dimensional case, one can proceed in exactly the same way as for the one-site problem. First of all, an analogous procedure to that carried out in 5.87–5.89 shows that the Liouvillian describing the evolution of $G(\underline{z}, t)$ is $\mathscr{L}^{\mathscr{G}} = \mathscr{L}(z_1, \frac{\partial}{\partial z_1}, ..., z_j, \frac{\partial}{\partial z_j}, ...)$ and hence that governing the evolution of $f(\underline{\alpha}, t)$ is $\mathscr{L}^{\mathscr{P}} = \mathscr{L}(1 - \frac{\partial}{\partial \alpha_1}, \alpha_1, ..., 1 - \frac{\partial}{\partial \alpha_j}, \alpha_j, ...)$. Secondly, the operator analogue of $\mathscr{L}^{\mathscr{P}}$ is immediately seen to be $\mathscr{L}^{\mathscr{O}} = \mathscr{L}(1 + i\hat{p}_1, \hat{\alpha}_1, ..., 1 + i\hat{p}_j, \hat{\alpha}_j, ...)$. Finally, path-integral representations for the solution of the Fokker–Planck equations can be constructed from $\mathscr{L}^{\mathscr{F}}$ and $\mathscr{L}^{\mathscr{O}}$ and shown to be the same, if the identification

$$\alpha_i \leftrightarrow z_i^*, \quad ip_i \leftrightarrow (z_i - 1) \tag{5.100}$$

is made.

Instead of considering the generalized Fokker–Planck equation obtained in the Poisson formalism, one can consider the associated generalized Langevin equation. There is a one to one correspondence between the two [179]. Let us consider again the simple case $A + A \rightarrow \emptyset$ [182]. The generalized Fokker–Planck equation is

$$\partial_t f(\{\alpha(n)\}_n, t) = -k \sum_{m=1}^N \left(\frac{\partial^2}{\partial \alpha^2(m)} - 2\frac{\partial}{\partial \alpha(m)} \right) (\alpha^2(m) f(\{\alpha(n)\}_n, t))$$

$$-d \sum_{m=1}^N \frac{\partial}{\partial \alpha(m)} ((\alpha(m-1) - 2\alpha(m) + \alpha(m+1)) f(\{\alpha(n)\}_n, t)) \tag{5.101}$$

The corresponding generalized Langevin equation is

$$d\alpha(r, t) = \left(D\frac{\partial^2}{\partial r^2}\alpha(r, t) - 2k\alpha(r, t)^2 \right) dt + \sqrt{-2k\alpha(r, t)^2} dW(r, t) \tag{5.102}$$

This equation should be considered as a pair of coupled stochastic equations for the real variables $\alpha_R(r, t)$ and $\alpha_I(r, t)$, respectively the real and imaginary parts of $\alpha(r, t)$

$$d\alpha_R(r, t) = \left(D\frac{\partial^2}{\partial r^2}\alpha_R(r, t) - 2k(\alpha_R^2(r, t) - \alpha_I^2(r, t)) \right) dt$$

$$-\sqrt{2k}\alpha_I(r, t) dW(r, t) \tag{5.103}$$

$$d\alpha_I(r,t) \;=\; \left(D\frac{\partial^2}{\partial r^2}\alpha_I(r,t) - 4k\alpha_R(r,t)\alpha_I(r,t) \right) dt$$

$$+\sqrt{2k}\alpha_R(r,t)dW(r,t) \tag{5.104}$$

As one sees from the above equations, the noise terms arising from the microscopic fluctuations are multiplicative with non-trivial correlation. In particular the noise–noise correlation is negative. This is an unusual feature related to the nonequilibrium aspect of such reaction-diffusion problems.

A Boltzmann-like approximation will overlook such complications and leads to erroneous results. These problems show clearly the advantage of working with cellular automata algorithms able to model important fluctuations in reaction diffusion problems.

5.9   Problems

5.1. Toffoli and Margolus [26] propose a cellular automaton (tube-worms rule) to model the Belousov–Zhabotinsky reaction. The state of each site is either 0 (refractory) or 1 (excited) and a local timer (whose value is 3, 2, 1 or 0) controls the refractory period. Each iteration of the rule can be expressed by the following sequence of operations: (i) where the timer is zero, the state is excited; (ii) the timer is decreased by 1 unless it is 0; (iii) a site becomes refractory whenever the timer is equal to 2; (iv) the timer is reset to 3 for the excited sites which have two, or more than four, excited sites in their Moore neighborhood. Simulate this automaton, starting from a random initial configuration of the timers and the excited states. Observe the formation of spiral pairs of excitations. Check that this rule is very sensitive to small modifications (in particular to the order of operations (i) to (iv)). What are the differences between this rule and the Greenberg–Hasting model discussed in section 5.2?

5.2. Modify the Greenberg–Hasting rule (or the Toffoli and Margolus tube-worms rule of the previous excercise) in order to model a one- or three-dimensional excitable media.

5.3. Contagion in epidemic model: a similar dynamics to that used for describing an excitable chemical media can be considered to model contagion in an epidemic disease [56]. Each site is occupied by an individual who can be either *suceptible*, *infectious* or *immune*. A healthy, suceptible individual is infected only if there is at least one

infectious person in its von Neumann neighborhood. An infectious individual remains so during m steps and then becomes immune. Immunity
stays for m steps: after that the immune individuals become suceptible
again. Consider a CA simulation of this contagion process, starting
with a random configuration of suceptible, infectious or immune individuals. What is the difference between this model and the Greenberg–
Hasting rule? Modify the epidemic rule to make a probabilistic CA.
What is then the difference from the forest fire models discussed in
chapter 2?

5.4. Propose a CA model of a prey–predator system where two species
(preys and predators) diffuse on a 2D lattice and interact according to
the following rules: new preys come to life at a rate proportional to
the current number of prey (e.g. one parent produces one offspring).
A prey is eaten with probability k when it encounters a predator.
Predators die with a given constant probability and come to life when
a parent predator eats a prey. Assuming that preys and predators are
homogeneously distributed, show that this system is roughly described
by the mean-field equations $\dot{p} = ap - kpP$ and $\dot{P} = -bP + kpP$ where
p stands for the total number of preys and P the total number of
predators. The quantities a, b and k are parameters of the model.
This system of equations is known as the Lotka-Volterra model and
yields an oscillatory behavior. Simulate the automaton model. Are
the species homogeneously distributed in space or do you observe
an oscillatory space–time pattern? Modify the rule as follows (i) each
predator has an internal "starving clock" which is decreased by 1 at
each time step. When a prey is eaten, the clock is reset to its maximum
value; when the clock runs down to zero, the predator dies. (ii) Add a
"pregnancy clock" which determines the time necessary for a predator
to give birth to a baby. How do these modification affect the global
behavior? Imagine other modifications that are sound from a biological
point of view. Can you easily write down the corresponding differential
equation?

5.5. Give a cellular automata rule for the $A \rightarrow B \rightarrow C \rightarrow A$ cyclic
reaction [182]. This model is characterized by three possible states
on each cell: A, B or C. There is a cyclic hierarchy of reaction: A is
transformed into B if at least one of the von Neumann neighbors is a
B. Similarly, B changes to C if and only if a C is around. And, finally,
a C turns into an A if an A is present in the neighborhood. Study the
behavior of this system in 1D or 2D lattices. Generalize this process
with more than three species. From the simulation, how many species
do you think are necessary to obtained a "frozen" configuration?

5.6. Write the microdynamics of the two-dimensional $A+A \to \emptyset$ reaction in which any two particles at the same site annihilate (not only those with opposite velocities). What is the reaction rate?

5.7. Write the microdynamics of the reversible reaction diffusion process $A+B \leftrightarrow C$ where the rate of the direct and inverse reactions are not necessarily identical.

5.8. Write the microdynamics of the two-dimensional $2A+B \to \emptyset$ reaction.

5.9. Propose a reaction rule corresponding to the so-called Selkov model: $A \leftrightarrow X$, $X + 2Y \leftrightarrow 3Y$ and $Y \leftrightarrow B$ where X and Y are the "active" species and A and B are held at fixed concentration by external mechanisms.

5.10. Measure the roughness of the interface between the A and B species in the $A + B \to \emptyset$ reaction diffusion process as a function of time. What are the effects of the diffusion constant and the reaction rate on this roughness?

5.11. Consider the reaction-diffusion process $A + B \to \emptyset$ in a two-dimensional box with the same number of A and B particles, but with correlated initial configurations. The initial state is prepared as follows. The A particles are randomly distributed in the box (Poisson process). In a first step, the same distribution is also used for the B particles. But then the B particles are allowed to diffuse for a few iterations, while the A remain at rest. In this way, the A and B distributions are made statistically dependent. The full reaction-diffusion process is then started. Compute the time dependance of the populations. Why does the long time behavior not agree with the prediction $N_A(t) = N_B(t) \sim t^{-1/2}$? Consider the same problem but now with the initial B configuration obtained from a translation of the A distribution. Does this change the time behavior?

5.12. One considers the reaction-diffusion process $A + B \to \emptyset$ where the two components are initially separated in space (the As are located in the domain $x < 0$ and the Bs in $x > 0$). Aussume there is now, at $x = 0$, a semi-permeable wall allowing the A particles to cross the wall but not the Bs. Discuss the dynamics of the reaction front.

5.13. Consider the one-species reaction-annihilation process $A+A \to \emptyset$ in a four-dimensional space. Initially, all the particles sit on a one-dimensional line. Write a CA algorithme which simulates this dynamics. Show that in the long time regime the number of A particles never vanishes. Why is this so?

5.14. Consider the reversible reaction-diffusion process $A + B \rightarrow C$ with rate k and $C \rightarrow A + B$ with rate g. The C species is also diffusing. If the two reactants A and B are initially separated in space (the As are located in the domain $x < 0$ and the Bs in $x > 0$) and no Cs are present at time $t = 0$, what will be the dynamics of the C reaction front as a function of k and g?

5.15. One of the best-known reaction-diffusion processes leading to bistability is the Schlögl model. The chemical reactions characterizing this model are:

$$A \underset{k_2}{\overset{k_1}{\rightleftharpoons}} X, \quad 2X + B \underset{k_4}{\overset{k_3}{\rightleftharpoons}} 3X$$

(i) Write a cellular automaton algorithm that simulates this process in a closed system. (ii) Consider the case of an open system in which A and B are created uniformly with a constant rate. Generalize the CA algorithm for this case.

5.16. Consider the decay of a point-like radioactive source. Write the corresponding master equation. Derive the Liouvillian in terms of the creation and annihilation operators. Compute the time dependance of the number of particles left.

5.17. Consider the following reaction–annihilation–coagulation processes. The A particles diffuse and react according to the processes: $A + A \rightarrow A$ with a reaction rate g and $A + A \rightarrow \emptyset$ with a reaction rate k. Write the corresponding action of the associated quantum field theory.

6

Nonequilibrium phase transitions

6.1 Introduction

The concept of phase transition plays a very important role in many domains of physics, chemistry, living matter and biology. The simplest case is the one in which a phase transition arises between two equilibrium states when varying a control parameter. Examples of control parameters are temperature, pressure or and concentration. The most familiar phase transitions are the paramagnetic–ferromagnetic transition, the liquid–gas transition and the supraconducting–normal phase transition in a metal. Equilibrium statistical mechanics gives a coherent description of the physics of these states. The determination of the presence of a phase transition and the study of its properties is not generally an easy problem, however there exists a well-defined framework to approach it. When, at the transition some quantities have a discontinuity, one speaks of a first-order phase transition. If the physical quantities vary continuously one speaks of second-order phase transitions [183]. This latter case is particularly tricky because one has to face a multiscale problem. Fluctuations of all wavelengths (from a few angström to the size of the system) play an equivalent important role at the transition. However, even in these complicated situations, such tools as the renormalization group method have been developed to solve these problems [171]. Although several analytical methods exist in equilibrium statistical mechanics, one often has to have recourse to numerical simulations. In particular, the Monte-Carlo method is widely used. As we have seen in chapter 2, Monte-Carlo algorithms can be viewed as probabilistic cellular automata providing that some care is taken with the evolution rules (checkerboard invariants).

The case of nonequilibrium phase transitions that we are now going to discuss is even more interesting for several reasons. In this case the phases transitions take place between different steady states of an open system.

Open system means that the system is in contact with reservoirs and that fluxes (of particles, energy, etc.) go through the system. The equilibrium situation is thus the particularly simple case in which the fluxes are zero.

Such open systems are often encountered and can have complex behaviors. Examples are provided by driven-diffusive systems [184], kinetic models with competition among different dynamical processes [185], autocatalytic systems which exhibit spatio-temporal patterns [167], or by living matter [130].

One remarkable feature of equilibrium phase transitions is their universal properties in the vicinity of the phase transition point. This fact has been noticed by the end of the nineteenth century. When expressed in reduced units, the coexistence curves of many different liquid–gas systems collapse onto a single curve [183]. Later it was realized that the study of the behavior of physical quantities near criticality was of great interest. Generically, in the vicinity of the critical temperature T_c of a continuous phase transition, the temperature dependence of a physical quantity Y can be written as:

$$Y(T) = Y_r(T) + Y_s(\tau) \tag{6.1}$$

where Y_r is an analytic function of T showing no singularity at $T = T_c$, while Y_s is a nonanalytic function of the reduced temperature $\tau = (T - T_c)/T_c$. This nonanalyticity usually takes the form of a power law:

$$Y_s(\tau) \sim \tau^y \tag{6.2}$$

and y is called the critical exponent for the variable Y. Several critical exponents can be defined. In the equilibrium case, hypothesis concerning the homogeneity of the thermodynamic potentials leads to relations among these exponents, called scaling relations.

To illustrate the problem, let us consider again a liquid–gas system. At a given critical density, pressure and temperature, the system changes continuously (without latent heat) from the gaseous to the liquid phase. Two quantities that can be experimentally determined are the isothermal compressibility and the specific heat at fixed density. It is found that both quantities diverge (with different critical exponents) as the critical point is approached. The universal character of the phenomenon is shown by the fact that the values of these exponents are the same for different liquids. More strikingly, these exponents turn out to have the same values for an order–disorder transition in a binary alloy. This universal behavior is nowadays well understood for equilibrium phase transitions. The so-called renormalization group theory [171,186] gives a clear explanation and classifies the different systems into universality classes. The different universality classes are characterized only by a small number of parameters, namely, for systems with short-range interactions, the dimensionality of the system d and the number of components n of the order parameter describing the change of symmetry at the transition.

The situation is not so clear in the framework of nonequilibrium systems. Nonequilibrium statistical mechanics is still in its infancy and there is no general theory available to describe such systems. Most of the known results are based on numerical simulations. However, here again the concept of universality classes appears to be relevant although we do not completely understand how the universality classes are characterized.

Accordingly, we shall discuss some simple examples to illustrate the basis ideas and possible approaches to study such nonequilibrium systems. First we shall sketch the theoretical approaches and their difficulties. Then we shall discuss the modeling in terms of cellular automata and review the problems and difficulties associated with this approach.

6.2   Simple interacting particle systems

A simple class of models exhibiting nonequilibrium phase transitions is the class of so-called *interacting particle systems* [187]. These systems are Markov processes on a lattice [188]. Each site can be in one of two states: vacant or occupied. The enumeration of the state of occupancy of all the lattice sites defines a configuration or a microstate of the system. Transitions between different configurations occur via elementary processes, related to creation, annihilation or hopping of particles. An example is provided by the problem of directed percolation discussed in chapter 2. We shall consider here two new models, the so-called *A model* (or *AM*) and *contact process model* (or *CPM*).

6.2.1   *The A model*

This model has been introduced by Dickman and Burschka [189] as a simple model describing poisoning transitions similar to the ones observed on catalytic surfaces. One considers a d-dimensional substratum covered by a regular hypercubic lattice. Each site has two possible states: empty or occupied by an A particle. The first step of the dynamical process is adsorption. The probability of a vacant site becoming occupied during a short time interval δt is $p\delta t$. The second step of the process is desorption. The probability of an occupied site x becoming vacant is $r\delta t$, provided that at least one of the nearest neighbors of x is vacant. During the time interval δt, one of the two processes occurs at each site, depending on its state. For simplicity we shall restrict ourselves to the case $r = (1-p)$; thus p is the only control parameter of the problem. Qualitatively speaking, one expects that if p is large enough, an initially empty substratum will be, after some time, completely covered by A particles. This is the poisoned phase or the adsorbing state. But, if p is small enough, the desorbing mechanism will be efficient enough to prevent such poisoning. Thus one may anticipate the existence of a threshold value p_c such that, in the stationary state, the covering fraction of A on the substratum X_A will be 1 for $p \geq p_c$ (poisoned phase) and smaller than 1 for $p < p_c$. If X_A varies continuously across p_c the transition will be of second order and its behavior near the threshold will be characterized by the critical exponent β:

$$1 - X_A(p) \sim (p_c - p)^\beta \qquad (6.3)$$

This model looks very simple and one may think that an exact analytical solution should be possible without too much effort. This is unfortunately not true as we shall see.

Analytical approaches. Several analytical approaches have been developed for this type of problem. They are based on more or less well-controlled approximations or perturbative-like expansions.

The first one is a mean-field type approximation. The simplest version is the one-site approximation in which one writes down the evolution equation for the probability that one cell is in a given state at a given time. Let us illustrate this method for a one-dimensional system.

In the mean-field spirit, all the sites play an equivalent role. Let $p_A(t)$ be the probability that a site is occupied by a particle A at time t, and $p_0(t)$ the probability that a site is empty. Assuming a discrete time evolution, we have

$$p_A(t+1) = p_A^3(t) + pp_A(t)(1 - p_A^2(t)) + pp_0(t) \qquad (6.4)$$

The first term on the right-hand side expresses the probability that a given site and its two neighbors are occupied by A particles. The second term is the probability that a particle A having at least one nearest neighbor site empty does not desorb during one time step. The final term is the probability that an empty site is filled during one time step. Similarly, one has

$$p_0(t+1) = (1 - p)p_A(t)(1 - p_A^2(t)) + (1 - p)p_0(t) \qquad (6.5)$$

The stationary state value p_A^s is a solution of the third-order equation $p_A^{s\,3}(1 - p) - p_A^s + p = 0$ and $p_0^s = 1 - p_A^s$. More elaborated approximations take into account the correlations between cells. For example, one can write the evolution equation for the joint probabilities p_{ij} that two nearest neighbor sites are respectively in the states i, j. In one dimension, these evolution equations are

$$p_{AA}(t+1) = p_{AA}(t)D(t)\left[p_{AA}^2(t) + \frac{p^2}{2}p_{0A}^2(t) + pp_{0A}(t)p_{AA}(t)\right]$$
$$+p^2 p_{0A}(t) + p^2 p_{00}(t) \tag{6.6}$$

$$p_{0A}(t+1) = p_{AA}(t)D(t)\left[\frac{p(1-p)}{2}p_{0A}^2(t) + (1-p)p_{0A}(t)p_{AA}(t)\right]$$
$$+2(1-p)p_{0A}(t) + 2p(1-p)p_{00}(t) \tag{6.7}$$

$$p_{00}(t+1) = p_{AA}(t)D(t)\frac{(1-p)^2}{4}p_{0A}^2(t) + (1-p)^2 p_{0A}(t) + p_{00}(1-p)^2 \tag{6.8}$$

where $D(t) = (p_{AA}(t) + \frac{1}{2}p_{0A}^2(t))$. By symmetry, $p_{0A} = p_{A0}$. Moreover, the one-site probabilities p_i are obtained from the joint probabilities p_{ij} by summing over all the possible values of j.

The results for the one-dimensional case are summarized in the typical phase diagram given in figure 6.1. The stationary values of the coverage of A particles is given as a function of the control parameter p, both for the one-site mean-field approximation and the direct simulation of the CA rules which will be introduced later. Near the critical point, all the different mean-field like approximations lead to $1 - X_A(p) \sim (p_c^a - p)$, i.e. to a critical exponent $\beta = 1$. The critical value of the probability p_c^a depends on the approximation considered and for the one-site mean-field approximation, $p_c = 0.676$. In fact, the value $\beta = 1$ is not correct when the dimension of the system is lower than an upper critical dimension $d_c = 4$. These mean-field theories often provide a reasonable qualitative description but cannot make quantitative predictions because they neglect fluctuations which play a very important role in the vicinity of the phase transition. Higher-order joint probabilities can be taken into account and each approximation scheme gives an approximate value for the critical probability p_c. However, the algebra soon becomes very tedious. Note that this approach is not restricted to one dimension; similar equations of motion can be derived in all dimensions.

A different analytical approach allowing one to take into account the fluctuations in a systematic way has been proposed by Dickman [190]. By analogy with what is done in equilibrium statistical mechanics, one performs a perturbative series expansion. The expansion is made around a model whose dynamics may be solved exactly and the development is performed directly in terms of the kinetic parameters rather than in terms of a power expansion of the interaction.

We shall not discuss this formalism in detail here but simply give the main ideas of this approach using again the one-dimensional A model as

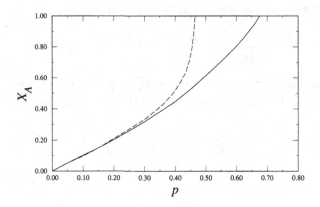

Fig. 6.1. Phase diagram for the one-dimensional A model. The full line correspond to the one-site mean-field approximation and the dotted line to the results of numerical simulations of the CA rules described later.

an example. As we have seen in chapter 5, it may be convenient to describe such a Markov process in terms of an operator formalism. However, if we do not allow multiple occupancy of the sites, the method developed in section 5.8 has to be modified and a different type of creation annihilation operators must be introduced.

Each site i can be in one of the two basic states $|\Phi_{0,i}\rangle$ and $|\Phi_{1,i}\rangle$ corresponding, respectively, to site i vacant or occupied. These vectors are orthonormal:

$$\langle\Phi_{j,i}|\Phi_{k,l}\rangle = \delta_{j,k} \tag{6.9}$$

A configuration of the system can be written as a product:

$$|\Psi\rangle = |\{j_1, j_2, ..., j_k, ...\}\rangle = \prod_i |\Phi_{j_i,i}\rangle \tag{6.10}$$

Creation and annihilation operators $a_i^\dagger, a_i$ are then defined as

$$a_i^\dagger|\Phi_{0,i}\rangle = |\Phi_{1,i}\rangle \tag{6.11}$$
$$a_i^\dagger|\Phi_{1,i}\rangle = 0 \tag{6.12}$$
$$a_i|\Phi_{1,i}\rangle = |\Phi_{0,i}\rangle \tag{6.13}$$
$$a_i|\Phi_{1,i}\rangle = 0 \tag{6.14}$$

Note that

$$a_i a_i^\dagger + a_i^\dagger a_i = I \tag{6.15}$$

and for $i \neq j$,

$$a_i a_j^\dagger - a_j^\dagger a_i = 0 \tag{6.16}$$

where I is the identity operator.

A physical state of the system at time t is given by

$$|\Psi\rangle(t) = \sum_{\{j_k\}} p(\{j_1, j_2, ..., j_k, ...\}, t)|\{j_1, j_2, ..., j_k, ...\}\rangle \qquad (6.17)$$

where the sum is over all the configurations and p is the probability distribution. Physical states $|\Psi\rangle$ must satisfy the positivity and normalization conditions:

$$\langle\{j_1, j_2, ..., j_k, ...\}|\Psi\rangle \geq 0, \forall \{j_1, j_2, ..., j_k, ...\} \qquad (6.18)$$

and

$$\langle\cdot|\Psi\rangle = 1 \qquad (6.19)$$

where

$$\langle\cdot| \equiv \prod_{\{j_k\}}(\langle\Phi_{0,j_k}| + \langle\Phi_{1,j_k}|) = \sum_{\{j_k\}}\langle\{j_1, j_2, ..., j_k, ...\}| \qquad (6.20)$$

Observables are represented by operators which are diagonal in the occupation representation and the expectation of an observable $\mathscr{A}$ in the state $|\Psi\rangle$ is given by

$$\langle\mathscr{A}\rangle_\Psi = \langle\cdot|\mathscr{A}|\Psi\rangle \qquad (6.21)$$

The evolution of the probability distribution can be written in the form of a Schrödinger-like equation:

$$\frac{d|\Psi\rangle}{dt} = S|\Psi\rangle \qquad (6.22)$$

where the evolution operator is defined in terms of the creation and annihilation operators. The conservation of normalization takes the simple form, $\langle\cdot|S = 0$.

The idea of the perturbation theory is to divide S into two parts, as $S = S^0 + V$, such that the evolution under S^0 is solvable and that $\langle\cdot|S^0 = \langle\cdot|V = 0$. Note that $\langle\Psi^0| = \langle\cdot|$; thus the stationary left eigenvector of S^0 is the normalization state. One would like to find a formal expression for the stationary state of the interacting problem, which is a solution of $S|\Psi\rangle = 0$, in terms of $|\Psi^0\rangle$. Let $[S^0]^{-1}$ be the inverse of $[S^0]$ in the non-zero subspace. Multiplying the equation $S|\Psi\rangle = 0$ by $(S^0)^{-1}$ and using $(S^0)^{-1}S^0 = 1 - |\Psi^0\rangle\langle\Psi^0|$, one finds

$$(1 + [S^0]^{-1}V)|\Psi\rangle = |\Psi^0\rangle \qquad (6.23)$$

This is the relation needed.

Let us illustrate the method on the model A in one dimension. The evolution operator S can be written in the form

$$S = S^0 + V = \sum_i S_i = \sum_i (S_i^0 + V_i^0) \tag{6.24}$$

where

$$S_i^0 = p(1 - a_i) + (1 - a_i^\dagger)a_i \tag{6.25}$$

and

$$V_i = -(1 - a_i^\dagger)a_i a_{i-1}^\dagger a_{i-1} a_{i+1}^\dagger a_{i+1} \tag{6.26}$$

The problem whose dynamics is given by S^0 is readily solvable. S_i^0 has two eigenvalues $0, -(1 + p)$, corresponding to the right eigenvectors:

$$|\psi_{0,i}\rangle = (1 - v)|\Phi_{0,i}\rangle + v|\Phi_{1,i}\rangle \tag{6.27}$$

$$|\psi_{1,i}\rangle = |\Phi_{0,i}\rangle - |\Phi_{1,i}\rangle \tag{6.28}$$

and the left eigenvectors:

$$\langle\psi_{0,i}| = \langle\Phi_{0,i}| + \langle\Phi_{1,i}| \tag{6.29}$$

$$\langle\psi_{1,i}| = v\langle\Phi_{0,i}|(1 - v)\langle\Phi_{1,i}| \tag{6.30}$$

where $v = p/(1 + p)$.

S^0 has a unique steady state $|\Psi^0\rangle = \prod_j |\psi_{0,i}\rangle$. The other eigenstates of S^0 are the states $|i_1, i_2, ..., i_m\rangle$ for which all the sites $i_1, i_2, ..., i_m$ are in the excited state $|\psi_{1,i}\rangle$, while all other sites are in their ground state $|\psi_{0,i}\rangle$. The corresponding eigenvalues are $\lambda_m = -m(1 + p)$ with $m = 1, 2, ...$.

Let P_m be the projection onto the subspace of states with exactly m excited sites. Then:

$$[S^0]^{-1} = \sum_{m=1}^{\infty} \lambda_m^{-1} P_m = -(1 - v) \sum_{m=1}^{\infty} m^{-1} P_m \tag{6.31}$$

It is necessary to introduce new creation and annihilation operators $b_i^\dagger$ and b_i defined as

$$b_i^\dagger |\psi_{0,i}\rangle = |\psi_{1,i}\rangle \tag{6.32}$$
$$b_i^\dagger |\psi_{1,i}\rangle = 0 \tag{6.33}$$
$$b_i |\psi_{1,i}\rangle = |\psi_{0,i}\rangle \tag{6.34}$$
$$b_i |\psi_{0,i}\rangle = 0 \tag{6.35}$$

They are related to the creation and annihilation operators $a_i^\dagger, a_i$ as follows:

$$a_i^\dagger = b_i(1 + (1 - v)b_i^\dagger) - (1 - v)b_i^\dagger((1 - v) + b_i) \tag{6.36}$$

$$a_i = b_i(vb_i^\dagger - 1) - vb_i^\dagger(b_i - v) \tag{6.37}$$

The steady-state occupation fraction can be written as

$$\bar{x} = \langle \Psi^0 | (vb_i b_i^\dagger - b_i) | \Psi \rangle \tag{6.38}$$

The interaction terms V_i can be expressed in terms of the bs and $b^\dagger$s. $[S^0]^{-1}V$ is then a six-order polynomial in v, while p appears to all order in this operator. One therefore adopts v as the expansion parameter. If the lowest-order calculations are easy to perform, the higher orders soon become very tedious. Moreover, it is not clear how the series converges, if it does.

This shows that, even more than for the equilibrium case, numerical simulations are very necessary to study such systems.

Cellular automata approaches.   The cellular automata version of this model is straightforward. One considers a d-dimensional lattice. Each cell of the lattice j has two possibles states: $|\Psi_j\rangle = |0\rangle$ or $|A\rangle$. The cellular automata probabilistic rules are
If $|\Psi_j\rangle(t) = |0\rangle$ then:

$$|\Psi_j\rangle(t+1) = \begin{cases} |0\rangle, & \text{with probability } (1-p), \\ |A\rangle, & \text{with probability } p. \end{cases} \tag{6.39}$$

If $|\Psi_j\rangle(t) = |A\rangle$ then:

$$|\Psi_j\rangle(t+1) = \begin{cases} |A\rangle, & \text{with probability } p \text{ if the site j has one nearest} \\ & \text{neighbor empty,} \\ |A\rangle, & \text{with probability 1, if all the neighbors are occupied,} \\ |0\rangle, & \text{with probability } (1-p) \text{ if the site j has at least one} \\ & \text{nearest neighbor empty.} \end{cases}$$
$$\tag{6.40}$$

Simulation of the above cellular automata rule has been performed for chains of lengths L between 30 and 65'536 [191]. As can be seen on figure 6.1, a second-order phase transition is obtained at $p_c = 0.465$. The order parameter critical exponent β extracted by fitting the data with the relation 6.3 over the range $0.001 \leq (p_c - p) \leq 0.02$ is $\beta = 0.280 \pm 0.010$. This shows that a mean-field-like approximation gives a qualitatively reasonable phase diagram, but overestimates the critical probability p_c and gives a poor value for the exponent β.

For the two-dimensional case, one-site and pair mean-field approximations predict a second-order phase transition for, respectively, $p_c = 0.800$ and 0.785. Here again the value of the order parameter critical exponent is $\beta = 1$. The results of numerical simulations of the CA rules lead to a phase diagram topologically similar to the one obtained for the one-dimensional case. The critical probability is $p_c = 0.747$. The order parameter critical exponent β extracted by fitting the data with the relation 6.3 over the range $0.001 \le (p_c - p) \le 0.02$ is $\beta = 0.52 \pm 0.01$.

6.2.2 The contact process model (CPM)

The contact process model is in many respects similar to the A model. However, the desorption mechanism is more subtle. In the A model, there is an equiprobable desorption each time that at least one of the nearest neighbors of a cell is empty. In the *CPM*, the probability of desorption decreases when the number of nearest neighbors increases. The cellular automata version of this model is defined by the following rules:
If $|\Psi_j\rangle(t) = |0\rangle$ then:

$$|\Psi_j\rangle(t+1) = \begin{cases} |0\rangle, & \text{with probability } (1-p), \\ |A\rangle, & \text{with probability } p. \end{cases} \qquad (6.41)$$

If $|\Psi_j\rangle(t) = |A\rangle$ then:

$$|\Psi_j\rangle(t+1) = \begin{cases} |0\rangle, & \text{with probability } (1 - n_j/z)q, \\ |A\rangle, & \text{with probability } (1 - (1 - n_j/z)q). \end{cases} \qquad (6.42)$$

where n_j is the number of occupied nearest-neighbor cells of the cell j, and z the coordination number of the lattice ($z = 2d$ for a d-dimensional hypercubic lattice). Here again to restrict the number of independent control parameters, we shall assume that $q = (1 - p)$.

The one-dimensional phase diagram is again topologically similar to the one obtained for the A model (see figure 6.1). A second-order phase transition occurs at $p_c = 0.281$ and the order parameter critical exponent is $\beta = 0.260 \pm 0.020$. In two dimensions, the second-order phase transition occurs at $p_c = 0.438$ and the order parameter critical exponent β extracted by fitting the data in the range $0.001 \le (p_c - p) \le 0.020$ is $\beta = 0.52 \pm 0.04$.

Several conclusions can be can drawn from the above results concerning the possible differences between a fully parallel dynamics and a sequential one. There are two categories of quantities. The first category contains the quantities which are "universal" i.e. independent of the type of dynamics used. This is the case of the critical exponents. The second category contains the "nonuniversal", quantities, i.e. ones depending on the chosen dynamics. The critical probabilities p_c are typical nonuniversal quantities and their values for a given model may be quite different in the cellular automata approach and the sequential updating. For example, for the A model in one dimension, one has $p_c = 0.465$ with parallel updating while sequential updating gives $p_c = 0.574$ [190].

6.3   Simple models of catalytic surfaces

In this section, we describe more complicated models related to the important problem of heterogeneous catalysis.

6.3.1   *The Ziff model*

The original Ziff model with sequential dynamics.   One of the simplest models describing the adsorption–dissociation–desorption process on a catalytic surface has been introduced by Ziff *et al.* [192]. This model is based upon some of the known steps of the reaction $A - B_2$ on a catalyst surface (for example $CO - O_2$). The basic steps in heterogeneous catalysis are the following:

- A gas mixture with concentrations X_{B_2} of B_2 and $X_A = (1 - X_{B_2})$ of A sits above the surface and can be adsorbed. The surface can be divided into elementary cells. Each cell can adsorb one atom only.

- The B species can be adsorbed only in the atomic form. A molecule B_2 approaching an empty cell will be dissociated into two B atoms only if another cell adjacent to the first one is empty. Otherwise it is rejected. The first two steps correspond to the reactions

$$A \rightarrow A(\text{ads}) \quad B_2 \rightarrow 2B(\text{ads}) \tag{6.43}$$

- If two nearest neighbor cells are occupied by different species they chemically react according to the reaction

$$A(\text{ads}) + B(\text{ads}) \rightarrow AB \tag{6.44}$$

and the product of the reaction is desorbed. In the example of the $CO - O_2$ reaction, the desorbed product is a CO_2 molecule.

This final desorption step is necessary for the product to be recovered and for the catalyst to be regenerated. However, the gas above the surface is assumed to be continually replenished by fresh material. Thus its composition is constant during the whole evolution. The original Ziff dynamics is sequential and defined as follows:

- A trial begins with the random collision of a gas molecule on a square lattice, which represents the surface. The colliding molecule is chosen to be A with probability X_A and B_2 with probability X_{B_2}. If the colliding molecule is A, then the following happens:

    (1) A site on the lattice is randomly chosen.

    (2) If the site is already occupied then the trail ends.

(3) Otherwise the *A* adsorbs and the four nearest neighbors are checked in random order.

(4) If a *B* is present in any of the nearest-neighbor sites, a reaction takes place between the central *A* particle and first *B* particle found. A a result, both sites become vacant.

- If the colliding molecule is B_2, then the following happens:

    (1) Two adjacent sites are randomly chosen on the surface.

    (2) If either site is occupied, the trial ends.

    (3) Otherwise, the B_2 dissociates and adsorbs on the two sites as two *B*s.

    (4) All the three nearest neighbors sites of *both B*s are then checked in random order for an *A* particle.

    (5) The first *A* particle found in any of these two neighborhood sites reacts with the corresponding *B* atom by vacating the two sites.

It is found by numerical simulation [192] that a reactive steady state occurs only in a window defined by

$$X_1 < X_A < X_2$$

where $X_1 = 0.389 \pm 0.005$ and $X_2 = 0.525 \pm 0.001$.

Outside this window, the steady state is a "poisoned" catalyst of pure A ($X_A > X_2$) or pure B ($X_A < X_1$). For $X_A > X_1$, the coverage fraction varies continuously with X_A and one can speak of a continuous (or second-order) nonequilibrium phase transition. At $X_A = X_2$, the coverage fraction varies discontinuously with X_A and one can speak of a discontinuous (or first-order) nonequilibrium phase transition.

The revisited model with parallel dynamics. It is not clear from a physical point of view if the dynamics of such a system is sequential or parallel. It is then legitimate to study a similar model with fully parallel, cellular automata dynamics.

In a simple approach [193], the cells of the automaton correspond to the elementary cells of the catalyst. In order to model the different processes, each cell j can be in four different states denoted $|\psi_j\rangle = |0\rangle$, $|A\rangle$, $|B\rangle$ or $|C\rangle$.

The state $|0\rangle$ corresponds to an empty cell, $|A\rangle$ to a cell occupied by an atom A, and $|B\rangle$ to a cell occupied by an atom B. The state $|C\rangle$ is artificial and represents a precursor state describing the conditional occupation of the cell by an atom B. Conditional means that during the next evolution

 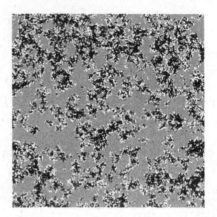

Fig. 6.2. Typical microscopic configuration in the stationary state of the CA Ziff model, where there is coexistence of the two species. This simulation corresponds to the generalized model described in equation 6.53. The gray and black dots represent, respectively, the A and B particles, while the empty sites are white. The control parameter X_A is larger in the right image than it is in the left.

step of the automaton, $|C\rangle$ will become $|B\rangle$ or $|0\rangle$ depending upon the fact that a nearest neighbor cell is empty and ready to receive the second B atom of the molecule B_2. This conditional state is necessary to describe the dissociation of B_2 molecules on the surface.

The time evolution of the CA is given by the following set of rules, fixing the state of the cell j at time $t + 1$, $|\psi_j\rangle(t + 1)$, as a function of the state of the cell j and its nearest neighbors (von Neumann neighborhood) at time t:

<u>R1</u> : If $|\psi_j\rangle(t) = |0\rangle$ then

$$|\psi_j\rangle(t+1) = \begin{cases} |A\rangle & \text{with probability } X_A \\ |C\rangle & \text{with probability } (1 - X_A) \end{cases} \qquad (6.45)$$

<u>R2</u> : If $|\psi_j\rangle(t) = |A\rangle$ then

$$|\psi_j\rangle(t+1) = \begin{cases} |0\rangle & \text{if at least one of the nearest neighbor cells} \\ & \text{of } j \text{ was in the state } |B\rangle \text{ at time } t \\ |A\rangle & \text{otherwise} \end{cases} \qquad (6.46)$$

<u>R3</u> : If $|\psi_j\rangle(t) = |B\rangle$ then

$$|\psi_j\rangle(t+1) = \begin{cases} |0\rangle & \text{if at least one of the nearest neighbor cells} \\ & \text{of } j \text{ was in the state } |A\rangle \text{ at time } t \\ |B\rangle & \text{otherwise} \end{cases} \qquad (6.47)$$

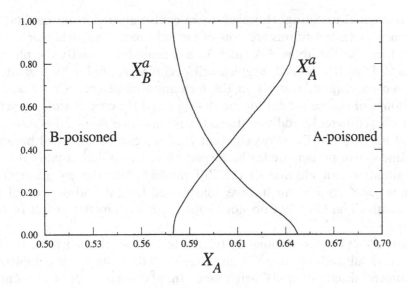

Fig. 6.3. Stationary state phase diagram in the simple CA model of the Ziff problem. X_A is the concentration of the A species injected on the surface. X_A^a and X_B^a are respectively the coverage fractions of A and B.

<u>R4</u> : If $|\psi_j\rangle(t) = |C\rangle$ then

$$|\psi_j\rangle(t+1) = \begin{cases} |0\rangle & \text{if none of the nearest neighbors was} \\ & \text{in the state } |C\rangle \text{ at time } t \\ |B\rangle & \text{otherwise} \end{cases} \qquad (6.48)$$

This last rule expresses the fact that the atoms of B_2 can be adsorbed only if they had been dissociated on two adjacent cells, that is if at least two adjacent cells were empty at time $t-1$. Rules R1, R4 describe the adsorption–dissociation mechanism while rules R2, R3 describe the reaction–desorption process.

Figure 6.2 shows typical stationary configurations obtained with a cellular automata version of the Ziff model. At time $t = 0$, all the cells are empty and a randomly prepared mixture of gases with fixed concentration of one species (X_A) sits on top of the surface. The rules are iterated until a stationary state is reached. The stationary state is a state for which the mean coverage fractions X_A^a and X_B^a of atoms of type A or B does not change in time, although microscopically the configurations of the surface changes.

As expected, it is found that the reactive steady state occurs only in a window limited by $X_1 < X_A < X_2$, where $X_1 = 0.5761 \pm 0.0004$ and $X_2 = 0.6515 \pm 0.0004$. The phase diagram is given in figure 6.3.

As anticipated, the limits of the window differ from those obtained with sequential updating. However, there is also an important qualitative difference. Both transitions are now of second order. This behavior is due to the fact that the above CA rules do not reproduce exactly the physics of the Ziff model. Indeed, a given cell occupied at time t by a B atom can take part simultaneously in the formation of several AB pairs; the situation is of course similar for the A atoms and the correct stoichiometry is not obeyed here. In addition, the adsorption phase $B_2 \rightarrow 2B$ is also not correct microscopically: three adjacent B atoms can be adsorbed because the same C can participate in the formation of two different pairs $B - B$. This situation not allowed in the Ziff model where the B_2 adsorption phase is geometrically much more constrained than the adsorption of an A molecule. The CA rules do not capture this asymmetric aspect of the reaction.

These difficulties are intimately related to the fact that all the cells are updated simultaneously in a CA and that a cell does not know completely the current situation of its neighbors. In particular it does not know whether or not these neighbors can also react with other particles in their own neighborhood.

This drawback can be cured [194] by adding a vector field to every site in the lattice, as shown in figure 6.4. A vector field is a collection of arrows, one at each lattice site, that can point in any of the four directions of the lattice. The directions of the arrows at each time step are assigned randomly. Thus, a two-site process is carried out only on those pairs of sites in which the arrows point toward each other (matching nearest-neighbor pairs (MNN)). This concept of *reacting matching pairs* is a general way to partition the parallel computation locally, very much in the same spirit as in lattice gas automata where the interaction takes place only among particles at the same site.

With reacting matching pairs, new CA rules can be defined as
<u>R1</u> : If $|\psi_j\rangle(t) = |0\rangle$ then

$$|\psi_j\rangle(t+1) = \begin{cases} |A\rangle & \text{with probability } X_A \\ \\ |C\rangle & \text{with probability } (1 - X_A) \end{cases} \tag{6.49}$$

<u>R2</u> : If $|\psi_j\rangle(t) = |A\rangle$ then

$$|\psi_j\rangle(t+1) = \begin{cases} |0\rangle & \text{if the MNN} \\ & \text{of } j \text{ was in the state } |B\rangle \text{ at time } t \\ |A\rangle & \text{otherwise} \end{cases} \tag{6.50}$$

<u>R3</u> : If $|\psi_j\rangle(t) = |B\rangle$ then

$$|\psi_j\rangle(t+1) = \begin{cases} |0\rangle & \text{if the MNN} \\ & \text{of } j \text{ was in the state } |A\rangle \text{ at time } t \\ |B\rangle & \text{otherwise} \end{cases} \tag{6.51}$$

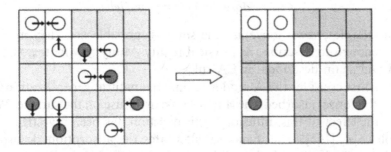

Fig. 6.4. Illustration of the modified rules R2 and R3. The arrows select which neigbor is considered for a reaction. Dark and white particles represent the *A* and *B* species, respectively. The shaded region correspond to cells that are not relevant to the present discussion such as, for instance, cells occupied by the intermediate *C* species.

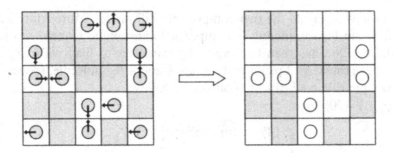

Fig. 6.5. Illustration of the modified rule R4. Light gray particles represent the intermediate *C* species and the arrows indicate how the matching pair is done. The shaded region corresponds to cells that are not relevant to the present discussion such as, for instance, cells occupied by *A* or *B* particles. White cells are empty cells.

<u>R4</u> : If $|\psi_j\rangle(t) = |C\rangle$ then

$$|\psi_j\rangle(t+1) = \begin{cases} |B\rangle & \text{if MNN is in the state } |C\rangle \text{ at time } t \\ |0\rangle & \text{otherwise} \end{cases} \tag{6.52}$$

The above rules *R2* and *R3* are illustrated in figures 6.4 and 6.5.

The simulations made on this model [194] reproduce the correct orders of the transitions (second and first order), but the limits of the reactive window are respectively 0.0495 and 0.0546, illustrating once more the nonuniversal character of these quantities.

6.3.2   More complicated models

The models described above are the simplest possible and they may not be rich enough to describe the physical reality. Many new features can be added and again described by CA rules.

Let us first consider the A model again. One natural generalization is to assume that once adsorbed, the A particles can diffuse on the surface. What will be the effect of this diffusion? This question has been investigated by Bagnoli *et al.* [195]. The dynamics alternates the adsorption–desorption rule discussed above with the diffusion rule described in section 4.2.

By changing the relative frequency of the diffusion and adsorption–desorption phases, the diffusion constant D can be tuned and different regimes can be investigated. As expected from physical grounds, the value of the critical probability p_c increases monotonically with D. In the limit $D \to \infty$, one recovers the mean-field value. The situation is more complicated for the critical exponents and will be discussed in the next section.

Let us now return to the two-components Ziff model. More complicated dynamics can be considered. An empty site can remain empty with some probability. One has then two control parameters to play with: X_A and X_{B_2} that are the arrival probability of an A and a B_2 molecule, repectively. For example rule 6.49 can be modified as follows:

If $|\psi_j\rangle(t) = |0\rangle$ then

$$
|\psi_j\rangle(t+1) = \begin{cases} |A\rangle & \text{with probability } X_A \\ |C\rangle & \text{with probability } X_{B_2} \\ |0\rangle & \text{with probability } 1 - X_A - X_{B_2} \end{cases} \tag{6.53}
$$

In addition, since empty sites may now be present on the lattice during the adsorption $B_2 \to 2B$ a generalization of rule 6.52 can be considered: a cell in the intermediate state C will give two adjacent B atoms if its matching arrow points to an empty site which is not pointed to by another C state. This new rule is illustrated in figure 6.6.

The phase diagram obtained for this generalized CA Ziff model is given in figure 6.7, with the value $X_{B_2} = 0.1$. Here again, the phase diagram is topologically similar to the case $X_{B_2} = 1 - X_A$, but the locations of the critical points are different.

The influence of several other mechanisms on the location of the active window have been investigated. Besides diffusion, the role of adatom–adatom interactions [196], desorption activation energy and precursor adsorption have been considered [197–199]. By tuning the parameters controlling the different effects, it is possible to obtain reasonable quantitative agreement with experimental situations.

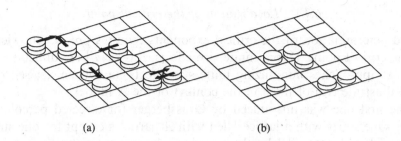

(a)   (b)

Fig. 6.6. A less constrained adsorption rule $B_2 \to 2B$ when some cells may be empty. Here the B_2 molecule (or C state) is represented as two disks on top of each other. Dissociation is possible if the upper disk can move to the site indicated by the arrow without conflict with other moves. Note that we still accept dissociation when a pair of matching Cs sit next to each other.

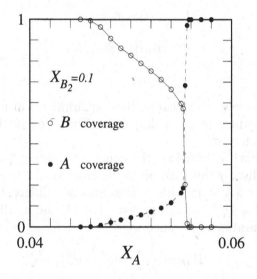

Fig. 6.7. Stationary state phase diagram corresponding to the CA Ziff model given by relation 6.53 and the generalized dissociation rule.

6.4 Critical behavior

As we have seen at the beginning of this chapter, a problem of great interest is the study of the critical behavior at a second-order phase transition. We shall briefly review some methods which can be used to extract the critical exponents for nonequilibrium systems from numerical simulations.

6.4.1   Localization of the critical point

Good determination of the critical exponents requires precise knowledge of the critical point. Due to finite size effects, this may be a difficult task. Several approaches minimizing finite size effects have been devised. We shall illustrate two of them in the context of the A model.

The first one was introduced by Grassberger for directed percolation [200]. One starts with a lattice filled with A particles except for one single site which is empty, and let the system evolve according to the CA rule. One measures the density of empty sites $n_0(t)$, the average cluster size $R^2(t)$ and the survival probability $P(t)$, that is the probability that the lattice is not poisoned by A particles after t iterations. If the system is sufficiently large that none of the clusters we simulate reaches the boundaries, the data are free of finite size effects. At the critical probability p_c, all the above quantities behave as a power law:

$$P(t) \sim t^{-\delta} \tag{6.54}$$

$$n_0(t) \sim t^{\eta} \tag{6.55}$$

$$R^2(t) \sim t^z \tag{6.56}$$

If one is not precisely at criticality, these quantities will have an upward (downward) curvature in a log–log plot. Thus this method leads to a precise determination of p_c.

Another approach is based on the concept of damage spreading [201]. Here again, we discuss this method in the context of the A model.

Let us consider a system with N sites and two different configurations $A = \{\psi_i^A\}$ and $B = \{\psi_i^B\}$ of this system. The Hamming distance between these two configurations is defined as

$$D_{AB} = \frac{1}{N} \sum_i |\psi_i^A - \psi_i^B| \tag{6.57}$$

At time $t = 0$, one prepares two copies of the system such that

$$\psi_j^A = \psi_j^B \quad \text{if} \quad j \neq 0 \tag{6.58}$$

and

$$\psi_0^A = 1 \quad \text{and} \quad \psi_0^B = 0 \tag{6.59}$$

Moreover, one imposes that this last condition remains true for all time. The initial Hamming distance is thus $D_{AB}(t = 0) = \frac{1}{N}$. One then starts the evolution in such a way that these two copies have strictly the same dynamics, defined by the cellular automata rules for the A model discussed previously. Locally, two types of damage can, in principle, occur.

At site i, $\psi_i^A = 1$ and $\psi_i^B = 0$; this may occur with a probability d_i^{10}.

At site i, $\psi_i^A = 0$ and $\psi_i^B = 1$; this may occur with a probability d_i^{01}.

Let us now restrict ourselves to the stationary state and denote by square brackets the time average in the stationary state. We have:

$$d_i^{10} = [\psi_i^A(1 - \psi_i^B)] \tag{6.60}$$

and

$$d_i^{01} = [(1 - \psi_i^A)\psi_i^B] \tag{6.61}$$

thus

$$\Gamma_i \equiv (d_i^{10} - d_i^{01}) = [\psi_i^A] - [\psi_i^B] \tag{6.62}$$

Let us now assume that we are not in the poisoned phase, i.e. $p < p_c$. The dynamical process is not frozen in. Assuming *ergodicity*, we can express the time averages in 6.62 by ensembles averages (denoted by $\langle .. \rangle$) on a nonconstrained nonequilibrium ensemble. One obtains:

$$[\psi_i^A] = \frac{\langle \psi_0 \psi_i \rangle}{\langle \psi_0 \rangle} \tag{6.63}$$

$$[\psi_i^B] = \frac{\langle (1 - \psi_0)\psi_i \rangle}{\langle 1 - \psi_0 \rangle} \tag{6.64}$$

Indeed, the presence of the ψ_0 insures that only configurations such that $\psi_0^A = 1$ and $\psi_0^B = 0$ are retained. Introducing the reduced correlation function

$$C_{0i} = \langle \psi_0 \psi_i \rangle - \langle \psi_0 \rangle \langle \psi_i \rangle \tag{6.65}$$

and assuming translational invariance,

$$X_A = \langle \psi_0 \rangle = \langle \psi_i \rangle \tag{6.66}$$

one has

$$\Gamma_i = \frac{C_{0i}}{X_A(1 - X_A)} \tag{6.67}$$

The probability of finding site i damaged is obviously $D_i = d_i^{10} + d_i^{01}$.

It is now important to notice that $d_i^{01} = 0$. Indeed, the probability of a site i being occupied is always larger for the A configuration than for the B one. Thus it is not possible that a given time $\psi_i^A = 0$ while $\psi_i^B = 1$, hence $D_i = \Gamma_i$. Introducing the total damage (which is just the Hamming distance) as

$$D_{AB} = \frac{1}{N} \sum_i D_i \tag{6.68}$$

one finds, in term of the nonequilibrium susceptibility defined as

$$\chi = \sum_i C_{0i} \tag{6.69}$$

$$D_{AB} = \frac{1}{N} \frac{\chi}{X_A(1 - X_A)} \tag{6.70}$$

Thus the Hamming distance is exactly related to the nonequilibrium one-point and two-point correlation functions, X_A and C_{0i}. In particular, when p approaches p_c, the Hamming distance diverges allowing a precise determination of p_c.

6.4.2 *Critical exponents and universality classes*

The determination of the critical exponents presents its own difficulties. Let us consider as an example the critical behavior of the A model with diffusion. The behavior of the order parameter exponent β as a function of the diffusion coefficient D has been studied by Bagnoli *et al.* [195]. Simulation shows an apparent dependence of β upon D. Two limit cases are clear. When $D = 0$, one finds the directed percolation value while, when $D \to \infty$, one finds the mean-field value, $\beta \to 1$. What is happening in between? The apparent D dependence does not mean that the true critical exponent is continuously varying with D. Indeed, the presence of a lattice of finite size prevents us going arbitrarily close to p_c. As D becomes larger, the width of the critical region may shrink, and it is more and more difficult to probe the real critical regime. In fact renormalization group arguments support the following scenario. For all finite values of D, the true critical behavior is one of directed percolation, while for $D = \infty$ one is in the mean-field regime. The continuously varying exponent measured reflects the crossover between the two limit cases. This situation can be described in terms of an effective exponent (see [195] for more details). This example shows that it may be difficult to extract an exponent from the numerical data if one does not have a clear theoretical idea of what is happening.

Similar studies can be made for the two-species models and in these cases one also finds that the critical exponents belong to the directed percolation universality class. This is not the place to discuss in detail the attributes of the directed percolation universality class. A thourough analysis of this question is given in the review article by Kinzel [202].

This universality class appears to be very large and robust and one can wonder if there exist some systems belonging to a different class.

Such models indeed exist and Grassberger *et al.* have shown [203] that simple Wolfram's rules with very special added noise belong to a different

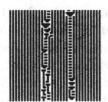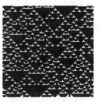

Fig. 6.8. Behavior of probabilistic rule 94 for p=0.0, 0.05, 0.25 and 1.0, respectively. A kink corresponds to two nearest neighbors black sites while an anti-kink corresponds to two nearest-neighbor white sites.

universality class from directed percolation. Let us consider for example the Wolfram's rule number 94

$$[(111);(101);(010);(100);(001);(000)] \rightarrow [0;0;1;1;1;0] \qquad (6.71)$$

and add the following noise:

$$[(011);(110)] \rightarrow \begin{cases} [0;0] & \text{with probability p} \\ [1;1] & \text{with probability } (1-p) \end{cases} \qquad (6.72)$$

For very small p, the system orders itself spontaneously: there are two symmetric adsorbing states. The corresponding patterns are, respectively, vertical stripes with $\psi_i = 0$, $\psi_{i+1} = 1$. Starting from a random initial state, after some short transients one finds small domains separated by doubly occupied neighbors called kinks. For $p = 0$, the kinks are frozen and between two kinks one observes periodic patterns with very short periods. For small p the kinks are no longer frozen. They move by annihilating random walks and the kink density decays in time as $(pt)^{1/2}$. If p grows further, in addition to enhancing the random walk it leads to a splitting of the kinks according to: kink→ kink+[kink+antikink]. For small p, the annihilation procedure dominates but if p is larger than a critical value $p_c = 0.13 \pm 0.02$, a single kink in the initial state is enough to create a completely disordered state. Note that in the limit $p = 1$, the rule is simply Wolfram's rule 22 which has a chaotic behavior. These different situations are illustrated on figure 6.8.

At $p = p_c$, the density of kinks decreases as

$$n_{kink} \sim t^{-\alpha}, \quad \alpha = 0.27 \pm 0.08 \qquad (6.73)$$

and for $p \rightarrow p_c$ from above, the density of kinks in the stationary state tends to zero as

$$n_{kink} \sim (p - p_c)^{\beta}, \quad \beta = 0.6 \pm 0.2 \qquad (6.74)$$

These exponents are not those of directed percolation.

This cellular automata model is in fact closely related to the more general problem of the *branching annihilating random walk (BARW)* [204].

In the BARW, particles on a lattice are subject to the following sequential dynamics. First, one chooses one particle randomly. With probability q this particle jumps to a nearest-neighbor site. If the new site is already occupied by one particle, both particles annihilate. Secondly, with probability $(1 - q)$ the particle gives birth to n offspring distributed randomly on the nearest-neighbor sites. In this problem there is a phase transition in the stationary states for $q = q_c$ between the empty phase and a phase with a non-zero particle density. Two cases have to be distinguished: if n is odd the exponents associated to the phase transition are those of directed percolation while if n is even, the exponents are those of the cellular automata described above [205].

The physical reason for this difference is simple. The even n case is special as the particle number is conserved modulo 2. This means that the absorbing state exists only if we start out with an even number of particles.

From the numerous models which have been studied, the following rule seems to be correct. Only systems possessing a unique absorbing state, or systems with many absorbing states such that, with probability one, the system falls into one of the many available absorbing states, belong to the directed percolation universality class [206]. This explains why the directed percolation universality class is large but also why there are exceptions.

6.5 Problems

6.1. Consider a variation of the A model introduced in section 6.2.1 In this new version, the particles are adsorbed in pairs at vacant nearest-neighbor sites. Study the phase diagram of this model in mean-field-like theory and by numerical CA simulations.

6.2. Consider the CA introduced in section 6.3.1 for the Ziff model. The rules are modified as follows: a site occupied by a B atom can also react with an A atom arriving on this site. Study the phase diagram of this model in mean-field-like theory and by cellular automata simulations.

6.3. Study the kinetics of the Ziff model (introduced in section 6.3.1) with a *sequential dynamics*. Compare with the site and pair mean-field approximation.

6.4. Consider an extension of the model A introduced in section 6.2.1. Once the A atoms are adsorbed on the surface, they can diffuse. Study the role of diffusion on the location of the critical threshold p_c and on the critical exponent β.

6.5. Driven-diffusive systems provide simple examples of nonequilibrium problems. Consider a one-dimensional lattice of N sites with periodic boundary conditions and such that nearest-neighbor particles are subject to an attractive interaction. At each site, one can have at most one particle. In equilibrium, the particles can hop to empty nearest-neighbor sites with a rate compatible with the fact that the particles are in thermal equilibrium with a heat bath at temperature T. The system can be put out of equilibrium by adding an external field favoring the hopping of the particles in the direction of the field.
(i) Write the master equation governing the probability $P(\Omega, t)$ that the configuration Ω is realized at time t (with and without the electric field).
(ii) Introduce a probabilistic CA which simulates the above dynamics.
(iii) Compute for different situations the Hamming distance $h_{AB}(t)$ at time t between two configurations Ω^A and Ω^B. The Hamming distance is defined as the sum of the local differences:

$$h_{AB}(t) = \frac{1}{N} \sum_{i=1}^{N} \eta_i^A(t) \oplus \eta_i^B(t)$$

where η_i represents the state of site i and $\oplus$ stands for the exclusive-or operation (hint: see [207]).

6.6. Voter model: consider an abitrary lattice and assume that each site is occupied by a voter who may have one of two opinions labeled $+$ and $-$. Each site keeps its opinion during some time interval, exponentially distributed, and then assumes the opinion of a randomly chosen neighboring site. Let $P_n(t)$ the fraction of voters who change their opinion n times up to time t. Compute $P_n(t)$ and $\langle n \rangle$ the average number of opinion changes in the mean-field approximation.

6.7. Write a CA rule which simulates the voter model described in the previous problem. Investigate the one-, two- and three-dimensional cases and measure $\langle n(t) \rangle$, the average number of opinion changes up to time t.

7

Other models and applications

This chapter presents a few more applications of the cellular automata and lattice Boltzmann techniques. We introduce some new ideas and models that have not been discussed in detail previously and which can give useful hints on how to address different problems.

We shall first discuss a lattice BGK model for wave propagation in a heterogeneous media and show how it can be applied to simulate a fracture process or make predictions for radio wave propagation inside a city. Second, we shall present how van der Waals and gravity forces can be included in an FHP cellular automata fluid in order to simulate the spreading of a liquid droplet on a wetting substrate. Then, we shall define a multiparticle fluid with a collision operator inspired by the lattice BGK method in order to avoid numerical instabilities and re-introduce fluctuations in a natural way. Finally, we shall explain how particles in suspension can be transported and eroded by a fluid flow and deposited on the ground. Snowdrift and sand dunes formation is a typical domain where this last model is applicable.

7.1 Wave propagation

7.1.1 One-dimensional waves

In this book, we have already encountered one-dimensional wave propagation. The chains of particles, or strings, discussed in section 2.2.9 move because of their internal shrinking and stretching. From a more physical point of view, this internal motion is mediated by longitudinal backward and forward deformation waves. This is easy to see from equation 2.46

$$q_1(t-1) - 2q_1(t) + q_1(t+1) = 2(q_2(t) - q_1(t) - a)$$
$$q_i(t-1) - 2q_i(t) + q_i(t+1) = q_{i-1}(t) - 2q_i(t) + q_{i+1}(t)$$
$$q_N(t-1) - 2q_N(t) + q_N(t+1) = 2(q_{N-1}(t) - q_N(t) + a)$$

Fig. 7.1. The string motion is due to forward and backward traveling waves f_1 and f_2.

where $q_i(t)$ is the location along the x-axis of particle i in the chain and a the equilibrium separation between consecutive particles. Now if we define Δ_i as the deformation from the equilibrium distance

$$\Delta_i = q_{i+1} - q_i - a$$

we obtain the following equations of motion

$$\Delta_1(t+1) - 2\delta_1(t) + \Delta_1(t-1) = \Delta_2(t) - 3\Delta_1(t) \tag{7.1}$$

$$\Delta_i(t+1) - 2\Delta_i(t) + \Delta_i(t-1) = \Delta_{i-1}(t) - 2\Delta_i(t) + \Delta_{i+1}(t) \tag{7.2}$$

$$\Delta_{N-1}(t+1) - 2\Delta_{N-1}(t) + \Delta_{N-1}(t-1) = \Delta_{N-2}(t) - 3\Delta_{N-1}(t) \tag{7.3}$$

Disregarding momentarily equations 7.1 and 7.3, it is interesting to note that equation 7.2 admits a very simple set of solutions

$$\Delta_i(t) = f(i-t) + g(i+t) \tag{7.4}$$

as can be seen by direct substitution. The functions f and g can be chosen arbitrarily and correspond to a backward and forward traveling wave. As we shall see, equations 7.1 and 7.3 impose reflection and transmission boundary conditions on these longitudinal deformation waves.

Now, let us return to the interpretation of the motion of our string particles, as presented in section 2.2.9. Black and white particles alternate along a chain and each type of particle moves at every other time step as if they were linked by a spring. Suppose we define $f_1(i,t)$ as the deformation on the *left* of a moving particle (say a black one) and $f_2(i,t)$ as the deformation on its *right*

$$f_1(i,t) = q_i - q_{i-1} - a$$
$$f_2(i,t) = q_{i+1} - q_i - a$$

for i odd or even depending on which kind of particle is going to move.

According to the rule of motion, our black particle moves so as to exchange f_1 and f_2, as shown in figure 7.1. Then, at the next iteration, f_1 and f_2 are no longer defined with respect to the black particle that just moved, but with respect to the white particles around it.

This choice of defining f_1 and f_2 as the left and right deformation, respectively, is less general than the solution 7.4. This is a particular case where f_1 is zero when f_2 is not and vice versa. But this choice corresponds to the actual motion of the string and leads to the same kind of formulation as found in lattice gas automata

$$\begin{pmatrix} f_1(i+1,t+1) \\ f_2(i-1,t+1) \end{pmatrix} = \begin{pmatrix} 1 & 0 \\ 0 & 1 \end{pmatrix} \begin{pmatrix} f_1(i,t) \\ f_2(i,t) \end{pmatrix}$$

With this formulation, it is obvious that the string deformation is mediated by a forward and a backward wave f_1 and f_2, respectively.

The boundary conditions 7.1 and 7.3 can be easily expressed in terms of the f_is. The particles at the right (or left) extremity of the string move so as to take a symmetrical position with respect to $q_{N-1} + a$ (or $q_2 - a$). Thus we have for the right end

$$q_N(t) = q_{N-1}(t) + a + f_1(N,t) \qquad \text{and} \qquad q_N(t+1) = q_{N-1}(t) + a - f_1(N,t)$$

with, also $q_{N-1}(t+1) = q_{N-1}(t)$ since the next-to-the-end particle does not move when the last one does. Therefore:

$$\begin{aligned} f_2(N-1,t+1) &= q_N(t+1) - q_{N-1}(t+1) - a \\ &= q_{N-1}(t) + a - f_1(N,t) - q_{N-1}(t) - a \\ &= -f_1(N,t) \end{aligned}$$

A similar result holds for the left end of the string. The interpretation is quite simple: at the extremity of the string, the forward wave is reflected back with opposite sign and so is the backward wave at the left string end.

Therefore, the evolution rule for the deformation of the string can be written as

$$\begin{pmatrix} f_1(i+1,t+1) \\ f_2(i-1,t+1) \end{pmatrix} = \begin{pmatrix} T & R \\ R & T \end{pmatrix} \begin{pmatrix} f_1(i,t) \\ f_2(i,t) \end{pmatrix} \tag{7.5}$$

where $T = 1$, $R = 0$ inside the string and $T = 0$, $R = -1$ at the free ends. Another interesting case is when the right string extremity is fixed and cannot move. Then, clearly $f_2(N-1,t+1) = f_1(N,t)$ and R should be set to $R = +1$.

7.1.2 Two-dimensional waves

The problem of modeling two-dimensional waves is more difficult because waves spread as they propagate in space. Thus, a representation of the deformation in terms of a field of integer values is in general not possible. However, the construction proposed in the previous section can be generalized provided that the f_is are real-valued.

Let us assume we have a two-dimensional lattice of black and white particles organized in a checkerboard fashion. Figure 7.2 shows a basic

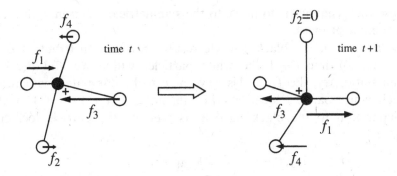

Fig. 7.2. Alternate motion of black and white particles in a two-dimensional system. The quantities f_1, f_2, f_3, and f_4 measure the distance, along the x-axis, separating the central particle from its four connected neighbors. The cross indicates the location of the center of mass of the white particles. Other quantities g_is can be defined to describe the separation along the vertical y-axis.

element of this lattice, made of a central black particle and its four white connecting neighbors.

As in the one-dimensional case, the black particles move at even time steps and the white ones at odd time steps. In figure 7.2, the motion is relative to the positions of the four connected white neighbors at rest. The rule of motion for the black particles is to jump to the symmetrical position with respect to the center of mass (the cross in the figure) of the four white surrounding particles.

Let us denote the location of the black particle by

$$\vec{r}_{i,j} = (x_{i,j}, y_{i,j})$$

The surrounding white particles will be at positions $\vec{r}_{i-1,j}$, $\vec{r}_{i+1,j}$, $\vec{r}_{i,j-1}$ and $\vec{r}_{i,j+1}$. We define their x-axis separation to the central black particle as

$$f_1(i,j,t) = x_{i-1,j}(t) - x_{i,j}(t) \qquad f_3(i,j,t) = x_{i+1,j}(t) - x_{i,j}(t)$$
$$f_2(i,j,t) = x_{i,j-1}(t) - x_{i,j}(t) \qquad f_4(i,j,t) = x_{i,j+1}(t) - x_{i,j}(t) \qquad (7.6)$$

Similarly, we can define g_1, g_2, g_3, and g_4 as the distance, along the y-axis, separating the white particles from the black ones.

If the origin of the coordinate system is the black particle (that is $\vec{r}_{i_1,j}(t) = 0$), the x-coordinate X_{CM} of the center of mass of the four white particles is

$$X_{CM} = \frac{1}{4} [f_1 + f_2 + f_3 + f_4]$$

The new position of the black particle is then

$$x_{i,j}(t+1) = 2X_{CM} = \frac{1}{2} [f_1 + f_2 + f_3 + f_4] \qquad (7.7)$$

because the dynamics is to move to the symmetrical position with respect to the center of mass.

The motion of the black particle results in a new distribution of the f_is. Since at time $t + 1$ the white particles will move and the black will be stationary, the $f_i(t + 1)$s have now to be computed with respect to each white particle location. For instance, $f_1(i + 1, j, t + 1)$ will be the separation to the black particle as seen by the particle located at $\vec{r}_{i+1,j}$

$$
\begin{aligned}
f_1(i + 1, j, t + 1) &= 2X_{CM} - x_{i+1,j} \\
&= \frac{1}{2}[f_1 + f_2 + f_3 + f_4] - f_3 \\
&= \frac{1}{2}[f_1 + f_2 - f_3 + f_4]
\end{aligned}
\tag{7.8}
$$

Similarly, $f_2(t+1)$, $f_3(t+1)$ and $f_4(t+1)$ can be computed for each white particle in figure 7.2. Thus we obtain the following evolution law for the x-separation between black and white particles

$$
\begin{pmatrix} f_1(i + 1, j, t + 1) \\ f_2(i, j + 1, t + 1) \\ f_3(i - 1, j, t + 1) \\ f_4(i, j + 1, t + 1) \end{pmatrix} = W \begin{pmatrix} f_1(i, j, t) \\ f_2(i, j, t) \\ f_3(i, j, t) \\ f_4(i, j, t) \end{pmatrix}
\tag{7.9}
$$

where the propagation matrix W reads

$$
W_{free} = \frac{1}{2} \begin{pmatrix} 1 & 1 & -1 & 1 \\ 1 & 1 & 1 & -1 \\ -1 & 1 & 1 & 1 \\ 1 & -1 & 1 & 1 \end{pmatrix}
\tag{7.10}
$$

where the subscript "free" indicates propagation in a medium with neither reflection nor absorption.

It is important to notice that the above dynamics requires to work with real-valued quantities f_i. In other words, by following our motion rule 7.7, the white and black particles will eventually jump off lattice sites. Thus, equation 7.9 is no longer a pure cellular automata rule as it was in the one-dimensional case but, rather, a lattice Boltzmann dynamics. Figure 7.3 illustrates this dynamics.

Equation 7.9 with W given by 7.10 leads to the propagation of two-dimensional waves. This will be proven below but, at this point, it should be clear from the physical interpretation we gave: masses are vibrating around their local equilibrium positions, and the deformation propagates like a wave. Note that equation 7.9 concerns only the deformations along the x-axis. This is enough to simulate *two-dimensional* wave propagation since an x-deformation also propagates along the y-axis.

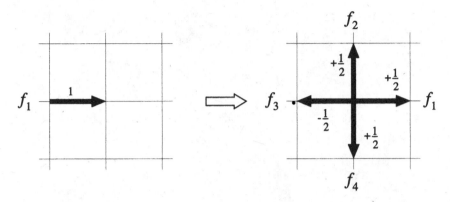

Fig. 7.3. Scattering of an incoming flux f_1 at a node. The value of each outgoing flux is indicated on the arrows.

However, to describe the overall motion of our system of white and black particles (the two-dimensional counterpart of the string rule discussed in section 2.2.9), it would be necessary to include the y-motion in terms of the other fields g_is. Then we could simulate the motion of a solid body made of many particles whose own internal movement mediates the large-scale motion of the whole system. Note that the interaction we have considered between black and white particles does not correspond exactly to that of a regular spring: we have here completely decoupled the x and y motions, which is not the case with a spring where the force depends on the Euclidean distance between the pairs of particles. However, our approach captures the essential features of the system, with fewer complications. Figure 7.4 shows an example of the motion of such a solid body bouncing off the wall of a container.

Using this interpretation of local internal deformation, a possible application of this model is to simulate dynamic fracture formation and cracks in a solid material. This problem is important because the way things break is still not well understood [208] and numerical simulations may help to develop a better intuition.

A typical experiment is to apply an initial stress to the system, for instance on the left and right sides of the sample. This gives non-zero initial values to the f_is and g_is at the boundaries. Then, a deformation wave will propagate and, if the value of a given $f_i(\vec{r}, t)$ exceeds some pre-assigned threshold at some site $\vec{r}$, we may assume that the corresponding bond breaks. The following dynamics of that site is then governed by three neighbors instead of four. In the present model, when a bond is broken, one adds a virtual particle at a distance $2r_0$ from the next-to-the-last particle [209] so that the same four-neighbor rule can be applied. The quantity r_0 is the equilibrium separation between adjacent "atoms".

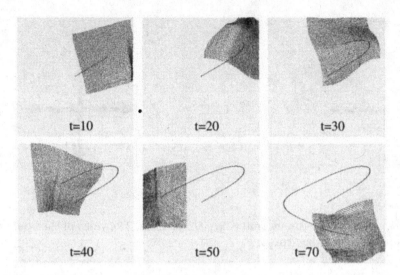

Fig. 7.4.   Motion of a deformable object made of many particles interacting according to the lattice Boltzmann wave rule. An initial momentum is given to the object by choosing the values of the quantities f_i and g_i. The propagation of this initial deformation produces an overall motion. The center of mass of the object follows the straight line shown on the pictures until its boundary reaches the limit of the simulation space. By forbidding the particles to cross the boundary, the entire object naturally bounces on the wall.

Fig. 7.5.   Dislocation of a two-dimensional media obeying the lattice Boltzmann wave model. The sample shown in this figure is subject to an external force pulling the left and right side in opposite directions. The breaking threshold is chosen so that it is more difficult to break a vertical bond than a horizontal one.

It differs for the horizontal and vertical coupling (it is zero in the latter case, for the x-motion).

A bond breaks if one of the f_i (or g_i) becomes too large. Each time a bond breaks a virtual particle is added. Figure 7.5 illustrates the various stages of this process. Of course, it is possible to distribute the breaking thresholds randomly among the bonds, so as to produce a disordered media.

Some modifications to the model can be considered. For instance, instead of comparing the distance f_i directly with the breaking threshold,

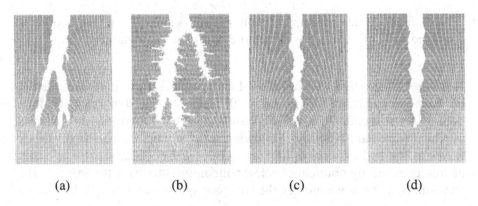

Fig. 7.6. Different patterns of fracture obtained with our model. Dissipation is an important parameter to obtain a smooth fracture, as in (d) rather than many micro-fractures and branching as in (b).

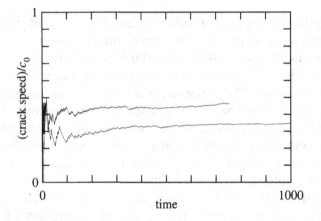

Fig. 7.7. Propagation speed of the factures shown in figures 7.6 (b) and (d) (upper and lower curve, respectively).

it is more natural to consider the deformation $f_i - r_0$. It is also possible to add some attenuation in the vibration (the way to do this is explained in the next paragraph). These parameters play an important role in the fracture behavior. Also the stress rate and the amount of disorder modify quantitatively the shape of the broken region. Figure 7.6 illustrates several patterns obtained with a breaking threshold based on $f_i - r_0$.

The fracture propagation speed, as measured from the simulation, is shown in figure 7.7. It may be seen that propagation is slower than sound speed $c_0 = 1/\sqrt{2}$ and that the fracture advances faster in the case where branching occurs. These facts are in agreement with experimental data.

Note finally that another version of this kind of fracture model is discussed in [210], though with a more complicated dynamics.

Reflection, absorption and source. Let us now return to the modeling of wave propagation. Equation 7.10 describes the dynamics inside a homogeneous media. In general, a system contains boundaries or obstacles at which waves are reflected. Following what we did for one-dimensional waves at the end of section 7.1.1, reflection at a boundary site can easily be implemented by bouncing back the incoming flux with the same or the opposite sign. Thus we modify the free propagation matrix 7.10 as follows

$$W_{ref}^{\pm} = \begin{pmatrix} 0 & 0 & \pm 1 & 0 \\ 0 & 0 & 0 & \pm 1 \\ \pm 1 & 0 & 0 & 0 \\ 0 & \pm 1 & 0 & 0 \end{pmatrix} \tag{7.11}$$

where the sign ± 1 depends whether or not there is a change of phase during the reflection step. A positive value corresponds to a free end while a negative one corresponds to an absorbing boundary.

With this approach, we can simulate free wave propagation and total reflection at obstacles. The lattice sites that represent obstacles are characterized by a "scattering" matrix $W_{ref}^{\pm}$, while free sites are evolved according to the scattering matrix W_{free}. In the next section, we will return to the general problem of modeling a change of medium and different speeds of propagation in each of them. For the time being let us explain how attenuated or damped waves can be modeled.

In the interpretation of black and white moving particles, attenuated waves may be seen as resulting from a particle movement with friction: the displacement is amortized and the moving particles jump closer to their initial position than they would otherwise do. Therefore, similarly to what we did in equation 7.7 we shall now write

$$x_{i,j}(t+1) = x_{i,j}(t) + \alpha \left(X_{CM} - x_{i,j}(t) \right) \tag{7.12}$$

where α is some real coefficient between 0 and 2 describing the amount of friction. Particular values are: $\alpha = 0$ meaning that the particle has not moved at all, $\alpha = 1$ meaning it has moved to the center of mass and $\alpha = 2$ corresponding to free motion. Thus, in an arbitrary coordinate system, we have

$$\begin{aligned} f_1(t+1) &= x_{i,j}(t+1) - x_{i+1,j} \\ &= \alpha X_{CM} + (1-\alpha)x_{i,j} - x_{i+1,j} \\ &= \frac{\alpha}{4}\left(x_{i-1,j} + x_{i+1,j} + x_{i,j-1} + x_{i,j+1}\right) + (1-\alpha)x_{i,j} - x_{i+1,j} \end{aligned}$$

$$= \left(\frac{\alpha}{4} - 1\right)(x_{i+1,j} - x_{i,j}) + \frac{\alpha}{4}(x_{i-1,j} - x_{i,j}) + \frac{\alpha}{4}(x_{i,j-1} - x_{i,j})$$
$$+ \frac{\alpha}{4}(x_{i,j+1} - x_{i,j})$$
$$= \frac{\alpha}{4}f_1 + \frac{\alpha}{4}f_2 + \left(\frac{\alpha}{4} - 1\right)f_3 + \frac{\alpha}{4}f_4$$

Repeating the same calculation for the other f_is yields the attenuation propagation matrix

$$W_{att}(\alpha) = \begin{pmatrix} \frac{\alpha}{4} & \frac{\alpha}{4} & \frac{\alpha}{4} - 1 & \frac{\alpha}{4} \\ \frac{\alpha}{4} & \frac{\alpha}{4} & \frac{\alpha}{4} & \frac{\alpha}{4} - 1 \\ \frac{\alpha}{4} - 1 & \frac{\alpha}{4} & \frac{\alpha}{4} & \frac{\alpha}{4} \\ \frac{\alpha}{4} & \frac{\alpha}{4} - 1 & \frac{\alpha}{4} & \frac{\alpha}{4} \end{pmatrix} \tag{7.13}$$

Clearly, when $\alpha = 0$, $W_{att} = W_{ref}^-$ and when $\alpha = 2$, $W_{att} = W_{free}$. In the general case, W_{att} can be written as

$$W_{att} = \frac{\alpha}{2}W_{free} + \left(1 - \frac{\alpha}{2}\right)W_{ref}^-$$

Damped waves are very useful to mimic an infinite system with a finite lattice: by smoothly absorbing the waves at the boundary of the lattice, it is possible to prevent them from coming back into the system as if they had freely continued their propagation beyond the limits of the simulation.

For the sake of completeness, we also give here the expression for W_{att} for a one-dimensional system, which can be obtained with a similar derivation.

$$1D: \quad W_{att}(\alpha) = \begin{pmatrix} \frac{\alpha}{2} & \frac{\alpha}{2} - 1 \\ \frac{\alpha}{2} - 1 & \frac{\alpha}{2} \end{pmatrix} \tag{7.14}$$

Another important ingredient to include in our lattice wave model is the presence of wave sources such as an emitting antenna. For instance a lattice site can excite a sine oscillation by forcing the time dependence of the f_i to follow some given function of time (a sine in this case). Similarly, plane waves can be produced by having an entire line of sites oscillating in phase.

As an illustration of our wave model, we show, in figure 7.8, the result of a simulation of a plane wave being reflected by a parabolic mirror. Observe that the wave naturally converges toward the focal point of the mirror, as a result of the collective effect of the reflector sites organized as a parabola.

Finally, let us mention that the propagation matrix 7.10 has been proposed and re-invented by several authors for different applications. Probably the first domain of application consisted of actual circuits organized as a two-dimensional lattice. Tension and current were used to

Fig. 7.8. Lattice Boltzmann wave model and focusing of a plane wave by a parabolic mirror.

study wave propagation in a kind of non-digital computer. This technique is widely known to the community studying wave guides and has received the name transmission line matrix (TLM) [211]. More recent works related to this wave propagation model include [212–214].

The dispersion relation. By using λ and τ as the lattice spacing and time step respectively, equation 7.9 can be rewritten as

$$f_i(\vec{r} + \tau \vec{c}_i, t + \tau) = \sum_{j=1}^{4} W_{i,j} f_j(\vec{r}, t)$$

where W is given by equation 7.10. The dispersion relation of this lattice Boltzmann wave model can be obtained by considering a solution of the form

$$f_i(r, t) = A_i e^{i(wt + \vec{k}\vec{r})}$$

After substitution, it is found that the above expression is a solution provided that w and k satisfy

$$e^{i\omega\tau} = \frac{1}{2}\left(\kappa \pm i\sqrt{4 - \kappa^2}\right) \tag{7.15}$$

where $\kappa = \cos(k_1\lambda) + \cos(k_2\lambda)$ and k_1 and k_2 are the two spatial components of $\vec{k}$. A Taylor expansion of 7.15 gives, with $k^2 = k_1^2 + k_2^2$

$$i\omega\tau - \frac{\omega^2\tau^2}{2} + O(\tau^3) = -\frac{\lambda^2 k^2}{4} + O(\lambda^4) \pm i\left[\frac{\lambda k}{\sqrt{2}} + O(\lambda^3)\right] \tag{7.16}$$

In view of taking the limit $\lambda \to 0$ and $\tau \to 0$ of this equation, it is interesting to consider the multiscale formalism introduced in section 3.2.3. The question is always to know how λ and τ are related to each other when they tend to zero. In principle, one assumes that both convective and diffusive phenomena are present. In other words τ is expected to have components of the order λ and also λ^2. Thus, we shall write

$$\tau = \epsilon\tau_1 + \epsilon^2\tau_2 \qquad \lambda = \epsilon\lambda_1$$

where ϵ is the parameter which will tend to zero.

We can now match the various terms of equation 7.16 according to their order of magnitude. We obtain

$$O(\epsilon): \qquad \omega = \pm\frac{1}{\sqrt{2}}\frac{\lambda_1}{\tau_1}k$$

which is exactly the dispersion relation expected for wave propagation with speed

$$c_0 = \frac{1}{\sqrt{2}}\lim_{\epsilon\to 0}\frac{\lambda}{\tau}$$

Similarly, for the second order in ϵ, we obtain

$$O(\epsilon^2): \qquad i\omega\tau_2 - \frac{\omega^2}{2}\tau_1^2 = -\frac{\lambda_1^2 k^2}{4}$$

Since we have just obtained that $\omega^2\tau_1^2(\lambda_1^2 k^2/2)$, we find that

$$i\omega\tau_2 = -\frac{\lambda_1^2 k^2}{4} + \frac{\lambda_1^2 k^2}{4} = 0$$

Hence, the dissipative term vanishes at the order ϵ^2 and we are left with pure wave propagation. Of course, in finite systems, corrections of higher order may show up.

7.1.3 The lattice BGK formulation of the wave model

Here we shall show that the lattice Boltzmann wave model introduced in the previous sections can be formulated as a lattice BGK equation, very much in the same spirit as the fluid model discussed in section 3.5.2. This fact is not so surprising since we have already noticed in section 3.2.3 that sound waves can propagate in a fluid. The key ingredients to obtaining a wave equation were to have mass and momentum conservation and also to neglect nonlinear velocity terms and viscous effects.

We shall now consider these ingredients again and formulate a lattice BGK model of colliding particles which privileges wave propagation over the normal fluid behavior. In a lattice BGK model, we write the evolution rule as

$$f_i(\vec{r} + \tau\vec{v}_i, t + \tau) = \frac{1}{\xi}f_i^{(0)}(\vec{r}, t) + \left(1 - \frac{1}{\xi}\right)f_i(\vec{r}, t) \qquad (7.17)$$

or, equivalently

$$f_i(\vec{r} + \tau \vec{v}_i, t + \tau) - f_i(\vec{r}, t) = \frac{1}{\zeta}\left(f_i^{(0)}(\vec{r}, t) - f_i(\vec{r}, t)\right) \tag{7.18}$$

where, here, $\vec{v}_i$ designates the velocity vector along the ith lattice direction. The quantities f_i are interpreted as the average number of particles traveling with velocity $\vec{v}_i$ at site $\vec{r}$ and time t. At this point the relaxation parameter ζ is a free quantity. We shall, later, see that only one specific value is acceptable to get clean wave propagation.

We now introduce the conserved quantities

$$\Psi = \sum m_i f_i \qquad \vec{J} = \sum m_i f_i \vec{v}_i \tag{7.19}$$

where m_i are free parameters corresponding to the mass of the particles traveling in lattice direction i.

The local equilibrium.　In order to be in a situation similar to that discussed in section 7.1.2, we now consider a two-dimensional case and a square lattice. However, for more generality (and to be able to adjust the propagation speed), we shall assume that rest particles are possible and accounted for by the field f_0. Thus, the index i runs from 0 to 4 and, of course, the speed $\vec{v}_0$ associated with the rest particles is null

$$\vec{v}_0 = 0$$

The physics involved in equation 7.18 is determined by the expression for the local equilibrium distribution $f_i^{(0)}$. To describe wave propagation we define

$$f_i^{(0)} = a_i \Psi + b_i \frac{\vec{v}_i \cdot \vec{J}}{2v^2} \tag{7.20}$$

where a_i and b_i are constants to be determined and $v^2 = (\lambda/\tau)^2$ is the modulus of the $\vec{v}_i$, for $i \neq 0$.

Compared with the hydrodynamical situation (see equation 3.246), we have here an expression for the local equilibrium which is *linear* in the particle current $\vec{J}$. This *a priori* choice can be understood from the fact that sound waves are seen in hydrodynamics when the velocity field is small and nonlinear terms neglected. We prove below that equation 7.20 is indeed the correct choice to produce waves instead of a hydrodynamical behavior.

This freedom in the choice of the local equilibrium distribution $f_i^{(0)}$ is actually a very interesting property of the lattice Boltzmann scheme: different physical behaviors can be captured by suitable choice of $f_i^{(0)}$, while still keeping the same evolution equation.

The value of the constants a_i and b_i can be calculated by imposing mass and momentum conservation. As usual, we require that the right-hand side of equation 7.18 vanishes when multiplied by m_i or $m_i\vec{v}_i$ and then summed over i. From equation 7.19, this amounts to asking that

$$\Psi = \sum_{i=0}^{4} m_i f_i^{(0)} \qquad \vec{J} = \sum_{i=0}^{4} m_i \vec{v}_i f_i^{(0)}$$

Using the expression 7.20 for $f_i^{(0)}$, we have

$$\Psi = \Psi \sum_{i=0}^{4} m_i a_i + \frac{1}{2v^2} \left(\sum_{i=0}^{4} m_i b_i \vec{v}_i \right) \cdot \vec{J}$$

and

$$\vec{J} = \Psi \sum_{i=0}^{4} m_i a_i \vec{v}_i + \frac{1}{2v^2} \left(\sum_{i=0}^{4} m_i b_i (\vec{v}_i \cdot \vec{J}) \vec{v}_i \right)$$

But we know from chapter 3 that $\sum_{i=1}^{4} \vec{v}_i = 0$ and $\sum_{i=1}^{4} v_{i\alpha} v_{i\beta} = 2v^2 \delta_{\alpha\beta}$, since the $\vec{v}_i$s point along the main directions of a square lattice. Thus, if $m_i b_i = 1$ we simply have (remember that $\vec{v}_0 = 0$)

$$\frac{1}{2v^2} \left(\sum_{i=0}^{4} m_i b_i (\vec{v}_i \cdot \vec{J}) \vec{v}_i \right) = \vec{J}$$

Therefore mass and momentum conservation can be enforced if

$$m_i = m \qquad a_i = a \qquad b_i = b \qquad \text{for } i = 1, 2, 3, 4$$

and

$$a_0 m_0 + 4ma = 1 \qquad b_0 = 0 \qquad b = \frac{1}{m} \qquad (7.21)$$

We can summarize these results as follows:

$$f_i^{(0)} = a\Psi + \frac{1}{m} \frac{\vec{v}_i \cdot \vec{J}}{2v^2} \qquad \text{if } i \neq 0$$
$$f_0^{(0)} = a_0 \Psi$$
$$a_0 m_0 + 4ma = 1 \qquad (7.22)$$

The propagation matrix. We can now rewrite the evolution law 7.18 using the above expression for $f_i^{(0)}$. We obtain

$$f_0(\vec{r}, t + \tau) = \frac{1}{\xi} [a_0 \Psi + (\xi - 1) f_0]$$

and, for $i \neq 0$,

$$f_i(\vec{r} + \tau \vec{v}_i, t + \tau) = \frac{1}{\xi} \left[a\Psi + \frac{1}{m} \frac{\vec{v}_i \cdot \vec{J}}{2v^2} + (\xi - 1)f_i \right]$$

By replacing Ψ and $\vec{J}$ with their expressions 7.19 in terms of $f_i^{(0)}$ we obtain

$$f_0(\vec{r}, t + \tau) = \frac{1}{\xi} \left[(a_0 m_0 + \xi - 1)f_0 + a_0 m \sum_{i=1}^{4} f_i \right]$$

$$f_i(\vec{r} + \tau \vec{v}_i, t + \tau) = \frac{1}{\xi} \left[\left(am + \xi - \frac{1}{2} \right) f_i + amf_{i+1} \right.$$

$$\left. + \left(am - \frac{1}{2} \right) f_{i+2} + amf_{i+3} + am_0 f_0 \right] \quad (7.23)$$

This can be written in a matrix form

$$\begin{pmatrix} f_0(\vec{r}, t + \tau) \\ f_1(\vec{r} + \tau \vec{v}_1, t + \tau) \\ f_2(\vec{r} + \tau \vec{v}_2, t + \tau) \\ f_3(\vec{r} + \tau \vec{v}_3, t + \tau) \\ f_4(\vec{r} + \tau \vec{v}_4, t + \tau) \end{pmatrix} = W \begin{pmatrix} f_0(\vec{r}, t) \\ f_1(\vec{r}, t) \\ f_2(\vec{r}, t) \\ f_3(\vec{r}, t) \\ f_4(\vec{r}, t) \end{pmatrix} \quad (7.24)$$

where

$$W = \frac{1}{\xi} \begin{pmatrix} a_0 m_0 + \xi - 1 & a_0 m & a_0 m & a_0 m & a_0 m \\ am_0 & am + \xi - \frac{1}{2} & am & am - \frac{1}{2} & am \\ am_0 & am & am + \xi - \frac{1}{2} & am & am - \frac{1}{2} \\ am_0 & am - \frac{1}{2} & am & am + \xi - \frac{1}{2} & am \\ am_0 & am & am - \frac{1}{2} & am & am + \xi - \frac{1}{2} \end{pmatrix}$$
$$(7.25)$$

By choosing the parameters in matrix W in an appropriate way, we find the matrix W_{free} given in equation 7.10. Indeed, in the case that there is no rest particle (i.e. $f_0 = 0$, $a_0 = 0$), we have

$$a = \frac{1}{4m}$$

Now, if ξ is chosen as

$$\xi = \frac{1}{2}$$

W reduces to

$$W = \begin{pmatrix} -1 & 0 & 0 & 0 & 0 \\ \frac{m_0}{2m} & \frac{1}{2} & \frac{1}{2} & -\frac{1}{2} & \frac{1}{2} \\ \frac{m_0}{2m} & \frac{1}{2} & \frac{1}{2} & \frac{1}{2} & -\frac{1}{2} \\ \frac{m_0}{2m} & -\frac{1}{2} & \frac{1}{2} & \frac{1}{2} & \frac{1}{2} \\ \frac{m_0}{2m} & \frac{1}{2} & -\frac{1}{2} & \frac{1}{2} & \frac{1}{2} \end{pmatrix}$$

which, except for the first row and column, is exactly the same as 7.10. Thus, with $f_0 = 0$ the lattice BGK formulation 7.18 is equivalent to the wave model developed in section 7.1.2.

Note also that the quantities f_i are not guaranteed to be positive. For some choice of ψ and $\vec{J}$, it is certainly possible to make $f_i^{(0)}$ negative as can be seen from the definition 7.20. Actually, it is clear that f_i will be negative since they represent an oscillating wave. This is not very satisfactory if the f_is are interpreted as an average number of particles. However, since the dynamics is linear, we can always think that f_i describes the fluctuations around some large positive constant value.

Adjusting the propagation speed. Equation 7.25 is a five by five matrix and the presence of rest particles allows us to change the wave propagation speed. In fact, each lattice site can have its own refraction index and it is then possible to model propagation through different media or in a disordered medium.

To obtain the resuting wave equation, we first compute the equation governing the evolution of the conserved quantities Ψ and $\vec{J}$. To this end, we shall use the same formalism (Chapman–Enskog expansion) as used in section 3.5.2. Since we have mass and momentum conservation we know that we have (see equations 3.87 and 3.88)

$$\partial_t \Psi + \partial_\beta J_\beta = 0 \qquad (7.26)$$

$$\partial_t J_\alpha + \partial_\beta \left[\Pi_{\alpha\beta} + \frac{\tau}{2} \left(\epsilon \partial_{t_1} \Pi_{\alpha\beta}^{(0)} + \partial_\gamma S_{\alpha\beta\gamma} \right) \right] = 0 \qquad (7.27)$$

where $\partial_t = \epsilon \partial_{t_1} + \epsilon^2 \partial_{t_2}$ is expressed in terms of the two relevant time scales and ϵ is the expansion parameter.

Using the usual definitions (see chapter 3), the momentum tensor is

$$\Pi = \Pi^{(0)} + \epsilon \Pi^{(1)}$$

and its the zeroth order is given by

$$
\begin{aligned}
\Pi_{\alpha\beta}^{(0)} &= \sum_i m_i f_i^{(0)} v_{i\alpha} v_{i\beta} \\
&= am\Psi \sum v_{i\alpha} v_{i\beta} + \frac{J_\gamma}{2v^2} \sum v_{i\alpha} v_{i\beta} v_{i\gamma} \\
&= 2amv^2 \Psi \delta_{\alpha\beta}
\end{aligned}
$$

Hence

$$\partial_\beta \Pi_{\alpha\beta}^{(0)} = 2amv^2 \partial_\alpha \Psi \qquad (7.28)$$

and

$$\epsilon \frac{\tau}{2} \partial_\beta \partial_{t_1} \Pi_{\alpha\beta}^{(0)} = am\tau v^2 \partial_\alpha \epsilon \partial_{t_1} \Psi = -am\tau v^2 \partial_\alpha \text{div} \vec{J} \qquad (7.29)$$

where we have used the continuity equation 3.77 for the last equality.

Now, the $O(\epsilon)$ correction to the momentum tensor is

$$\Pi_{\alpha\beta}^{(1)} = \sum_i m_i f_i^{(1)} v_{i\alpha} v_{i\beta}$$

As obtained in section 3.5.2, $f_i^{(1)}$ has the solution

$$f_i^{(1)} = \tau\xi \left(a_i \delta_{\alpha\beta} - \frac{b_i}{2v^2} v_{i\alpha} v_{i\beta} \right) \partial_{1\alpha} J_\beta \qquad (7.30)$$

where $\partial_{1\alpha}$ is the order ϵ of the spatial derivative with respect to coordinate α, i.e. $\partial_\alpha = \epsilon \partial_{1\alpha}$. Thus,

$$
\begin{aligned}
\epsilon \Pi_{\alpha\beta}^{(1)} &= \tau\xi am \, \mathrm{div}\vec{J} \sum v_{i\alpha} v_{i\beta} - \frac{\tau\xi}{2v^2} \partial_\gamma J_\delta \sum v_{i\alpha} v_{i\beta} v_{i\gamma} v_{i\delta} \\
&= \tau\xi 2amv^2 \delta_{\alpha\beta} \mathrm{div}\vec{J} - \frac{\tau\xi}{2v^2} \partial_\gamma J_\delta T_{\alpha\beta\gamma\delta}
\end{aligned}
$$

where $T_{\alpha\beta\gamma\delta}$ is, by definition, the fourth-order tensor

$$T_{\alpha\beta\gamma\delta} = \sum v_{i\alpha} v_{i\beta} v_{i\gamma} v_{i\delta}$$

Remember that, since we are here working on a square lattice, without the diagonals, $T_{\alpha\beta\gamma\delta}$ is not isotropic (it depends on the particular orientation of the vectors $\vec{v}_i$). This is of course not desirable and we shall see below how this problem is solved. For the time being we have

$$\partial_\beta \epsilon \Pi_{\alpha\beta}^{(1)} = 2\xi amv^2 \tau \partial_\alpha \mathrm{div}\vec{J} - \frac{\tau\xi}{2v^2} T_{\alpha\beta\gamma\delta} \partial_\beta \partial_\gamma J_\delta \qquad (7.31)$$

Finally, we also have to compute

$$
\begin{aligned}
S_{\alpha\beta\gamma}^{(0)} &= \sum_i m_i v_{i\alpha} v_{i\beta} v_{i\gamma} f_i^{(0)} \\
&= \frac{1}{2v^2} \left(\sum v_{i\alpha} v_{i\beta} v_{i\gamma} v_{i\delta} \right) J_\delta
\end{aligned}
$$

The term involving Ψ vanishes since it contains the product of three $\vec{v}_i$s. Therefore

$$\frac{\tau}{2} \partial_\beta \partial_\gamma S_{\alpha\beta\gamma}^{(0)} = \frac{\tau}{4v^2} T_{\alpha\beta\gamma\delta} \partial_\beta \partial_\gamma J_\delta \qquad (7.32)$$

We can now collect relations 7.28, 7.29, 7.31 and 7.32, and plug them into equation 7.27. After grouping the terms, we have

$$\partial_t J_\alpha + 2amv^2 \partial_\alpha \Psi + (2\xi - 1) am\tau v^2 \partial_\alpha \mathrm{div}\vec{J} - (2\xi - 1)\frac{\tau}{4v^2} T_{\alpha\beta\gamma\delta} \partial_\beta \partial_\gamma J_\delta = 0 \qquad (7.33)$$

As we can see, by choosing $\xi = 1/2$, we cancel the anisotropic term containg $T_{\alpha\beta\gamma\delta}$ as well as the dissipative term $(2\xi - 1) am\tau v^2 \partial_\alpha \mathrm{div}\vec{J}$.

In the fluid model of section 3.5.2, ξ plays a similar role in the sense that $\xi = 1/2$ corresponds to the limit of zero viscosity. But there, this limit is numerically unstable, due to nonlinear instabilities. In the wave model, on the other hand, this limit is stable and is the only one which is acceptable due to isotropy (note that for $\xi > 1/2$, the fluid model is still isotropic since the lattice is not the same as it is here).

With $\xi = 1/2$ we can now rewrite the momentum conservation equation 7.33 as

$$\partial_t J_\alpha + 2amv^2 \partial_\alpha \Psi = 0 \tag{7.34}$$

The wave equation can now be obtained easily. We first differentiate the continuity equation 7.26 with respect to time

$$\partial_t^2 \Psi + \partial_t \partial_\beta J_\beta = 0$$

and we take the spatial derivative ∂_α of 7.34 and sum over α

$$\partial_t \partial_\alpha J_\alpha + 2amv^2 \nabla^2 \Psi = 0$$

After substitution of the second equation into the first one, we obtain

$$\partial_t^2 \Psi - 2amv^2 \nabla^2 \Psi = 0 \tag{7.35}$$

Thus, as expected, Ψ obeys the wave equation with propagation speed c, where

$$c = v\sqrt{2am}$$

As opposed to the situation described in section 7.1.2 (without "rest particles"), this propagation speed c is now adjustable. Indeed, in conditions 7.21 we still have the freedom to choose the quantity $a_0 m_0$.

Equation 7.18 is linear in the f_i. Thus, in principle there is no need to consider a sophisticated method (the Chapman–Enskok expansion) to derive the wave equation. However, in the LBGK framework, this method is straightforward to apply and makes it unnecessary to compute the eigenvalues of a 5×5 matrix.

In summary, with $\xi = 1/2$ the wave model with ajustable speed is given by the propagation matrix

$$W(n) = \frac{1}{2n^2} \begin{pmatrix} 2n^2 - 4 & \frac{m}{m_0}(4n^2 - 4) & \frac{m}{m_0}(4n^2 - 4) & \frac{m}{m_0}(4n^2 - 4) & \frac{m}{m_0}(4n^2 - 4) \\ \frac{m_0}{m} & 1 & 1 & 1 - 2n^2 & 1 \\ \frac{m_0}{m} & 1 & 1 & 1 & 1 - 2n^2 \\ \frac{m_0}{m} & 1 - 2n^2 & 1 & 1 & 1 \\ \frac{m_0}{m} & 1 & 1 - 2n^2 & 1 & 1 \end{pmatrix} \tag{7.36}$$

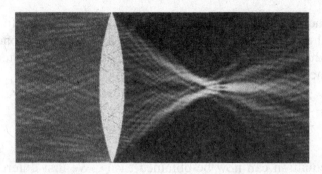

Fig. 7.9. Lattice Bolztmann wave model and focusing of a plane wave by a convex lens with a refraction index of 1.7. The source is on the leftmost side of the system. Energy is shown in this simulation. Light gray denotes high energy.

where n is the refraction index defined as the ratio of the free propagation speed $v/\sqrt{2}$ to the propagation speed $c = 2amv^2$

$$n^2 = \frac{v^2}{2c^2}$$

Figure 7.9 illustrates the propagation of a plane wave through a lens shaped medium. The lattice points outside the lens are characterized by a matrix $W(1)$ and $f_0(\vec{r}, t) = 0$ while the internal sites have $W(n)$ with $n \geq 1$. Due to the change of speed between the two media and the shape of the interface, we observe a focusing of the incident plane wave.

Time reversal invariance. In the previous section, we saw that the relaxation time ξ should be set to $1/2$ in order to obtain a good wave model. There is another interesting way to see this, using time reversal symmetry. Time reversal is a fundamental symmetry of microscopic physics and is clearly satisfied by the wave equation which has only second-order time derivative. We shall see here that time reversal implies

$$\xi = \frac{1}{2}$$

For this purpose, let us write equation 7.18 as

$$f'_i = \left(1 - \frac{1}{\xi}\right) f_i + \frac{1}{\xi} f_i^{(0)} = \sum_{j=0}^{4} W_{ij} f_j \tag{7.37}$$

where f'_i designates the outgoing flux and W the propagation matrix.

As explained in section 2.2.7, a dynamics is reversible in time if, after reversing the direction of the flux, the system traces back its own past. In our case, reversing the flux is achieved with a reversal matrix R such that

$$(Rf)_i = f_{i+2} \ (i \neq 0) \qquad \text{and} \qquad (Rf)_0 = f_0 \tag{7.38}$$

where f stands for the five-element column vector whose ith component is f_i. Thus, in order for equation 7.37 to be invariant under time reversal, one should have (see section 2.2.7)

$$WRf' = Rf$$

Using that (see equation 7.20)

$$f_i^{(0)} = a_i \Psi(f) + \frac{b_i}{2v^2} \vec{v}_i \cdot \vec{J}(f)$$

we have

$$(WRf')_i = \left(1 - \frac{1}{\xi}\right)(Rf')_i + \frac{1}{\xi}\left[a_i \Psi(Rf') + \frac{b_i}{2v^2}\vec{v}_i \cdot \vec{J}(Rf')\right] \qquad (7.39)$$

Let us now compute $\Psi(Rf')$ and $\vec{J}(Rf')$

$$
\begin{aligned}
\Psi(Rf') &= \sum_{j=0}^{4} m_j (Rf')_j = m_0(Rf')_0 + \sum_{j\neq 0} m(Rf')_j \\
&= m_0 f_0' + \sum_{j\neq 0} m f_{j+2} = \sum_{j=0}^{4} m_j f_j' \\
&= \sum_{j=0}^{4} m_j f_j \qquad \text{(because of mass conservation)} \\
&= \Psi(f)
\end{aligned}
$$

Similarly,

$$
\begin{aligned}
\vec{J}(Rf') &= \sum_{j=0}^{4} m_j \vec{v}_j (Rf')_j = \sum_{j\neq 0} \vec{v}_j m(Rf')_j \\
&= -\sum_{j\neq 0} m\vec{v}_{j+2} f_{j+2} = -\sum_{j=0}^{4} m_j \vec{v}_j f_j' \\
&= \sum_{j=0}^{4} m_j \vec{v}_j f_j \qquad \text{(because of momentum conservation)} \\
&= -\vec{J}(f)
\end{aligned}
$$

Substituting these expressions for $\Psi(Rf')$ and $\vec{J}(Rf')$ into 7.39 yields (for $i \neq 0$)

$$(WRf')_i = \left(1 - \frac{1}{\xi}\right) f_{i+2}' + \frac{1}{\xi}\left[a\Psi(f) - \frac{b}{2v^2}\vec{v}_i \cdot \vec{J}(f)\right]$$

$$= \left(1 - \frac{1}{\xi}\right) f'_{i+2} + \frac{1}{\xi} \left[a\Psi(f) + \frac{b}{2v^2}\vec{v}_{i+2}\cdot\vec{J}(f)\right]$$

$$= \left(1 - \frac{1}{\xi}\right)\left[\left(1 - \frac{1}{\xi}\right)f_{i+2} + \frac{1}{\xi}f^{(0)}_{i+2}\right] + \frac{1}{\xi}f^{(0)}_{i+2}$$

$$= \left(1 - \frac{1}{\xi}\right)^2 f_{i+2} + \frac{1}{\xi}\left(2 - \frac{1}{\xi}\right)f^{(0)}_{i+2}$$

Since we require that

$$(WRf')_i = (Rf)_i = f_{i+2}$$

we have the conditions

$$\left(1 - \frac{1}{\xi}\right)^2 = 1 \qquad \left(2 - \frac{1}{\xi}\right) = 0$$

whose solution is

$$\xi = 1/2$$

Similarly, the condition $(WRf')_0 = (Rf)_0 = f_0$ also imposes this value for ξ.

In conclusion, with $\xi = 1/2$, the LBGK formulation of the wave model becomes

$$f_i(\vec{r} + \tau\vec{v}_i, t + \tau) = 2f^{(0)}_i(\vec{r}, t) - f_i(\vec{r}, t) \tag{7.40}$$

This expression gives a natural way to extend the model to a three-dimensional cubic lattice (see also [212]) since our derivation did not use any specific properties of the two-dimensional space.

In the previous section, we saw that $\xi = 1/2$ makes the viscosity vanish, that is requiring time reversal invariance is equivalent to removing dissipation (actually breaking of time reversal symmetry is the standard definition of dissipation). This observation shows that the LBGK fluid models of section 3.5.2 is not reversible in time. Thus, this symmetry is lost when compared with the cellular automata fluids such as FHP. In the cellular automata approach, the microdynamics is time reversal invariant. This symmetry breaks only when the average behavior is considered to obtain the Navier–Stokes equation.

Energy conservation. Our first interpretation of the wave model was to say that the f_i represent local elastic deformations in a solid medium made of alternating black and white particles connected by springs. From this physical picture, we may argue that $\sum f_i^2$ corresponds to some potential energy. This is the way we defined energy in the string model of section 2.2.9. Now, since the black and white particles move between time t and time $t + \tau$ in such a way that the springs are always in maximal

deformation state when one looks at them, we may also argue that kinetic
energy vanishes at the observation time steps.

In this section we want to obtain the condition that the evolution matrix
W must satisfy in order to conserve the energy

$$E(\vec{r},t) = \sum_{i=0}^{4} f_i^2 = \langle f|f \rangle$$

where $\langle f|f \rangle$ denotes the scalar product.

It is easy to check that the matrix W given by 7.25 obeys

$$RWR = W$$

where R is the flux reversal operator introduced in 7.38

$$R = \begin{pmatrix} 1 & 0 & 0 & 0 & 0 \\ 0 & 0 & 0 & 1 & 0 \\ 0 & 0 & 0 & 0 & 1 \\ 0 & 1 & 0 & 0 & 0 \\ 0 & 0 & 1 & 0 & 0 \end{pmatrix}$$

The above property means that W and R commute (note that $R^2 = 1$) and
reflect a natural spatial symmetry of the dynamics: if an incoming flux f_i
enters a site it produces outgoing flux $f_j' = W_{ji}f_i$. Now the incoming flux
is $f_{i+2} = (Rf)_i$ we shall get the same outgoing flux, but pointing in the
reverse direction, namely $(Rf')_j$.

Energy conservation requires that the energy E' after scattering is equal
to the energy before scattering

$$E' = \langle f'|f' \rangle = \langle Wf|Wf \rangle = \langle W^\perp Wf|f \rangle$$

where $W^\perp$ is the transpose of matrix W. Thus energy will be conserved if

$$W^\perp W = 1$$

In the case that the dynamics is invariant under time reversal (and that
is really what we want here), we learned in the previous section that

$$WRW = R$$

Since we have also shown that $RWR = W$ we obtain that

$$W^2 = 1$$

Therefore the condition of energy conservation demands

$$W^\perp = W$$

or, in other words that W be symmetric. We can return to equation 7.36
and use the extra freedom in the choice of $\frac{m_0}{m}$ to make W symmetric. The

condition is

$$\frac{m_0}{m} = \frac{m}{m_0}(4n^2 - 4) \qquad \text{or} \qquad \frac{m_0}{m} = 2\sqrt{n^2 - 1}$$

Thus, with energy conservation and time reversal invariance, the wave propagation matrix in a medium of refraction index $n \geq 1$ is

$$W(n) = \frac{1}{2n^2} \begin{pmatrix} 2n^2 - 4 & 2\sqrt{n^2 - 1} & 2\sqrt{n^2 - 1} & 2\sqrt{n^2 - 1} & 2\sqrt{n^2 - 1} \\ 2\sqrt{n^2 - 1} & 1 & 1 & 1 - 2n^2 & 1 \\ 2\sqrt{n^2 - 1} & 1 & 1 & 1 & 1 - 2n^2 \\ 2\sqrt{n^2 - 1} & 1 - 2n^2 & 1 & 1 & 1 \\ 2\sqrt{n^2 - 1} & 1 & 1 - 2n^2 & 1 & 1 \end{pmatrix}$$

$$(7.41)$$

Note that here the limit $n = 1$ corresponds to taking $m_0 = a_0 = 0$ in equation 7.22.

7.1.4   *An application to wave propagation in urban environments*

As an illutration of our wave propagation model we consider, in this section, the problem of simulating how a radio wave propagates in an urban environment. The reason for studying this question stems from the rapid development of mobile, personal systems such as cellular phones.

Large cities constitute an interesting situation because radio waves are absorbed, reflected and scattered in a complicated way by the buildings and the amplitude pattern of a wave emitted by any antenna surrounded by such obstacles may be very complicated and beyond analytical calculation. In addition to the physical problem of solving the wave equation in a disordered environment, there is the problem of resource optimization: some regions of a city will experience a high mobile communications traffic and the base stations in charge of relaying these calls should be sized so that they can service everyone. The solution is to place more base stations in regions of dense traffic. Thus, a city is divided into cells (defined as the region covered by a base station) whose size is chosen so that the potential number of calls within this cell matches the limited number of channels available at the corresponding base station. For continuity reasons during transmission, adjacent cells must have overlapping boundaries.

An optimized set of base stations should ensure complete coverage with a minimum number of cells of appropriate area. Therefore the starting point for the planning or maintenance of any mobile communication network is based on a prediction of the radio wave attenuation around the transmitter (antenna). In this problem, it is important to note that the antenna may well be located at an elevation which is less than the building height.

Fig. 7.10. Wave propagation in a region of the city of Bern, as predicted by the lattice Boltzmann model. The gray levels show, from white to black, the intensity of the wave. The source antenna is located at the dot mark. The white straight line indicates the path along which intensity measurements have been compared with real measurements.

In our simulation we consider the map of a city and its discretization on a square lattice. Two different choices can be considered for the lattice sites corresponding to the boundary of buildings. First, they can be governed by a propagation matrix αW_{ref} where $\alpha < 1$ is an empirical constant which accounts for the energy absorbed by the bulidings. Another choice is to take the matrix $W_{att}(\alpha)$ on the building walls.

The other lattice sites (corresponding to streets) obey the W_{free} propagation matrix. Finally, a source can be located anywhere along the streets to simulate the presence of an antenna.

The simulation of such wave propagation in a city is given in figure 7.10.

When the transmitting antenna is below the rooftop, the essential features of a building layout are assumed to be captured by a two-dimensional simulation (which is equivalent to a line source and infinitely high buildings). A real antenna, however, is considered as a point source in a three-dimensional space. Quantitative comparisons of simulations with real measurements become possible only after an appropriate renormalization is performed.

In addition to this 2D propagation, the simulation shown in figure 7.10 has another problem. It has not been performed at the frequency used in personal mobile communication systems (900 and 1800 MHz). The reason is that the discreteness of the lattice is not fine enough to work at such a high frequency. The renormalization proposed is the following [215].

We suppose that the measured (real) amplitude A_{3D} of the wave can be approximated by the outcome a_{2D} of our simulation using the following relation:

$$A_{3D}(\vec{r}) \approx a_{2D}\Lambda(\lambda\delta(\vec{r}))^{-1/2} \tag{7.42}$$

where Λ is the actual wavelength used in the measurement and λ the simulated one. The quantity δ represents the distance *around* the buildings separating the transmitter from the receiver.

The rationale for this expression is that it is true for wave propagation in vacuum. It is then assumed that the effects of multiple reflections and scattering on buildings is already captured in a_{2D}: these are the geometrical contributions which are expected to be the same in two and three dimensions. Finally, since the distance traveled by the wave is not accounted for in the same way in 2D and 3D, the rest of the renormalization is assumed to be included in the function $\delta(\vec{r})$. Here we have used for $\delta(\vec{r})$ the so-called Manhattan distance which is the shortest distance following the lattice edges representing the discrete city map.

Figure 7.11 gives a comparison between the predictions of our model and real measurements. The renormalization given by equation 7.42 has been used to compute the wave intensity along the path shown in figure 7.10. We observe good qualitative agreement between the two curves and also acceptable quantitative agreement knowing that our simulation neglects all small-scale details of the real city and that each building is assumed to produce the same absorption and reflection. Finally, it should be noted that, in the simulations, some streets are too narrow to permit wave penetration which may explain why some intensity peaks are not correctly reproduced by the simulations.

7.2 Wetting, spreading and two-phase fluids

7.2.1 Multiphase flows

Sytems in which different species of fluids coexist, move and interact are frequently encountered in science. Such systems are often termed multiphase flows and represent an important domain of application of the lattice gas approach. Models for miscible, immiscible and reactive two-phase flows have been proposed [35,216,97,217,46]. Initially cellular automata models were developed, and more recently, lattice Boltzmann models have been proposed.

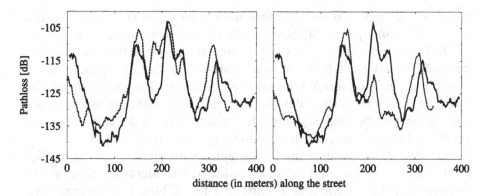

Fig. 7.11. Lattice Boltzmann predictions of wave intensity (dashed line) and *in situ* measurements (thick lines) along a street in the city of Bern (courtesy of Swiss Telecom PTT). The left panel corresponds to a simulation with a propagation αW_{ref}^{-} on building sites, whereas the right panel shows the result obtained with an attenuation propagation matrix $W_{att}(\alpha)$ at each building wall.

Multiphase flows are necessary to model, for instance, a Rayleigh–Taylor instability where two immiscible fluids lie on top of each other, so that the heavy one is above the light one. Small perturbations cause the heavy fluid to move down, giving rise to a spectacular mushroom-like unstable interface. Chemical processes such as combustion or front flame [218,219] can also be investigated within this framework.

Multispecies fluids can be modeled in much the same way as the reaction-diffusion systems discussed in chapter 5, except that now, particles obey hydrodynamics instead of a random walk. Thus, an FHP model or an isoptropic lattice Boltzmann model with momentum conservation is necessary to represent the basic dynamics of each species. For practical reasons, one often associates a different "color" to each species involved, in order to mark and distinguish them.

Interactions between the species are the key ingredient of multiphase flows and, depending on the nature of the physical processes, different interactions will be considered.

For instance, an immiscible two-phase flow can be simulated with a local attraction between particles of the same type and a local repulsion between particles of different type. For this purpose, a "color" gradient is computed at each lattice site, using the color densities on the neighboring grid points. The direction of this gradient actually determines the local interface between the species. The output of particle collisions is then chosen so that a particle of a given color preferentially moves in the direction where its own species resides. Of course in such a collision, mass (color) is conserved separately for each type of fluid, whereas momentum can be exchanged. A realistic surface tension naturally emerges from such a rule, as a consequence of these microscopic attractions/repulsions.

The lattice gas approach has the important advantage that the interface between the different fluids appears naturally as a consequence of the way the fluids are modeled in terms of particles. The evolution of this interface results from the collective behavior of all particles and does not need to be incorporated explicitly in the model (as is the case with more traditional approaches, which introduces many complications).

For more detail on how the lattice Boltzmann method can be applied to multiphase flows, we refer the reader to the literature cited at the beginning of this section and in particular to ref. [46]. For the remainder of this discussion, we present two specific applications where body forces or surface tension are included. The first one is a cellular automata fluid with inter-molecular attracting forces to simulate the wetting and spreading of a liquid droplet on a solid substrate. The second application is an Ising fluid model. Our main purpose is to illustrate how lattice gas techniques can be extended to include more complex physical processes.

7.2.2   *The problem of wetting*

The phenomena of wetting occurs, for instance, when a fluid is in contact with a solid interface, such as the wall of a container. It is well known that, at equilibrium, the free surface of the fluid joins the solid with some given contact angle, determined by the respective interaction between the molecules in the solid, those in the fluid and those in the surrounding air.

For instance (see figure 7.12), a liquid droplet sitting on a flat solid surface rests in equilibrium with a static contact angle θ given by Young's relation

$$\gamma_{lv} \cos \theta = \gamma_{sv} - \gamma_{sl}$$

where γ_{lv}, γ_{sv} and γ_{sl} are the liquid/vapor, solid/vapor and solid/liquid surface tensions, respectively.

When such a balance between the surface tensions is achieved, one has a situation of *partial* wetting. However there are some situations where γ_{lv}, γ_{sv} and γ_{sl} are such that θ goes to zero. There is a transition between partial wetting to a situation of *complete* wetting where the liquid droplet spreads over the solid surface. Complete wetting can be observed in simple every day life experiments: when a droplet of oil is deposited on the surface of water, it spreads slowly to form a thin layer which may eventually break if there is free space left on the water surface.

The spreading of volatile and non-volatile liquids is different. In the first case there is equilibrium between the liquid and its vapor and molecules of the liquid can be efficiently transported through the vapor phase and

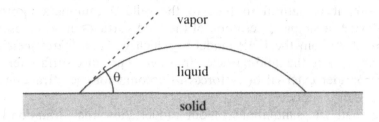

Fig. 7.12.   A liquid droplet in equilibrium with its vapor.

rapidly recondense into a thin layer over the substrate. On the other hand a non-volatile liquid covers the substrate much more slowly than a volatile substance, because it has to flow.

It has long been observed that the spreading of non-volatile liquids are often accompanied by the formation of a so-called precursor film at the interface between the substrate and the liquid [220]. This film spreads faster than the bulk of the droplet and is typically a monolayer of molecular thickness [221]. Sophisticated experimental techniques [222] have also shown that the fluid spreads as a series of superposed molecular layers.

As a result of these features, the spreading of liquid is a very interesting phenomena for which much is still to be understood from first principles. The Navier–Stokes equation, commonly used to described fluid motion, is not the appropriate approach in this case. The usual boundary condition in the Navier–Stokes equation is a "no slip" condition at the liquid–solid interface. A zero velocity field at the boundary is clearly incompatible with droplet spreading [223] and the existence of a precursor film. Furthermore, the Navier–Stokes equation assumes that each fluid element is composed of many molecules. This hypothesis is not true in our case since the layering which is observed occurs at a molecular scale.

Several theoretical models have been proposed to describe the spreading of a droplet [224, 225]. They assume the existence of two-dimensional layers of fluid, moving on top of each other with some viscous force. These models give predictions of the speed of the precursor film, the droplet profile and the partial/complete wetting transition.

Wetting and spreading phenomena are of great importance in industrial processes used in the manufacture of lubricant and paints. From a numerical point of view, spreading can be investigated in terms of molecular dynamics simulations, since it results from inter-molecular attractive or repelling forces. However, molecular dynamics requires massive computational resources and we are interested here in the long time regime for which this approach may not be appropriate.

Therefore, it is natural to turn to the cellular automata approach which provides a simple description of the fluid particles at a microscopic level. We start from the FHP model to which we add nearest-neighbor interactions among the fluid particles, in order to produce surface tension. We also consider external body forces to account for the attraction due to a solid wall or gravity.

As we shall see, simulations of our model show the formation of a precursor film, provided suitable boundary conditions are used at the liquid–solid interface. The motion of the upper fluid layers (above the precursor film) are found to obey power laws. The droplet profile is found to be parabolic.

7.2.3   An FHP model with surface tension

The plain FHP was introduced in section 3.2. Here we shall use the same terminology and denote by $n_i(\vec{r}, t) = 0$ (or $n_i(\vec{r}, t) = 1$) the absence (or presence) of a particle entering site $\vec{r}$ at time t with velocity $\vec{v}_i$.

In order to simulate wetting phenomena, we need first to include a surface tension in the fluid. Second, since spreading is due to Lennard–Jones interactions between the fluid particles and the solid substrate, we need to include external forces acting on the fluid. The resulting force $\vec{F}(\vec{r}, t)$ acting on each particles produces a new velocity distribution $n_i'(\vec{r}, t)$ so that

$$\sum_{i=1,6} n_i'(\vec{r}, t) = \sum_{i=1,6} n_i(\vec{r}, t) = n(\vec{r}, t) \tag{7.43}$$

and

$$\vec{P}'(t, \vec{r}) - \vec{P}(t, \vec{r}) = n(\vec{r}, t)\vec{F}(r, t) \tag{7.44}$$

The quantities $\vec{P}(t, \vec{r})$ and $n(\vec{r}, t)$ are the total momentum and the total number of particles at site $\vec{r}$ and time t, respectively. $\vec{P}'(t, \vec{r})$ is the total momentum just after collision. The momentum is defined as

$$\vec{P}(\vec{r}, t) = \sum_{i=1,6} n_i(\vec{r}, t)\vec{v}_i \tag{7.45}$$

Equation 7.44 states that particle number is conserved such that the variation of momentum for each particle is equal to the total force (Newton's law of mechanics). Equation 7.43 states that no particles are lost or created during the interaction. After the interaction has taken place, the particles move to a nearest-neighbor site, according to the direction of their new velocity.

Clearly, in a cellular automata model, $\vec{P}$ can take only a finite number of values and it is not expected that equation 7.44 will always have a solution. We may even say that usually it has no solution and the new particle distribution should be computed so as to minimize the error. As a consequence, if $\vec{F}$ is the mutual attraction between the fluid particles, the quantity $\sum_{\vec{r}} \vec{P}(\vec{r}, t)$ may not be exactly conserved during time evolution.

A better way to ensure momentum conservation while still having the existence of mutual attraction would be to follow the approach of Rothman [216]: two fluids are considered which could be interpreted here as the liquid and the surrouding gas, respectively. Momentum can be exchanged between the two fluids so that global conservation is guaranteed. However, within each fluid, momentum may not be conserved. Our approach may be seen as a simplified version of this approach, in which the second fluid is disregarded. Empty lattice sites may be considered to act as reservoirs of momentum.

Surface tension.   Let us now discuss the way to compute the force $\vec{F}$ in our model. Surface tension is produced by short-range mutual attraction. Keeping only interactions with nearest-neighbor particles, we define

$$\vec{F}(\vec{r}, t) = f \sum_{i,=1,6} \vec{c}_i \sum_{j=1,6} n_j(\vec{r} + \vec{c}_i, t) \qquad (7.46)$$

, where, as usual, $\vec{c}_j$ denotes the unit vector joining two nearest-neighbor sites (j is labeling the direction) and f is some constant describing the magnitude of the elementary attraction.

Figure 7.13 illustrates the result of such an interaction. The force $\vec{F}$ given in equation 7.46 (with $f = 1$) causes the particles at the middle site to change their velocities according to equation 7.44. This example shows a situation where equation 7.44 can be fulfilled exactly. This is usually not the case and one has to find the "best" approximation to equation 7.44 by inspection of all possible distributions of the n_is. Of course, a lattice Boltzmann version of this model would give much more flexibility in the implementation of equation 7.44 because the n_i would no longer be restricted to the values 0 or 1.

In order to adjust the surface tension, a threshold quantity can be introduced in the rule: the mutual interaction $\vec{F}$ between fluid particles is retained only when it is larger than this threshold value. In particular, this prevents a single particle from capturing another one and forming with it a stable oscillating pair,

In the same spirit, we can also introduced in the rule a "temperature effect," that is some probability that surface tension does not operate at a site. When this happen, the particles are submitted only to the action of the external force. The purpose of this probability is to cause evaporation, i.e to let particles escape from the attraction of the liquid bulk.

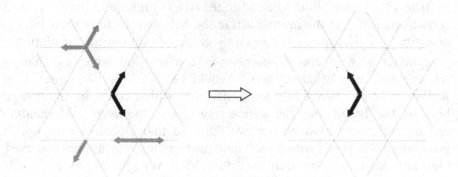

Fig. 7.13. Example of surface tension forces. We show the result of the attraction of the gray particles on the two black ones. The resulting force is a horizontal vector pointing to the left which causes the black particles to change their momentum along the horizontal direction.

External forces.   External forces, like those produced by gravity or by the resulting attraction due to the substrate, sum up to the internal forces. This external force field is modeled as follows. At each lattice site, one associates the angle and the amplitude of the external force. For instance, gravity is constant in amplitude and direction at each site. Attractions due to a solid wall give a force normal to the surface, with a $1/\ell^3$ decay, where ℓ is the distance to the wall. This value is obtained as the sum of all Lennard-Jones contributions from every solid particles at the interface.

Usually, it is not appropriate to have the external force act at every lattice site: a small fraction of them are sufficient but not always the same ones. For this reason, a probabilistic force is considered. The force is taken into account only when its amplitude is larger than a random number which is generated at each time step and each lattice site.

Boundary conditions.   Finally, we have to introduce the rule for the particles reaching the solid substrate or the limit of the system. We can consider three different possible behaviors when a particle encounter a solid wall, as shown in figure 7.14. We call them: (a) specular reflection; (b) bouncing back; and (c) trapping wall.

The first two conditions are standard boundary conditions in lattice gas models. We have introduced condition (c) in order to take into account the strong attraction that may exist for particles right on the wall. The idea is that, once a particle has hit such a wall, strong short-range forces trap it. The particle is then restricted to move parallel to the wall. Of course, due to the exclusion principle, such trapping is only possible when

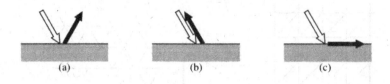

Fig. 7.14. The various boundary conditions that have been used when a particle meets a solid wall. (a) Specular reflection, (b) bounce back condition and (c) trapping wall condition. Condition (c) is necessary to produce a precursor film.

there is no particle already traveling in the considered direction. If this is the case, the particle will bounce back from where it came. As we shall see, the trapping boundary condition appears essential to produce the precursor film.

7.2.4 Mapping of the hexagonal lattice on a square lattice

Simulating a hexagonal lattice on a computer requires some mapping, since the natural data structures on a computer are Cartesian grid (arrays). There are two simple ways to map the hexagonal lattice on a grid, as shown in figures 7.15 and 7.16.

In the first case (figure 7.15), the square lattice undergoes a deformation which amounts to tilting the vertical lines by 60 degrees and shrinking the vertical scale by a factor of $\sqrt{3}/2$. If each lattice site is connected to its north, east, west, south, north-east and south-west neighbors, then in the transformed system a hexagonal topology is obtained.

If (i, j) is the location of lattice site $\vec{r}$ in the square lattice computer topology, the corresponding coordinates (x, y) in the physical topology (hexagaonal) are

$$x = \lambda\left(i + \frac{j}{2}\right)$$

$$y = \lambda\frac{\sqrt{3}}{2}j$$

where λ is the length of the lattice spacing.

In the second mapping technique (see figure 7.16), odd and even horizontal lines are treated differently. Odd lines are shifted half a lattice spacing with respect to even lines. In addition to north, east, west and south neighbors, lattice sites on an even line are also connected to their north-east and south-east neighbors. On the other hand, lattice sites residing on an odd line are connected to their north-west and south-west

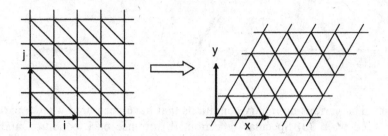

Fig. 7.15.   Mapping of a hexagonal lattice onto a square lattice, by means of a 60 degree deformation.

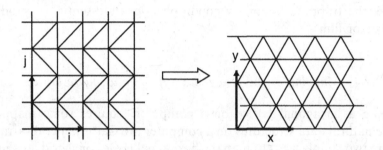

Fig. 7.16.   Mapping of a hexagonal lattice onto a square lattice, by means of a translation of half a lattice spacing for every second horizontal lines.

neighbors. Thus the mapping can be expressed as

$$\text{Even } j: \qquad x = \lambda i \qquad y = \lambda \frac{\sqrt{3}}{2} j$$

$$\text{Odd } j: \qquad x = \lambda(i + \frac{1}{2}) \qquad y = \lambda \frac{\sqrt{3}}{2} j$$

The advantage of the first technique over the second one is that the transformation is the same for all j and that the connectivity of each lattice site is identical. On the other hand, boundaries are more easily dealt with in the second method. Periodic boundary conditions in the vertical direction are preserved in the mapping of figure 7.16 while they are not in figure 7.15. Also, a vertical wall is inclined in the computer topology of the first mapping, while, in the second case it can be implemented as a straight vertical line. Thus, despite its more difficult implementation, the second method is probably safer.

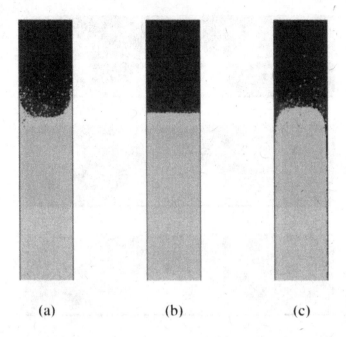

(a)   (b)   (c)

Fig. 7.17. Free surface of a liquid in a capillary tube: (a) a wetting situation; (b) inert walls; and (c) a non-wetting situation. Gravity is acting downwards.

7.2.5 Simulations of wetting phenomena

Capillary tube. First, we consider a "capillary tube" simulation in order to see the formation of a meniscus. The result is shown in figure 7.17, in which three kinds of solid–liquid interaction are considered. Situation (a) corresponds to an attracting $1/\ell^3$ force, where ℓ is the distance to the wall of the container. In (b), we have a situation with no liquid–solid interactions. Situation (c) is equivalent to (a) except for the sign of the force, which is repulsive.

Spreading. Now, we consider the situation of spreading. A drop is deposited on a horizontal attracting substrate. Its surface is ideally flat and produces a $1/\ell^3$ force above it. Gravity is neglected. As the simulation goes on, the droplet spreads over the substrate, as illustrated in figure 7.18. Note that particles can be exchanged between adjacent layers. This motion takes place mostly downwards, due to the direction of the substrate attraction and, as a result, the fluid layers must flow horizontally.

    Boundary conditions are periodic along the horizontal axis (direction of spreading). The upper limit of the system is a reflecting wall. On the substrate, we have tried the three types of boundary conditions discussed in the previous section.

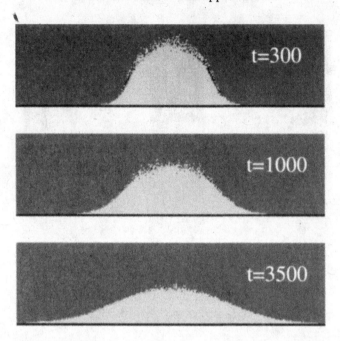

Fig. 7.18. Spreading of a liquid droplet on an attractive substrate. Bouncing back condition is used at the substrate. These snapshots are shown after 300, 1000 and 3500 iterations. Evaporation can occur, as shown by dot particles.

With both specular reflection and bouncing back conditions, the droplet is seen to spread away. However, no precursor film are observed. The fluid may be considered as being at the transition between partial and complete wetting. We can measure the position $x_0(t)$ of the first fluid layer as a function of time (the existence of layers of molecular size is obvious here, due to space discreteness). We observe, in figure 7.19(a), a $t^{1/2}$ behavior

$$x_0(t) \sim t^{1/2}$$

in agreement with the results of [226,227]. Upper layers are found to advance also with power law, but with different exponents.

The droplet profile can be measured. It turns out to fit relatively well a parabola, at least for the lower part of the droplet, as indicated on figure 7.19(b). This result agrees with the prediction of [227] for the solid-on-solid model described in [225,226].

Finally, if we consider our "trapping wall" boundary condition on the substrate, a rapidly advancing precursor film is formed at the liquid–solid interface. As in the model [226], it moves at constant speed ($x_0(t) \sim t$), much faster than the rest of the fluid. This film is fed by particles flowing

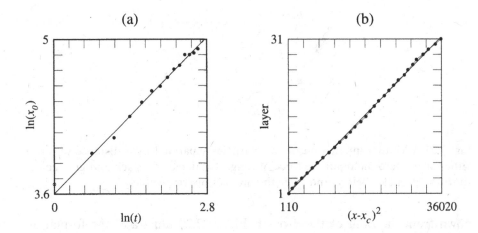

Fig. 7.19. (a) Measurement of the time behavior of the first fluid layer at the liquid–solid interface (horizontal advance versus time in a log–log scale); the slope gives $x \sim t^{0.5}$. (b) Measurement of the droplet profile at a given time, with an appropriate change of scale; only the first 31 layers are taken into account (one-third of the total height); this plot indicates a parabolic profile, at least for the foot of the droplet.

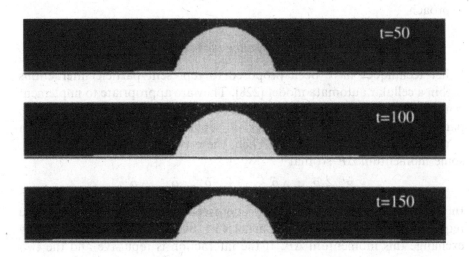

Fig. 7.20. Snapshots of a spreading droplet, as given by our model with the "trapping" boundary condition. A rapidly advancing precursor film is present at the liquid–solid interface.

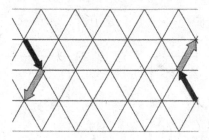

Fig. 7.21. Mutual interaction of two particles separated by a distance of $r = 4$ lattice sites. Here the repulsive force of range 4 is depicted. Black and gray colors show the particles before and after the interaction, respectively.

down from the bulk of the droplet. Figure 7.20 illustrates the formation of the precursor film.

Although the way we introduced the interaction between the particle is not exact from the point of view of momentum conservation, we see that this microscopic approach exhibits several key features of the wetting and spreading phenomena: the formation of a meniscus in capillary tubes and the existence of a precursor film during the spreading of a droplet. The trapping boundary condition provides a very intuitive explanation of the formation of the precursor film. More complex situations, such as spreading over a rough surface, could be studied easily with the present approach.

7.2.6   *Another force rule*

Other techniques have been proposed to represent particle interactions within a cellular automata model [228]. They are appropriate to implement both short and long range forces. The rule is the following: suppose two particles (of mometum $\vec{P}_1$ and $\vec{P}_2$) are separated by a distance r, which is the exact range of the force. Then, these two particles may exchange some momentum $\Delta \vec{P}$ so that

$$\vec{P}_1' = \vec{P}_1 + \Delta \vec{P} \qquad \vec{P}_2' = \vec{P}_2 - \Delta \vec{P}$$

In this way, momentum is exactly conserved. Practically, this force rule requires matching the pairs of particles at distance r that can effectively exchange this momentum $\Delta \vec{P}$. If the interaction is repulsive and the two particles move toward each others, we may say that the particle having a positive component of momentum loses some of it at the expense of the other particle having a negative component of momentum. If the interaction is attractive, it then applies to particles moving away from each other. Figure 7.21 illustrates this force rule. Finally, note that a crystal-like interaction can be produced by switching from a repulsive to an attractive force depending on the distance separating the two interacting particles.

7.2.7 *An Ising cellular automata fluid*

In this section we consider a two-phase cellular automata fluid. Each particle of the fluid can be in two possible states, say $s = 1$ or $s = -1$. If we call this extra degree of freedom a *spin*, this fluid can be compared with the Ising system we defined in section 2.2.3 but in which, now, the spins can move according to the rules used in lattice gas models for hydrodynamics.

In order to produce a surface tension effect at the interface between the two phases, we shall also introduce an interaction between nearest-neighbor spins using the same approach as found in classical dynamical Ising models.

This interaction will be defined through real-valued fields and, thus, the present model can be viewed as a mixture of a cellular automata rule and a coupled map lattice dynamics.

It is interesting to remark that a temperature can be defined in a natural way through the Ising interaction. In addition to being a binary fluid model, this system has some of the ingredients of a ferrofluid [229] if the spin is interpreted as the magnetization carried by the particles. Note that a lattice Boltzmann model for magnetic fluids has also been proposed in the literature [230].

The collision rule. Before we define more precisely the spin interaction, let us return to the particle motion. A collision rule which conserves mass, momentum *and* spin can be defined in analogy with the FHP rule described in section 3.2.

We denote by $s_i(\vec{r}, t) \in \{-1, 0, 1\}$ the state of the automaton along lattice direction i at site $\vec{r}$ and time t ($s_i = 0$ means the absence of a particle). Clearly the presence of a particle is characterized by $s_i^2 = 1$, regardless of its spin. Thus, the collision term can be obtained by using s_i^2 as an occupation number.

When a collision takes place, the particles are redistributed among the lattice directions but the same number of spin $+1$ and -1 particles should be present in the output state as there were in the input state. A way to guarantee this spin conservation is to assume that the particles are distinguishable, at least as far as their spin is concerned.

In figure 7.22 we show the case of a collision between two spin 1 particles and one spin -1 particle. We assume that all particles bounce back to where they came from (this is not the only possibility to conserve momentum and any permutation of the three particles could be considered

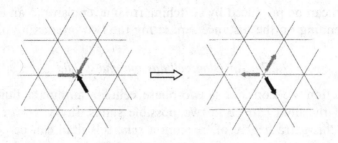

Fig. 7.22. Three-particle collision in a FHP Ising fluid. The black arrow indicates a particle with spin $+1$ while the gray arrows show particles with spin -1.

too). Thus, for such a three-body collision, the state after collision is given by

$$s_i(\vec{r} + \lambda \vec{c}_i, t + \tau) \; = \; s_i - s_i s_{i+2}^2 s_{i+4}^2 (1 - s_{i+1}^2)(1 - s_{i+3}^2)(1 - s_{i+5}^2)$$
$$+ s_{i+3} s_{i+1}^2 s_{i+5}^2 (1 - s_i^2)(1 - s_{i+2}^2)(1 - s_{i+4}^2)$$

The identity of the particle appearing or disappearing in direction i is expressed by the variables s_i and s_{i+3}, whereas the presence or absence of other particles in direction j is indicated by s_j^2.

Similarly, by including the two-body collisions, we can express the full collision term as

$$s_i(\vec{r} + \lambda \vec{c}_i, t + \tau) = s_i$$
$$- s_i s_{i+2}^2 s_{i+4}^2 (1 - s_{i+1}^2)(1 - s_{i+3}^2)(1 - s_{i+5}^2)$$
$$+ s_{i+3} s_{i+1}^2 s_{i+5}^2 (1 - s_i^2)(1 - s_{i+2}^2)(1 - s_{i+4}^2)$$
$$- s_i s_{i+3}^2 (1 - s_{i+1}^2)(1 - s_{i+2}^2)(1 - s_{i+4}^2)(1 - s_{i+5}^2)$$
$$+ pq s_{i+1} s_{i+4}^2 (1 - s_i^2)(1 - s_{i+2}^2)(1 - s_{i+3}^2)(1 - s_{i+5}^2)$$
$$+ p(1 - q) s_{i+4} s_{i+1}^2 (1 - s_i^2)(1 - s_{i+2}^2)(1 - s_{i+3}^2)(1 - s_{i+5}^2)$$
$$+ (1 - p)(1 - q) s_{i+2} s_{i+5}^2 (1 - s_i^2)(1 - s_{i+1}^2)(1 - s_{i+3}^2)(1 - s_{i+4}^2)$$
$$+ (1 - p)q s_{i+5} s_{i+2}^2 (1 - s_i^2)(1 - s_{i+1}^2)(1 - s_{i+3}^2)(1 - s_{i+4}^2)$$
$$(7.47)$$

where p and q are random Boolean variables that are 1 with probability $1/2$, independently at each site and time step. These quantities are necessary because, with two-body collisions, several collision outputs are possible and one has to select randomly one of them.

Spin interaction. An important part of this Ising fluid model is the interaction between spins at the same sites and spins on adjacent lattice sites. This interaction produces surface tension and can be adjusted through a parameter which corresponds to the temperature of the system (which is asumed to be uniform here).

The interaction we propose here does not conserve the number of spins of each sign. It conserves only the number of particles and, for this reason, does not represent two different fluids but two possible state of the same fluid. Of course, the miscibility or immiscibility of the two phases can be tuned through the temperature.

The updating rule for the spin dynamics is taken from the Monte-Carlo method [231], using the so-called Glauber transition rule. The main idea is that a spin flips (changes sign) if it can lower the local energy of the system. The energy of the pair of spins s_i and s_j is computed as $E = -J_1 s_i s_j$ if the two spins are nearest neighbors on the hexagonal lattice and $E = -J_0 s_i s_j$ if they are both on the same site (remember that up to six particles can populate a given site).

However, a spin can flip even if this results in a local increase of energy. But, then, the change is accepted only with a probability $W(s \to -s)$ which depends on the temperature. In the Glauber dynamics, this probability is given by [231]

$$W(s_i \to -s_i) = \frac{1}{2}(1 - s_i \tanh(E_i))$$

where E_i is the energy before the flip

$$E_i = \frac{1}{k_B T}(J_0 m_i + J_1 M_i) s_i$$

and $m_i = \sum_{j \neq i} s_j$ is the on-site "magnetization" seen by spin s_i and $M_i = \sum_{<ji>} s_j$ is the "magnetization" carried by all the particles j on the neighboring sites of spin i. The quantity T is the temperature and k_B the Boltzmann constant that we can set to 1 when working with an appropriate temperature scale. When more than one particle are present at a site, only one of them, chosen at random, is checked for such a spin flip.

The above transition rule is obtained from the detailed balance condition, namely

$$\frac{W(s_i \to -s_i)}{W(-s_i \to s_i)} = \frac{\exp(-E(-s_i)/(k_B T))}{\exp(-E(s_i)/(k_B T))}$$

where $E(\pm s_i)$ denotes the Ising energy as a function of s_i and has the properties to drive an ergodic system to thermodynamic equilibrium.

As opposed to the standard the Monte-Carlo approach, where the lattice sites are visited sequentially and in a random way, here we update synchronously all the sites belonging to a given sub-lattice. Indeed, for coherence of the dynamics it is important not to update simultaneously

```
2   0   1   2   0   1   2   0   1   2   0   1
  0   1   2   0   1   2   0   1   2   0   1   2
2   0   1   2   0   1   2   0   1   2   0   1
  0   1   2   0   1   2   0   1   2   0   1   2
```

Fig. 7.23. The three sub-lattices on the hexagonal lattice used for the syn-chrounous spin update. The values 0, 1, 2 label the sites according to the sub-lattice to which they belong.

any two spins that are neighbors on the lattice. This is for the same reason as explained in section 2.2.3 when we discussed the Q2R rule.

In a hexagonal lattice, it is easy to see that the space can be partitioned into three sub-lattices so that all the neighbors of one sub-lattice always belong to the other two (see figure 7.23).

Therefore, the spin interaction rule described above cycles over these three sub-lattices and alternates with the FHP particle motion discussed in the previous section.

It is of course possible to vary the relative frequency of the two rules (Glauber and FHP). For instance we can perform n successive FHP steps followed by m successive steps of the Ising rule in order to give more or less importance to the particle motion with respect to the spin flip. When $n = 0$ we have a pure Ising model on a hexagonal lattice but with possibly a different number of spins per site.

If the temperature is large enough and a periodic boundary imposed, the system evolves to a configuration where, on average, there are the same number of particles with spins up and down. Of course, the situation is not frozen and the particles keep moving and spins continuously flip. As in a regular Ising systems, there is a critical temperature below which we can observe a collective effect like a global magnetization (an excess of particles of a given spin) and the growth of domains containing one type of spin. This situation is illustrated in figure 7.24 and corresponds to the case $n = m = 1$, namely one spin update cycle followed by one step of FHP motion. It is observed that the critical temperature depends on the frequency updating n and m.

Another interesting situation corresponds to the simulation of a Raleigh–Taylor instability. Two immiscible fluids are on top of each other and the heavier is above the lighter. Due to gravity, the upper fluid wants to penetrate through the lower one. Since the two fluids are immiscible, the interface between them becomes unstable and, as time goes on, gives rise to a mushroom-like pattern.

Gravity can be added to our model in the same way as explained in

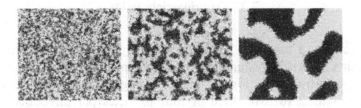

Fig. 7.24.   Three snapshots of the evolution of the Ising FHP model below the critical temperature. Particles with spin +1 are shown in black while gray points show particles with −1. White cells indicate empty sites.

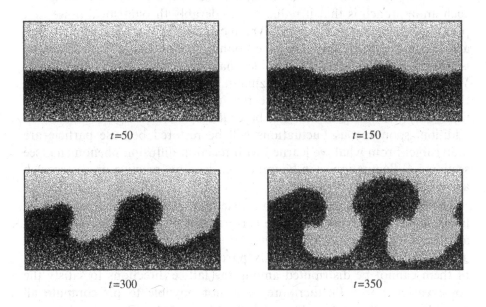

Fig. 7.25.   Rayleigh–Taylor instability of the interface between two immiscible fluids. Particles with spin +1 are shown in black and are "lighter" than gray particles having spin −1. An approximate immisciblity is obtained by choosing a low temperature in the model.

section 7.2.3. Two immiscible fluids can be modeled by having a low temperature T in the Glauber dynamics and producing the necessary surface tension. The upper fluid layer is initialized with only particles of spin −1, whereas the lower layer contains only spins +1 (see figure 7.25). Gravity is adjusted so that "light" particles go up and heavy particles go down. After a few iterations, the flat interface destabilizes as shown in the final panel of figure 7.25.

7.3   Multiparticle fluids

In this section we present a method to go beyond the cellular automata
fluid models which have an exclusion principle limiting the number of
particles per lattice direction to one. Although it may be interesting to
increase this limit to two, three or four particles per direction, here we
would like to define a lattice hydrodynamics with an arbitrary number
of particles on each site. Hence, the particle density can vary from zero
to infinity (actually infinity stands for the largest number that can be
represented in the computer).

Our purpose is very similar to that in section 5.7. The advantage of
such an approach is that it reduces considerably the statistical noise that
plagues cellular automata hydrodynamics (such as the FHP model) and,
also, more effectively fine tunes the boundary conditions. More flexibility
is offered because the model variables may vary over a much larger range.
With respect to the lattice Boltzmann fluids discussed in section 3.5,
the present approach will remove the numerical instabilities because it
deals only with integers that can be exactly processed in a computer. In
addition, spontaneous fluctuations will be restored because particle are
indivisible. From what we learned with reaction-diffusion phenomena (see
section 5.4) fluctuations can be important in some reactive processes such
as combustion [218,219].

The main drawback of a multiparticle fluid model, compared with the
lattice Boltzmann approach, is the performance of the algorithm. Defining
a collision between an arbitrary number of particles which conserve mass
and momentum is not an easy task: particles are indivisible and fractions
of them cannot be distributed among the lattice directions to satisfy the
conservation laws. Furthermore, it is not possible to pre-compute all
possible collisions (as we do in a cellular automaton) because there are
an infinite number of configurations. Thus, more sophisticated algorithms
should be devised which slow down computation of collision output.

There are several requirements we would like to impose on such a
multiparticle model. First, its algorithmic complexity should grow more
slowly than linearly with the number of particles. Otherwise it would be
as fast to run several instances of a standard cellular automata fluid to
obtain the same statistical accuracy on the physical quantities of interest.
Second, we would like to be able to exploit the fact that more types of
collisions are possible when many particles collide, and define a collision
rule that lowers the intrinsic model viscosity. As an example of the new
possiblities to define particle interactions, figure 7.26 shows two examples
of collisions involving several particles traveling along the same link.

Other aspects of multiparticle fluid models can be found in [105,232]

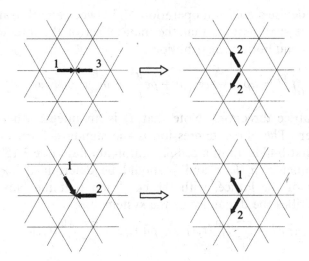

Fig. 7.26. Two examples of multiparticle collisions which conserve mass and momentum. The values indicated on the arrows give the number of particles traveling on the corresponding link. It is interesting to note that these collisions have a non-zero global momentum, as opposed to the usual FHP model.

(note that these models are also referred to as Integer Lattice Gases). Here we present a multiparticle model for which the algorithmic complexity grows as the the square root of the number of particles present at a given site. For this reason, it is expected to eventually win over a basic cellular automata model as far as the CPU time required to obtained a given accuracy is concerned. In addition, the rule we propose is rather simple and allows us to adjust the viscosity with an external parameter.

7.3.1 The multiparticle collision rule

As earlier (see chapter 3), we shall describe the multiparticle microdynamics in terms of quantities $f_i(\vec{r}, t)$ describing the number of particles entering site $\vec{r}$ at time t along lattice direction i and which now take their value in $\{0, 1, 2, ..., \infty\}$. The evolution equation will be, as usual

$$f_i(\vec{r} + \tau \vec{v}_i, t + \tau) = \mathscr{F}_i(f(\vec{r}, t))$$

where $\vec{v}_i$ denotes the elementary particle velocities, τ the time between two consecutive iterations and $\mathscr{F}_i$ the collision term. The physical quantities such as the particle density ρ and velocity field $\vec{u}$ are defined as

$$\rho(\vec{r}, t) = \sum_i f_i(\vec{r}, t) \qquad \rho \vec{u}(\vec{r}, t) = \sum_i f_i(\vec{r}, t) \vec{v}_i$$

where index i runs over the lattice directions.

In order to devise a collision operation $\mathscr{F}_i$ between an arbitrary number of particles, we shall assume that the main effect of the interaction is to restore a local equilibrium distribution

$$f_i^{(0)} = a\rho + \frac{b}{v^2}\rho\vec{v}_i \cdot \vec{u} + \rho e\frac{u^2}{v^2} + \rho\frac{h}{v^4}v_{i\alpha}v_{i\beta}u_\alpha u_\beta$$

along each lattice direction. Note that f_i is an integer whereas $f_i^{(0)}$ is a real number. The above expression is the standard form of the local equilibrium that builds up in a cellular automata fluid (see 3.152 or 3.246) and the parameters a, b, e and h should be determined according to the geometry of the lattice, with the condition that the Navier–Stokes equation describes the dynamics of the system, and

$$\rho(\vec{r},t) = \sum_i f_i^{(0)}(\vec{r},t) \qquad \rho\vec{u}(\vec{r},t) = \sum_i f_i^{(0)}(\vec{r},t)\vec{v}_i$$

The relaxation to the local equilibrium is governed by a parameter ξ. Thus, as in the LBGK situation, we shall require that the number of particles f_i' leaving (after collision) a given site along direction i be given by

$$f_i' = \frac{1}{\xi}f_i^{(0)} + \left(1 - \frac{1}{\xi}\right)f_i + \Delta f_i \tag{7.48}$$

where Δf_i is a random quantity accounting for the fact that (after collision) the actual particle distribution may depart from its ideal value.

In practice f_i' can be obtained as follows. Let $N = \sum_i f_i$ the total number of particles at the given site. Let us denote by p_i the probability that a particle leaves the site along direction i. We define

$$p_i = \max\left(0, \left[\frac{1}{\xi}f_i^{(0)} + \left(1 - \frac{1}{\xi}\right)f_i\right]/M\right)$$

where M is a normalization constant adjusted so that $\sum_i p_i = 1$. Clearly, if none of the p_i is zero, $\sum_i p_i = (\sum_i f_i)/M$ (since $\sum_i f_i^{(0)} = \sum_i f_i$) and $M = N$ the total number of particles on the site. From a numerical analysis, the risk that

$$\frac{1}{\xi}f_i^{(0)} + \left(1 - \frac{1}{\xi}\right)f_i < 0$$

decreases rapidly as the relaxation time ξ deviates from the zero viscosity limit $\xi = 1/2$ (see section 3.5). On the other hand, it increases when the number of particles decreases.

To compute the collision output, we run through each of the N particles and place them in direction i with probability p_i. This will give us a temporary particle distribution $\tilde{f}_i$ which then must be corrected to obtain f_i', in order to ensure exact momentum conservation.

Practically, $\tilde{f}_i$ is obtained by choosing a random number $s \in [0, 1[$ and comparing it with the quantity

$$s_0 = 0 \qquad s_i \equiv \sum_{j=1}^{i} p_i$$

If $s_{i-1} \leq s < s_i$, the particle is placed in direction i (by construction, the probability that this happens is indeed $p_i = s_i - s_{i-1}$). Thus we write

$$\tilde{f}_i = \sum_{h=1}^{N} (s_{i-1} \leq s_h < s_i) \tag{7.49}$$

where $(s_{i-1} \leq s_h < s_i)$ is to be taken as a Boolean value which is 1 when the condition is true and zero otherwise.

Since s_h is a random number uniformly distributed in the interval $[0, 1[$, $\tilde{f}_i$ is also a random quantity. If none of the p_i is zero, its expectation value is

$$\langle \tilde{f}_i \rangle = \sum_{h=1}^{N} p_i = \frac{1}{\xi} f_i^{(0)} + \left(1 - \frac{1}{\xi}\right) f_i \tag{7.50}$$

Note that when N is large enough, equation 7.49 can be computed using the same Gaussian approximation as used in section 5.7. In this way, the algorithmic complexity of the operation is independent of the number of particles N.

While the distribution $\tilde{f}_i$ of output particles obviously conserves the number of particle, equation 7.50 shows that it only conserves momentum on the average. Some particles must be pushed to other directions. Momentum tuning is performed iteratively, according to the following steps

- At each site where momentum is not correctly given by $\sum_j \tilde{f}_j \vec{v}_j$, choose at random one lattice direction i.

- If $\tilde{f}_j > 0$ move one particle randomly to an adjacent direction.

- Accept the change if it does not worsen the momentum.

- Iterate this procedure until a particle distribution f_i' is obtained which satisfies momentum conservation $\sum_j f_j' \vec{v}_j = \sum_j f_j \vec{v}_j$.

From the way the particles are distributed with the probabilities p_i, we expect that roughly $\sqrt{(N)}$ of them are misplaced. This gives an estimate of the number of iterations necessary to re-adjust the particles direction.

According to the above discussion it is reasonable to argue that, on average, the quantity Δf_i defined in equation 7.48 vanishes. This fact is confirmed numerically. Consequently, we have

$$f_i(\vec{r} + \tau\vec{v}_i, t + \tau) \approx \frac{1}{\xi} f_i^{(0)}(\vec{r}, t) + \left(1 - \frac{1}{\xi}\right) f_i(\vec{r}, t) \qquad (7.51)$$

where we have used $f_i(\vec{r} + \tau\vec{v}_i, t + \tau) = f_i'(\vec{r}, t)$, due to the definition of particle motion.

Equation 7.51 is identical to the BGK microdynamics, except that now it approximates a multiparticle dynamics, that is f_i are integer quantities. In the limit where the correlations beween the f_i can be neglected (remember that $f_i^{(0)}$ is a nonlinear function of all f_j) we may take the average of the above equation to obtain

$$\langle f_i(\vec{r} + \tau\vec{v}_i, t + \tau) \rangle = \frac{1}{\xi} f_i^{(0)}(\langle\rho\rangle, \langle\rho\vec{u}\rangle) + \left(1 - \frac{1}{\xi}\right) \langle f_i \rangle \qquad (7.52)$$

If the average distribution $\langle f_i(\vec{r}, t) \rangle$ is smooth in space and time, it can be differentiated. Then, the same derivation as that given in section 5.7 can be repeated to show that a correct hydrodynamical behavior emerges from our multiparticle model: equation 7.52 is equivalent to the Navier–Stokes equation with viscosity

$$\nu = c_s^2 \left(\xi - \frac{1}{2}\right)$$

where c_s^2 is the sound speed whose value is model dependent (different for hexagonal, square or cubic lattices) [233].

The present muliparticle scheme is intrinsically stable because it is based on integer numbers and probabilities. No small fluctuation will be amplified unphysically to make the arithmetic to blow up as is the case with the LGBK model. In principle, any value of the relaxation parameter ξ can be considered. Of course if $\xi \leq 1/2$ it is not clear whether or not a hydrodynamical behavior is reproduced.

7.3.2 *Multiparticle fluid simulations*

In this subsection we present some simple results of our multiparticle lattice gas fluid. We have considered a two-dimensional hexagonal lattice, including the possibility of having rest particles. Using the standard LBGK expression for such a situation [233] we can write the appropriate local equilibrium function $f_i^{(0)}$ as

$$\begin{cases} f_i^{(0)} = \dfrac{\rho}{12}\left[1 + \dfrac{4}{v^2}\vec{v}_i \cdot \vec{u} + 8\dfrac{u_\alpha u_\beta}{v^4}\left(v_{i\alpha}v_{i\beta} - \dfrac{v^2}{4}\delta_{\alpha\beta}\right)\right] \\ f_0^{(0)} = \dfrac{\rho}{2}\left[1 - 2\dfrac{u_\alpha u_\beta}{v^2}\right] \end{cases} \qquad (7.53)$$

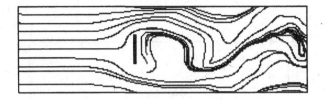

Fig. 7.27. Flow instability past a plate. The system size is 80×500 (although only half of it is shown here). The parameters of the simulations are: $u_\infty = 0.15$, $\xi = 0.55$ and $\rho = 3500$ on average.

The first simulation is shown in figure 7.27. It pictures the flow instability which develops in a fluid flow past an obstacle (von Karman street).

A second experiment (figure 7.28) shows the velocity time autocorrelation in an equilibrium multiparticle fluid. We consider a periodic system of size 400×400, with a random initial configuration of zero average velocity and 500 particles per site. The quantity of interest is the velocity–velocity time correlation

$$C(t) = \langle \vec{v}(\vec{r}, t)\vec{v}(\vec{r}, 0)\rangle - \langle \vec{v}(\vec{r}, 0)\rangle^2$$

where $\vec{v}(\vec{r}, t)$ is defined as

$$\vec{v}(\vec{r}, t) = \sum_i \vec{v}_i f_i(\vec{r}, t)$$

This quantity is known to exhibit the so-called long time tails, that is it obeys a power law [87,234,86] instead of the fast decay exponential predicted by a *linearized* Boltzmann equation [235]. The plot in figure 7.28 shows the behavior of

$$c(t) = \langle f_i(\vec{r}, t)f_i(\vec{r}, 0)\rangle - \langle f_i(\vec{r}, 0)\rangle^2$$

which is one of the basic components of $C(t)$. Here, since the system in at equilibrium and homogeneous in space, the averaging process is performed by summing the contribution of all lattice points $\vec{r}$. By choosing $\langle f_i \rangle = 0.0837$ (instead of the theoretical value $\langle f_i \rangle = 1/12 = 0.0835$ given by equation 7.53 with $\vec{u} = 0$), we obtain that $c(t)$ behaves as

$$c(t) \sim t^{-1}$$

as expected for a two-dimensional fluid [234,86].

Finally, we consider a multiparticle simulation of a Poiseuille flow. Fluid particles are injected on the left side of a channel of length L and width W with a rightward velocity. On the upper and lower channel limits, the usual no-slip condition is imposed, by bouncing back incoming particles in order to produce a zero speed flow at the boundary. According to

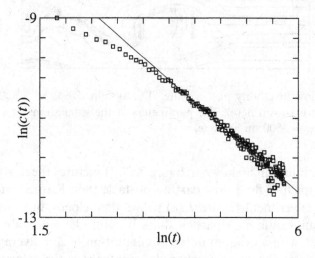

Fig. 7.28. Long time tail in the multiparticle fluid model. Simulations are performed on a 400×400 lattice, with an average density of 500 and a relaxation time $\xi = 1$. The slope of the straight line is -1.

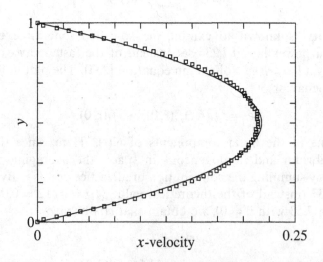

Fig. 7.29. Velocity profile in a multiparticle Poiseuille flow. The plot shows the horizontal average velocity $\langle u_x(y) \rangle$ as a function of y the vertical position between the upper and lower boundaries. The solid line corresponds to the best parabola fitting the data.

hydodynamics (see [80] for instance), we expect to observe a parabolic velocity profile. Figure 7.29 shows the measured velocity profile.

7.4 Modeling snow transport by wind

As mentioned several times in this book, the lattice gas approach makes it quite natural to introduce complex boundary conditions. It is, for instance, easy to deal with a dynamically changing boundary. This kind of problem appears in particular in sedimentation processes where particles in suspension in a fluid are transported by the flow and, under the effect of gravity, deposit on the ground. As a result, the shape of the ground continually changes according to its own dynamics, like for instance the piling and toppling of a granular material. The fluid flow has of course to self-adjust to this moving boundary. This change of flow will, in turn, change the way deposition takes place.

Thus, we see that the problem can be quite intricate and difficult to formulate in terms of differential equations. Moreover, in addition to deposition, the whole process also includes erosion phenomena: the fluid may pick up some of the particles located at the top of the deposit and transport them downstream again according to various possible mechanisms such as creeping, saltation or suspension.

Such processes take place, for instance, in sand dune formation, sedimentation by water and snowdrift formation. Although complete formulation of the sedimentation problem is difficult, the lattice gas approach give a very natural framework to propose intuitive rules accounting for the flow motion, particle transport, erosion and deposition. One of the major difficulties is to deal (on the same lattice) with the very different orders of magnitude that exist between the characteristic scales governing the fluid dynamics and the suspension. But, as usual, the hope is that, provided a scale separation exist, the correct phenomena can be captured even though the real scale separation is much larger than the one in the model.

In this section we present such a lattice gas model to describe snow transport by wind and the mechanisms of deposition and erosion.

7.4.1 The wind model

The problem of snow transport by wind is important in exposed areas such as mountain environments. Predicting the location of snow deposits after a windy period is crucial to protect roads against snowdrifts or control the formation of wind-slab (responsible for about 80% of accidentally caused avalanches). Placing snow fences at appropriate locations can substantially modify the shape of the deposit, by screening out the wind. This technique is widely used in practice to control and prevent overaccumulation of snow at critical spots or, in contrast, to store snow on ski trails.

From the simulation point of view, such a problem has two main components. First, there is the fluid (the wind here) and, second, the particles (the snow flakes) that are transported. We shall consider here a two-dimensional system, for the sake of simplicity.

From what we have learned so far, a lattice Boltzmann model like the one discussed in section 3.5 (that is a LBGK model on a square lattice with diagonals) is the natural choice to simulate wind flow. Indeed, our aim is to simulate high Reynolds number flows such as $R > 10^4$ for small-scale modeling and up to 10^6 for large-scale situations such as a mountain area.

The main difference from the situation described in section 3.5 is that now we shall include rest particles. As a consequence, the lattice directions will be labelled from $i = 0$ to $i = 8$, where $i = 0$ refers to the rest particle population (in particular the associated velocity $\vec{v}_0$ is zero). This model is often referred to as the *D2Q9* model [233] (2 dimensions and 9 particle populations).

The physical quantities of interest are

- the density $\rho = \sum_{i=0}^{8} f_i$,

- and the velocity field $\vec{u} = \frac{1}{\rho} \sum_{i=1}^{8} f_i \vec{v}_i$.

The lattice Boltzmann dynamics is expressed by the usual equation

$$f_i(\vec{r} + \tau \vec{v}_i, t + \tau) = f_i(\vec{r}, t) + \frac{1}{\xi}(f_i^{eq}(\vec{u}(\vec{r}, t), \rho(\vec{r}, t)) - f_i(\vec{r}, t))$$

where ξ is the relaxation time and $f_i^{eq}(\vec{u}, \rho)$ is the equilibrium distribution [233]

$$f_i^{eq}(\vec{u}, \rho) = t_i \rho \left[1 + \frac{v_{i\alpha} u_\alpha}{c_s^2} + \frac{1}{2} \left(\frac{v_{i\alpha} u_\alpha}{c_s^2} \right)^2 - \frac{\vec{u} \cdot \vec{u}}{2c_s^2} \right]$$

where

$$c_s^2 = \frac{1}{3}$$

is the speed of sound and the quantity t_i are

$$t_0 = \frac{4}{9}$$

for the rest particles,

$$t_1 = t_3 = t_5 = t_7 = \frac{1}{9}$$

for the horizontal and vertical particles and, finally

$$t_2 = t_4 = t_6 = t_8 = \frac{1}{36}$$

for the diagonal particles.

The shear viscosity v is given in terms of the relaxation parameter ξ, as in section 3.5, according to the relation

$$v = \frac{2\xi - 1}{6}$$

Decreasing the relaxation time too much leads to instability and the limit of high Reynolds number (i.e $v \approx 0$) cannot be attained. The solution is to consider what is called a subgrid model. This is a frequent approach in standard CFD resolution techniques. The idea is to assume that an effective viscosity results from the unresolved scales, that is scales below the lattice spacing λ.

Our goal is not to give a theoretical discussion of subgrid models but, rather, to adopt a pragmatic approach and show how to introduce eddy viscosity in a lattice Boltzmann model, following S. Hou *et al.*'s work [104].

Practically, in a subgrid model, one adapts the relaxation time locally, according to the value of local strain tensor

$$S_{\alpha\beta} = \frac{1}{2} \left(\partial_\beta u_\alpha + \partial_\alpha u_\beta \right)$$

Instability appears where the magnitude $|S_F| = \sqrt{2S_{\alpha\beta}S_{\alpha\beta}}$ of this tensor is large.

One common subgrid model is the Smagorinski model. It amounts to correcting the relaxation time as

$$\xi' = \xi + 3C_{smago}^2 \lambda^2 |S_F|$$

where $C_{smago} > 0$ is the so-called Smagorinsky constant (in practice, a parameter of the model set, for instance, to 0.3).

It turns out that the magnitude of the tensor $S_{\alpha\beta}$ can be computed locally, without taking extra derivatives, just by considering the nonequilibrium momentum tensor

$$\Pi_{\alpha\beta}^{(1)} = \sum_i v_{i\alpha}v_{i\beta}(f_i - f_i^{eq})$$

Then, the quantity $|S_F|$ is obtained directly as

$$|S_F| = \frac{-\xi + \sqrt{\xi^2 + 18\lambda^2 C_{smago}^2 \sqrt{\Pi_{\alpha\beta}^{(1)}\Pi_{\alpha\beta}^{(1)}}}}{6\lambda^2 C_{smago}^2}$$

Therefore the bare viscosity v is transformed to

$$v' = v + v_t$$

with v_t the Smagorinski eddy viscosity, given by

$$v_t = C_{smago}^2 \lambda^2 |S_F|$$

7.4.2　The snow model

On the same lattice described in the previous subsection, we now define another kind of particle $S_i(r,t)$ (S stands for snow, sand or solid). These particles interact with the wind particles and have a set of local evolution rules to account for deposition, collisions, cohesion, erosion and gravity effects.

All the processes pertaining to sedimentation and erosion are not fully understood, at least for the case of snow. To model these phenomena, we follow the cellular automata philosophy of keeping only the key microscopic interactions, hoping that they will be sufficient to capture the macroscopic behavior of the system.

On each lattice site, snow distribution is represented by $S_i(r,t)$, $i \in 0,\dots,8$. S_i are integer values equal to the number of particles moving in the ith direction (S_0 for the particles remaining motionless). Therefore, snow is described with a multiparticle-like dynamics.

In real life, the influence of wind during the snowfall is important, but transport after the fall is essential for re-shaping of the deposit. It is assumed that the snow transport can be split into three main processes:

- *creeping* of particles along the ground, consisting of very small jumps of the snow particles;

- *saltation*, which occurs mainly in the first 30 centimeters above the surface. This process is essential for snow transport;

- *suspension* or *turbulent diffusion/dispersion* of particles. Particles are caught in the saltation layer by strong eddies and can be transported over long distances. This process is often seen as "smoke" above the mountain crests.

Unlike other numerical models [236,237], the aim of our lattice model is not to treat these processes (mainly saltation and suspension) separately. By setting reasonable microscopic rules we expect to naturally recover the macroscopic reality as the emergence of a complex behavior.

These rules governing the snow particles and their interaction with wind particles have several components: transport mechanisms, including gravity, erosion and, finally deposition rules. They are now described in more detail.

Transport by wind. Neglecting the inertial effects, we may assume that a snow particle follows the wind velocity field $\vec{u}(\vec{r}, t)$ and experiences gravity. Due to friction, the main effect of gravity is to impose vertical falling speed $-\vec{u}_g$ to the particle in suspension. Thus, to a first approximation, a flake at site $\vec{r}$ will move, after τ units of time, to

$$\vec{r}(t + \tau) = \vec{r}(t) + (\vec{u} - \vec{u}_g)\tau$$

Unfortunately, the quantity $\vec{r}(t + \tau)$ defined in this way no longer has any reason to correspond to a lattice site. Thus, it is no longer possible to use a lattice gas approach to describe the motion of snow particles if we do not use a trick.

One possibility, however, is to view the particle transport as a probabilistic process. We write

$$\vec{r}(t + \tau) = \vec{r}(t) + \vec{v}_i \tau$$

where $\vec{v}_i$ is one of the nine velocity vectors associated with the *D2Q9* square lattice (see figure 3.15 and remember that $\vec{v}_0 = 0$). For each particle motion, one of the $\vec{v}_i$ is chosen so that, on average

$$\langle \vec{r}(t + \tau) - \vec{r}(t) \rangle = (\vec{u} - \vec{u}_g)\tau$$

Practically, we can proceed as follows. We first assume that $|\vec{u} - \vec{u}_g| < v$ where $v = (\lambda/\tau)$ is the maximum speed pemitted on the lattice. Then we introduce four probabilities of motion p_x, p_{-x}, p_y, p_{-y} corresponding to motion along the x- or y-axis, in the plus or minus direction. We define

$$p_x = \max\left(0, \frac{\vec{v}_1 \cdot (\vec{u} - \vec{u}_g)}{v^2}\right) \qquad p_{-x} = \max\left(0, \frac{\vec{v}_5 \cdot (\vec{u} - \vec{u}_g)}{v^2}\right)$$

$$p_y = \max\left(0, \frac{\vec{v}_3 \cdot (\vec{u} - \vec{u}_g)}{v^2}\right) \qquad p_{-y} = \max\left(0, \frac{\vec{v}_7 \cdot (\vec{u} - \vec{u}_g)}{v^2}\right)$$

Since $\vec{v}_1 = -\vec{v}_5$ and $v_3 = -\vec{v}_7$ one cannot have p_x and p_{-x} simultaneously non-zero (and similarly for p_y and p_{-y}). On the other hand, we may have any combination of x- and y-motion.

To select the direction of motion of a particle with this approach, two random numbers q_x and q_y, uniformly distributed in the interval $[0, 1]$, are chosen. If $q_x < p_x$ (or $q_x < p_{-x}$) the particle will move along the positive x-direction (or the negative x-direction). Otherwise, if q_x is larger than both p_x and p_{-x}, no motion along the horizontal axis is permitted.

The same algorithm is applied to q_y for the y-axis motion. When both the x- and y-axis are selected, the particle moves with velocity $\vec{v}_2$, $\vec{v}_4$, $\vec{v}_6$ or $\vec{v}_8$, according to the sense of motion. The average velocity $\langle \vec{v} \rangle$ of the particle is then

$$\langle \vec{v} \rangle = p_x \vec{v}_1 + p_{-x} \vec{v}_5 + p_y \vec{v}_3 + p_{-y} \vec{v}_7$$

Fig. 7.30. Example of particle transport in a fixed, circular velocity field. The lattice contains 128×128 sites and a particle source is set in the middle of the small circle. The particles follow their path with a dispersion due to the probabilistic algorithm.

Since p_x and p_{-x} are incompatible (and so are p_y, p_{-y}) and $\vec{v}_1 = -\vec{v}_5$, $\vec{v}_3 = -\vec{v}_7$, we may as well write

$$\langle \vec{v} \rangle = \left(\frac{\vec{v}_1 \cdot (\vec{u} - \vec{u}_g)}{v^2} \right) \vec{v}_1 + \left(\frac{\vec{v}_3 \cdot (\vec{u} - \vec{u}_g)}{v^2} \right) \vec{v}_3 = \vec{u} - \vec{u}_g$$

since $\vec{v}_1$ and $\vec{v}_3$ are orthogonal. This last relation shows that, although the particles in suspension move on the lattice, their average position follows the desired velocity flow $\vec{u} - \vec{u}_g$.

As we may have many snow particles at a site, the above procedure should be repeated for each of them. Of course, it is possible to speed up this phase by using the Gaussian approximation described in section 5.7, and know in a few steps how many particles will move with each of the possible velocities $\vec{v}_i$.

An example of such a particle transport over a lattice is shown in figure 7.30. We observe an intrinsic particle dispersion due to the probabilitic nature of the algorithm. This dispersion, which is significant over a long distance, can be given a physical content such as, for instance, the effects of microscopic fluctuations present in the system.

Deposition.   Deposition is a crucial ingredient of our model. Particles in suspension eventually reach the bottom of the system, which follows the ground profile. As they fall down, particles pile up and modify the shape of the terrain. In our model, a lattice site can either be solid (like "hard" soil or deposited snow) or free (where fluid and snow particles can move). When a snow particle should move to a neighbor solid site, it remains in position as a result of being solidified. Since we deal with a

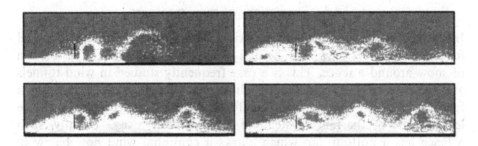

Fig. 7.31. Four sequential stages in a snow deposit simulation. The wind is blowing from left to right and the presence of a fence (with a ground clearance) is shown by the vertical line segment. Light gray regions indicate solid snow (the deposit) and white dots show the flying snow (particles in suspension). In order to feed the system with snow, a source of snow is placed at the lower left corner of the system.

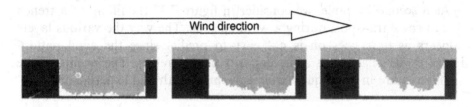

Fig. 7.32. Three sequential stages of snow deposit in a trench. A source of snow is placed on the left side of the system.

multiparticle snow model, a site will become a solid site only after it has received more than a given threshold number of particles. This technique, which takes into account some cohesion between snow particles, makes the use of complex piling–toppling rules (such as described in section 2.2.6) unnecessary.

Erosion. The erosion process is very complex and not yet fully understood. Deposited particles can be picked up again by the wind, but the physical conditions that are necessary to produce this and the interactions that are involved are not easy to identify [238]. Some empirical laws exist that express the erosion rate in terms of the local properties of the system, such as the wind speed at the erosion surface. Several rules can be considered in the lattice gas approach to reproduce qualitatively the experimental observations. The basic erosion mechanism we use is that a solid snow particle is ejected vertically with a random speed when the appropriate erosion conditions are met.

7.4.3 Simulations of snow transport

In this section we present two numerical simulations of our lattice model for snow transport by wind. The first situation is to obtain the deposition of snow around a fence. This is a case frequently studied in wind tunnels and which has many practical applications. The purpose of placing a fence is to screen out the wind and force the snow to accumulate on the leeward site of the fence instead of flying further away. Usually, such a fence has a bottom gap which allows a continual wind flow below it. This prevents the snow from accumulating at this place and thus rapidly burying the fence, which would make it completely ineffective.

Figure 7.31 shows several stages of the deposit formation. The snow deposit grows in a non-regular way and the wind pattern evolves as the accumulation becomes larger. It is also interesting to note the presence of a smaller deposit in front of the fence, due to eddies forming windward of the fence. These simulations are very close to wind tunnel experiments.

As a second example, we consider in figure 7.32 the filling of a trench due to snow transport during a wind episode. The way the various layers appears as time goes on is not easy to predict since the wind pattern in the cavity changes as snow deposits in the trench. The results of the simulation are in good qualitative agreement with real experiments.

References

[1]   R. Livi, S. Ruffo, S. Ciliberto, and M. Buiatti, editors. *Chaos and Complexity*. World Scientific, 1988.

[2]   I.S.I and R. Monaco, editors. *Discrete Kinetic Theory, Lattice Gas Dynamics and Foundations of Hydrodynamics*. World Scientific, 1989.

[3]   G. Doolen, editor. *Lattice Gas Method for Partial Differential Equations*. Addison-Wesley, 1990.

[4]   P. Manneville, N. Boccara, G.Y. Vichniac, and R. Bideau, editors. *Cellular Automata and Modeling of Complex Physical Systems*. Springer Verlag, 1989. Proceedings in Physics 46.

[5]   A. Pires, D.P. Landau, and H. Herrmann, editors. *Computational Physics and Cellular Automata*. World Scientific, 1990.

[6]   J.M. Perdang and A. Lejeune, editors. *Cellular Automata: Prospect in Astrophysical Applications*. World Scientific, 1993.

[7]   Minnesota IMA cellular automata bibliography http://www.ima.umn.edu/bibtex/ca.bib.

[8]   Santa-Fe cellular automata bibliography. ftp://alife.santafe.edu/pub/topics/cas/ca-faq.bib.

[9]   T. Toffoli D. Farmer and S. Wolfram, editors. *Cellular Automata, Proceedings of an Interdisciplinary Workshop*, volume 10. Physica D, North-Holland, 1984.

[10]   J.-P. Boon, editor. *Advanced Research Workshop on Lattice Gas Automata Theory, Implementations, and Simulation, J. Stat. Phys*, **68** (3/4), 1992.

[11]   S. Ulam. Random processes and transformations. *Proc. Int. Congr. Math.*, 2:264–275, 1952.

[12]   A.W. Burks. Von Neumann's self-reproducing automata. In A.W. Burks, editor, *Essays on Cellular Automata*, pp. 3–64. University of Illinois Press, 1970.

[13]   U. Pesavento. An implementation of von Neumann's self-reproducing machine. *Artificial Life*, **2**:337–354, 1995.

[14] A. Reggia, S.L. Armentrout, H.-H. Chou, and Y. Peng. Simple systems that exhibit self-directed replication. *Science*, **259**:1282, 1993.

[15] E.F. Codd. *Cellular Automata*. Academic Press, 1968.

[16] C.G. Langton. Self-reproduction in cellular automata. *Physica D*, **10**:135–144, 1984.

[17] J. Byl. Self-reproduction in small cellular automata. *Physica D*, **34**:259–299, 1989.

[18] C.G. Langton, C. Taylor, J.D. Farmer, and S. Rasmussen, editors. *Artificial Life II*. Addison-Wesley, 1992.

[19] C.G. Langton. Editor's Introduction. *Artificial Life*, **1**:v–viii, 1994.

[20] M. Gardner. The fantastic combinations of John Conway's new solitaire game life. *Scientific American*, **220(4)**:120, 1970.

[21] E.R. Berlekamp, J.O. Conway, and R.K. Guy. *Winning Ways for your Mathematical Plays*, volume 2. Academic Press, 1982. Chapter 25.

[22] M. Gardner. *Wheels, Life and Other Mathematical Amusements*. Freeman, 1983.

[23] K. Preston and M. Duff. *Modern Cellular Automata: Theory and Applications*. Plenum Press, 1984.

[24] S. Wolfram. *Theory and Application of Cellular Automata*. World Scientific, 1986.

[25] S. Wolfram. *Cellular Automata and Complexity*. Addison-Wesley, Reading MA, 1994.

[26] T. Toffoli and N. Margolus. *Cellular Automata Machines: a New Environment for Modeling*. The MIT Press, 1987.

[27] Information Mechanics Group. CAM8: a parallel, uniform, scalable architecture for CA Experimentation. http://www.im.lcs.mit.edu/cam8.

[28] J. Hardy, Y. Pomeau, and O. de Pazzis. Molecular Dynamics of a classical lattice gas: Transport properties and time correlation functions. *Phys. Rev. A*, **13**:1949–60, 1976.

[29] J.C. Maxwell. *Scientific Paper II*. Cambridge University Press, 1890.

[30] H. Cornille. Exact (2+1)-dimensional solutions for two discrete velocity Boltzmann models with four independent densities. *J. Phys. A*, **20**:L1063–67, 1987.

[31] J.E. Broadwell. Shock structure in a simple discrete velocity gas. *Phys. Fluids*, **7**:1243, 1964.

[32] U. Frisch, B. Hasslacher, and Y. Pomeau. Lattice-gas automata for the Navier–Stokes equation. *Phys. Rev. Lett.*, **56**:1505, 1986.

[33] S. Wolfram. Cellular automaton fluid: basic theory. *J. Stat. Phys.*, **45**:471, 1986.

[34] D. d'Humieres, Y. Pomeau, and P. Lallemand. Simulation of 2-D von Karman streets using a lattice gas. *C.R. Acad. Sci. Paris, II*, **301**:1391, 1985. Reprinted in *Lattice Gas Methods for Partial Differential Equations*, ed. G. Doolen, p.423, Addison-Wesley, 1990.

[35] P. Calvin, D. d'Humières, P. Lallemand, and Y. Pomeau. Cellular automata for hydrodynamics with free boundaries in two and three dimensions. *C.R. Acad. Sci. Paris, II*, **303**:1169, 1986. Reprinted in *Lattice Gas Methods for Partial Differential Equations*, ed. G. Doolen, p. 415, Addison-Wesley, 1990.

[36] D. d'Humières, P. Lallemand, and U. Frisch. Lattice gas models for 3D hydrodynamics. *Europhys. Lett.*, **2**:291, 1986.

[37] D. d'Humières and P. Lallemand. Lattice gas models for 3D hydrodynamics. *Europhys. Lett.*, **2**:291, 1986.

[38] S.A. Orszag and V. Yakhot. Reynolds Number Scaling of Cellular Automata Hydrodynamics. *Phys. Rev. Lett*, **56**:1691–1693, 1986.

[39] S. Chen, K. Diemer, G.D. Doolen, K. Eggert, C. Fu, S. Gutman, and B.J. Travis. Lattice Gas Automata For Flow Through Porous Media. *Physica D*, **47**:72–84, 1991.

[40] E. Aharonov and D. Rothman. Non-Newtonian flow (through porous media): a lattice Boltzmann method. *Geophys. Res. Lett.*, **20**:679–682, 1993.

[41] D.W. Grunau, T. Lookman, S.Y. Chen, and A.S. Lapedes. Domain growth, wetting and scaling in porous media. *Phys. Rev. Lett.*, **71**:4198–4201, 1993.

[42] D.H. Rothman. Immiscible lattice gases: new results, new models. In P. Manneville, N. Boccara, G.Y. Vichniac, and R. Bideau, editors, *Cellular Automata and Modeling of Complex Physical Systems*, pp. 206–231. Springer Verlag, 1989. Proceedings in Physics 46.

[43] M. Bonetti, A. Noullez, and J.-P. Boon. Lattice gas simulation of 2-d viscous fingering. In P. Manneville, N. Boccara, G.Y. Vichniac, and R. Bideau, editors, *Cellular Automata and Modeling of Complex Physical Systems*, p. 239–241. Springer Verlag, 1989. Proceedings in Physics 46.

[44] A.K. Gunstensen, D.H. Rothman, S. Zaleski, and G. Zanetti. Lattice Boltzmann Model of Immiscible Fluids. *Phys. Rev. A*, **43**:4320–4327, 1991.

[45] D. Grunau, Shiyi Chen, and K. Eggert. A Lattice Boltzmann Model for Multi-phase Fluid Flows. *Phys. Fluids A*, **5**:2557–2562, 1993.

[46] D. Rothman and S. Zaleski. *Lattice-Gas Cellular Automata: Simple Models of Complex Hydrodynamics*. Collection Aléa. Cambridge University Press, 1997.

[47]   U. D'Ortona, M. Cieplak, R.B. Rybka, and J.R. Banavar. Two-color nonlinear cellular automata: surface tension and wetting. *Phys. Rev. E*, **51**:3718–28, 1995.

[48]   B. Boghosian, P. Coveney, and A. Emerton. A lattice-gas model of microemulsions. *Proceedings of the Royal Society of London*, **452**:1221–1250, 1996.

[49]   J.T. Wells, D.R. Janecky, and B.J. Travis. A lattice gas automata model for heterogeneous chemical reaction at mineral surfaces and in pores network. *Physica D*, **47**:115–123, 1991.

[50]   J.-P. Boon, D. Dab, R. Kapral, and A. Lawniczak. Lattice Gas Automata for Reactive Systems. *Phys. Rep.*, **273**:55–148, 1996.

[51]   G.G. McNamara and G. Zanetti. Use of the Boltzmann Equation to Simulate Lattice-Gas Automata. *Phys. Rev. Lett.*, **61**:2332–2335, 1988.

[52]   F. Higuera, J. Jimenez, and S. Succi. Boltzmann approach to lattice gas simulations. *Europhys. Lett*, **9**:663, 1989.

[53]   D.B. Bahr and J.B. Rundle. Theory of lattice Boltzmann simulation of glacier flow. *J. Glaciology*, **41**(139):634–40, 1995.

[54]   E. Banks. Information processing and transmission in cellular automata. Technical report, MIT, 1971. MAC TR-81.

[55]   Thinking Machines Corp. *CM Fortran Libraries Reference Manual*, January 1994. Version 2.1, p.28.

[56]   R. J. Gaylord and K. Nishidate. *Modeling Nature with Cellular Automata using Mathematica*. Springer-Verlag, 1996.

[57]   K. Culick II and F. Yu. Undecidability of CA classification scheme. *Complex System*, **2**:177–190, 1988.

[58]   J. Hemmingsson. A totalistic three-dimensional cellular automaton with quasiperiodic behavior. *Physica A*, **183**:255, 1992.

[59]   H. Chaté and P. Manneville. Collective behaviors in spatially extended systems with local interactions and synchronous updating. *Prog. Theor. Phys.*, **87**:1–67, 1992.

[60]   J. Krug and H. Spohn. Universality classes for deterministic surface growth. *Phys. Rev. A*, **38**:4271–83, 1988.

[61]   K. Chen P. Bak and C. Tang. A forest-fire model and some thoughts on turbulence. *Phys. Lett. A*, **147**:297, 1990.

[62]   B. Drossel and F. Schwabl. Self-Organized Critical Forest-Fire Model. *Phys. Rev. Lett.*, **69**:1629, 1992.

[63]   P. Rujàn. Cellular Automata and Statistical Mechanical Models. *J. Stat. Phys.*, **49**:139, 1987.

[64]   G. Vichniac. Simulating physics with cellular automata. *Physica D*, **10**:96–115, 1984.

[65] J.D. Gunton and M. Droz. *Introduction to the Theory of Metastable and Unstable States*. Springer Verlag, 1983.

[66] J. Rajechenbach, E. Clément, J. Duran, and T. Mazozi. Experiments on bidimensional Models of sand: study of the dynamics. In J. Vannimenius A. McKane, M. Droz and D. Wolf, editors, *Scale Invariance and Non-equilibrium Dynamics*, pp. 313–27. NATO ASI Series, Plenum Press, 1995.

[67] E. Guyon and J.-P. Troadec. *Du sac de bille au tas de sable*. Edition Odile Jacob, 1994.

[68] H.M. Jaeger and S.P. Nagel. La physique de l'état granulaire. *La Recherche*, **23**(249):1380–87, December 1992.

[69] D.E. Wolf, M. Schreckenberg, and A. Bachem, editors. *Traffic and Granular Flow*. World Scientific, 1996.

[70] A. Károlyi and J. Kertész. Hydrodynamics cellular automata for granular media. In R. Gruber and M. Tomassini, editors, *Proceeding of the 6th Joint EPS-APS International Conference on Physics Computing: PC '94*, pp. 675–681, 1994.

[71] I. Stewart. The ultimate in anty-particle. *Scientific American*, pp. 88–91, July 1994.

[72] J. Propp. Trajectory of generalized ants. *Math. Intelligencer*, **16**(1):37–42, 1994.

[73] O. Biham, A.A. Middleton, and D. Levine. Self-organization and dynamical transition in traffic-flow models. *Phys. Rev. A*, **46**:R6124, 1992.

[74] K. Nagel and H.J. Herrmann. Deterministic models for traffic jams. *Physica A*, **199**:254, 1993.

[75] M. Kikuchi S. Yukawa and S. Tadaki. Dynamical phase transition in one-dimensional traffic flow model with blockage. *J. Phys. Soc. Jpn*, **63**(10):3609–3618, 1994.

[76] B. Chopard. A Cellular Automata Model of Large Scale Moving Objects. *J. Phys. A*, **23**:1671–1687, 1990.

[77] M. Sipper. Co-evolving Non-Uniform Cellular Automata to Perform Computations. *Physica D*, **92**:193–208, 1996.

[78] B. Chopard and M. Droz. Cellular automata model for heat conduction in a fluid. *Phys. Lett. A*, **126**:476, 1988.

[79] P. Grosfils, J.-P. Boon, and P. Lallemand. Spontaneous fluctuation correlation in thermal lattice gas automata. *Phys. Rev. Lett.*, **68**:1077, 1992.

[80] D.J. Triton. *Physical Fluid Dynamics*. Clarendon Press, 1988.

[81] K. Molvig, P. Donis, R. Miller, and G. Vichniac. Multi-species Lattice-Gas Automata for Realistic Fluid Dynamics. In P. Manneville, N. Boccara, G.Y. Vichniac, and R. Bideau, editors, *Cellular Automata and*

Modeling of Complex Physical Systems, pp. 206–231. Springer Verlag, 1989. Proceedings in Physics 46.

[82]   M. Hénon. Viscosity of a lattice gas. *Complex Systems*, 1:763–789, 1987. Reprinted in *Lattice Gas Methods for Partial Differential Equations*, ed. G. Doolen, pp. 179, Addison-Wesley, 1990.

[83]   K. Diemer, K. Hunt, S. Chen, T. Shimomura, and G. Doolen. Density and velocity dependence of Reynolds numbers for several lattice gas models. In G. Doolen, editor, *Lattice Gas Methods for Partial Differential Equations*, pp. 137–177. Addison-Wesley, 1990.

[84]   L.P. Kadanoff, G.R. McNamara, and G. Zanetti. From automata to fluid flow: comparison of simulation and theory. *Phys. Rev. A*, **40**:4527–4541, 1989.

[85]   N. Margolus, T. Toffoli, and G. Vichniac. Cellular automata supercomputers for fluid-dynamics modeling. *Phys. Rev. Lett.*, **56**:1694–1696, 1986.

[86]   D. Frenkel. Long-time decay of velocity autocorrelation of two-dimensional lattice gas cellular automata. In P. Manneville, N. Boccara, G.Y. Vichniac, and R. Bideau, editors, *Cellular Automata and Modeling of Complex Physical Systems*, pp. 144–154. Springer Verlag, 1989. Proceedings in Physics 46.

[87]   R. Balescu. *Equilibrium and Nonequilibrium Statistical Mechanics*. John Wiley & Sons, 1975.

[88]   U. Frisch, D. d'Humières, B. Hasslacher, P. Lallemand, Y. Pomeau, and J.-P. Rivet. Lattice Gas Hydrodynamics in two and Three Dimension. *Complex Systems*, 1:649–707, 1987. Reprinted in *Lattice Gas Methods for Partial Differential Equations*, ed. G. Doolen, p. 77, Addison-Wesley, 1990.

[89]   M. A. van der Hoef. Introduction to lattice gas cellular automata as a model for simple fluids. In J.M. Pergang and A. Lejeune, editors, *Cellular Automata: Prospects in Astrophysical Applications*, pp. 87–118. World Scientific, 1993.

[90]   B. Chopard and M. Droz. Cellular automata approach to nonequilibrium correlation functions in a fluid. *Helvetica Physica Acta*, **61**:893–6, 1988.

[91]   P. Résibois and M. Leener. *Classical Kinetic Theory of Fluids*. John Wiley, 1977.

[92]   G. Zanetti. Hydrodynamics of lattice-gas automata. *Phys. Rev. A*, **40**:1539–1548, 1989.

[93]   T. Naitoh, M.H. Ernst, and J.M. Dufty. Long-time tails in two-dimensional cellular automata fluids. *Phys. Rev. A*, **42**:7187, 1990.

[94]   R. Brito and M.H. Ernst. Propagating Staggered Wave in Cellular Automata Fluids. *J. Phys. A*, **24**:3331, 1991.

[95] D. d'Humières, Y.H. Quian, and P. Lallemand. Invariants in Lattice gas Models. In I.S.I and R. Monaco, editors, *Discrete Kinetic Theory, Lattice Gas Dynamics and Foundations of Hydrodynamics*, pp. 102–113. World Scientific, 1989.

[96] R. Benzi, S. Succi, and M. Vergassola. The lattice Boltzmann equation: theory and application. *Phys. Rep.*, **222**(3):145–197, 1992.

[97] A. Gustensen, D.H. Rothman, S. Zaleski, and G. Zanetti. Lattice Boltzmann model of immiscible fluids. *Phys. Rev. A*, **43**:4320–4327, 1991.

[98] S. Succi, M. Vergassola, and R. Benzi. Lattice-Boltzmann scheme for two-dimensional magnetohydrodynamics. *Phys. Rev. A*, **43**:4521, 1991.

[99] F.J. Alexander, S. Chen, and J.D. Sterling. Lattice Boltzmann thermohydrodynamics. *Phys. Rev. E*, **47**:2249–2252, 1993.

[100] F. Higuera, J. Jimenez, and S. Succi. Lattice gas dynamics with enhanced collision. *Europhys. Lett*, **9**:345, 1989.

[101] Hudong Chen, Shiyi Chen, and W.H. Matthaeus. Recovery of Navier–Stokes equations using a lattice-gas Boltzmann method. *Phys. Rev. A*, **45**:R5339–42, 1992.

[102] Y.H. Quian, D. d'Humières, and P. Lallemand. Lattice BGK Models for Navier–Stokes Equation. *Europhys. Lett*, **17**(6):470–84, 1992.

[103] P. Bhatnager, E.P. Gross, and M.K. Krook. A model for collision process in gases. *Phys. Rev.*, **94**:511, 1954.

[104] S. Hou, J. Sterling, S. Chen, and G.D. Doolen. A lattice Boltzmann subgrid model for high Reynolds number flows. *Fields Institute Communications*, **6**:151–166, 1996.

[105] R. Chatagny and B. Chopard. Multiparticle lattice gas models for hydrodynamics. In J.-M. Alimi, A. Serna, and H. Scholl, editors, *Science on the Connection Machine System: Proceedings of the Second European CM Users Meeting 1993*, pp. 239–47. Thinking Machines Corporation, 1995.

[106] G.H. Weiss. Random walks and their applications. *American Scientist*, **71**:65, 1983.

[107] G.H. Weiss, editor. *Contemporary Problems in Statistical Physics*. SIAM, 1994.

[108] G.I Taylor. Diffusion by continuous movements. *Proc. Lond. Math. Soc*, **20**:196–212, 1921.

[109] E.H. Hauge. What can we learn from Lorentz models? In G. Kirczenow and J. Marro, editors, *Transport Phenomena*, p. 337. Springer, 1974.

[110] N.G. van Kampen. *Stochastic Processes in Physics and Chemistry*. North-Holland, 1981.

[111] X.P. Kong and E.G.D. Cohen. Anomalous diffusion in lattice-gas wind-tree model. *Phys. Rev. B*, **40**:4838, 1989.

[112] B. Chopard and M. Droz. Cellular automata model for diffusion processes. *J. Stat. Phys.*, **64**:859–892, 1991.

[113] J.D. Jackson. *Classical Electrodynamics.* John Wiley, 1975.

[114] A.S. Nowick and J.J. Burton. *Diffusion in Solids.* Academic Press, 1978.

[115] J. Philibert. *Diffusion et Transport de Matière dans les Solides.* Les Editions de Physique, Les Ulis, 1985.

[116] W. Dieterich, P. Fulde, and I. Peschel. Theoretical models for superionic conductors. *Adv. Phys.*, **29**:527, 1980.

[117] J.P. Hulin, E. Clément, C. Baudet, J.-F. Gouyet, and M. Rosso. Qualitative analysis of an invading-fluid invasion front under gravity. *Phys. Rev. Lett.*, **61**:333, 1988.

[118] B. Chopard, M. Droz, and M. Kolb. Cellular automata approach to nonequilibrium diffusion and gradient percolation. *J. Phys. A*, **22**:1609, 1989.

[119] T. Vicsek. *Fractal Growth Phenomena.* World Scientific, 1989.

[120] T.A. Witten and L.M. Sander. Diffusion-limited aggregation. *Phys. Rev. B*, **27**:5686, 1983.

[121] S. Tolman and P. Meakin. Off-lattice and hypercubic-lattice models for diffusion-limited aggregation in dimension 2–8. *Phys. Rev. A*, **40**:428–37, 1989.

[122] J.W. Evans. Random and cooperative sequential adsorption. *Reviews of Modern Physics*, **65**:1281–1329, 1993.

[123] I. Langmuir. Adsorption and gases on glass, mica and platinium. *J. Am. Chem. Soc.*, **40**:1361, 1918.

[124] P. Schaaf, A. Johner, and J. Talbot. Asymptotic behavior of particle deposition. *Phys. Rev. Lett.*, **66**:1603–1605, 1991.

[125] G. Tarjus and P. Viot. Generalized Car-Parking Problem as a Model Particle Deposition with Entropy-Activated Rate Process. *Phys. Rev. Lett.*, **68**:2354, 1992.

[126] B. Senger, P. Schaaf, J.-C. Voegel, A. Johner, A. Schmitt, and J. Talbot. Influence of bulk diffusion on the adsorption of hard spheres on a flat surface. *J. Chem. Phys.*, **97**:3813–3820, 1992.

[127] B. Senger, J. Talbot, P. Schaaf, A. Schmitt, and J.-C. Voegel. Effect of the Bulk Diffusion on the Jamming Limit Configurations for Irreversible Adsorption. *Europhys. Lett.*, **21**:135–140, 1993.

[128] B. Chopard, P. Luthi, and M. Droz. Microscopic approach to the formation of Liesegang patterns. *J. Stat. Phys.*, **76**:661–677, 1994.

[129] J. E. Pearson. Complex Patterns in a Simple System. *Science*, **261**:189–192, July 1993.

[130] J.D. Muray. *Mathematical Biology.* Springer-Verlag, 1990.

[131] L.M. Brieger and E. Bonomi. A stochastic cellular automaton model of nonlinear diffusion and diffusion with reaction. *J. Comp. Phys.*, **94**:467–486, 1991.

[132] E.J. Garboczi. Permeability, Diffusivity and Microstructural Parameters: a Critical Review. *Cement and Concrete Res.*, **20**:591–601, 1990.

[133] R. Kapral and K. Showalter, editors. *Chemical Waves and Patterns.* Kluwer Academic, 1995.

[134] J.P. Keener and J.J. Tyson. The dynamics of scroll waves in excitable media. *SIAM Rev.*, **34**:1–39, 1992.

[135] E.E. Selkov. Self-oscillation in glycolysis: a simple kinetic model. *Eur. J. Biochem.*, **4**:79, 1968.

[136] R. Fisch, J. Gravner, and D. Griffeath. Threshold-range scaling of excitable cellular automata. *Statistics and Computing*, **1**:23, 1991.

[137] J. Gravner and D. Griffeath. Threshold grouse dynamics. *Trans. Amer. Math. Soc.*, **340**:837, 1993.

[138] A.A. Ovchinnikov and Y.B. Zeldovich. Role of density fluctuation in bimolecular reaction kinetics. *Chem. Phys.*, **28**:215, 1978.

[139] D. Toussain and F. Wilczek. Particle–antiparticle annihilation in diffusive motion. *J. Chem. Phys.*, **78**:2642, 1983.

[140] K. Kang and S. Redner. Scaling approach for the kinetics of recombination processes. *Phys. Rev. Lett.*, **52**:955, 1984.

[141] H. Schnörer, I.M. Sokolov, and A. Blumen. Fluctuations statistics in the diffusion-limited $A + B \rightarrow 0$ reaction. *Phys. Rev. A*, **42**:7075, 1990.

[142] K. Lindenberg, B.J. West, and R. Kopelman. Diffusion-limited $A + B \rightarrow 0$ reaction: Correlated initial condition. *Phys. Rev. A*, **42**:890, 1990.

[143] S. Cornell, M. Droz, and B. Chopard. Some properties of the diffusion-limited reaction $nA + mB \rightarrow C$ with homogeneous and inhomogeneous initial conditions. *Physica A*, **188**:322–336, 1992.

[144] R. Kopelman. Rate processes on fractals: theory, simulations and experiments. *J. Stat. Phys.*, **42**:185, 1986.

[145] B. Chopard, L. Frachebourg, and M. Droz. Multiparticle lattice gas automata for reaction-diffusion systems. *Int. J. Mod. Phys. C*, **5**:47–63, 1994.

[146] L. Gálfi and Z. Rácz. Properties of the reaction front in an $A + B \rightarrow C$ type reaction-diffusion process. *Phys. Rev. A*, **38**:3151, 1988.

[147] B. Chopard and M. Droz. Microscopic study of the properties of the reaction front in an $A + B \rightarrow C$ reaction-diffusion process. *Europhys. Lett.*, **15**:459–464, 1991.

[148] Y.E. Koo, L. Li, and R. Kopelman. Reaction front dynamics in diffusion-controlled particle antiparticle annihilation: experiments and simulations. *Mol. Cryst. Liquid Cryst.*, pp. 187–192, 1990.

[149] Z. Jiang and C. Ebner. Simulation study of reaction fronts. *Phys. Rev. A*, **42**:7483, 1990.

[150] S. Cornell, M. Droz, and B. Chopard. Role of fluctuations for inhomogeneous reaction-diffusion phenomena. *Phys. Rev. A*, **44**:4826–32, 1991.

[151] S. Cornell and M. Droz. Steady-state reaction-diffusion front scaling for $mA + nB \rightarrow$ [inert]. *Phys. Rev. Lett.*, **70**:3824–27, 1993.

[152] H. K. Henisch. *Crystals in Gels and Liesegang Rings*. Cambridge University Press, 1988.

[153] K. Jablczynski. La formation rythmique des pécipités: Les anneaux de Liesegang. *Bull. Soc. Chim. France*, **33**:1592, 1923.

[154] R.E. Liesegang. Über einige Eigenschaften von Gallerten. *Naturwiss. Wochenschr.*, **11**:353, 1896.

[155] S. Prager. Periodic precipitation. *J. Chem. Phys.*, **25**:279, 1956.

[156] Y.B. Zeldovitch, G.I. Barrenblatt, and R.L. Salganik. The quasi-periodical formation of precipitates occuring when two substances diffuse into each other. *Sov. Phys. Dokl.*, **6**:869, 1962.

[157] D.A. Smith. On Ostwald's supersaturation theory of rhythmic precipitation (Liesegang rings). *J. Phys. Chem.*, **81**:3102, 1984.

[158] W. Ostwald. *Lehrbuch der allgemeinen Chemie*. Engelman, 1897.

[159] G.T. Dee. Patterns Produced by Precipitation at a moving Reaction Front. *Phys. Rev. Lett.*, **57**:275–78, 1986.

[160] B. Mathur and S. Ghosh. Liesegang Rings-Part I: Revert system of Liesegang Rings. *Kolloid-Zeitschrift*, **159**:143, 1958.

[161] E. Kárpáti-Smidróczki, A. Bki, and M. Zrinyi. Pattern forming precipitation in gels due to coupling of chemical reaction with diffusion. *Colloid. Polym. Sci.*, **273**:857–865, 1995.

[162] H. K. Henisch. *Periodic Precipitation*. Pergamon Press, 1991.

[163] B. Chopard, H.J. Herrmann, and T. Vicsek. Structure and growth mechanism of mineral dendrites. *Nature*, **353**:409–412, October 1991.

[164] K.M. Pillai, V.K. Vaidyan, and M.A. Ittyachan. On the theory of Liesegang phenomena. *Colloid Polym. Sci.*, **258**:831–38, 1980.

[165] W.H. Press, B.P. Flannery, S.A. Teukolsky, and W.T. Vetterling. *Numerical Recipes: The Art of Scientific Computing*. Cambridge University Press, 1989.

[166] T. Karapiperis and B. Blankleider. Cellular automata model of reaction-transport process. *Physica D*, **78**:30–64, 1994.

[167] A. Turing. The chemical basis of morphogenesis. *Phil. Trans. Roy. Soc. London*, **B237**:37, 1952.

[168] V. Dufiet and J. Boissonnade. Conventional and unconventional Turing patterns. *J. Chem. Phys.*, **96**:664, 1991.

[169] J. Schnackenberg. Simple chemical reaction systems with limit cycle behaviour. *J. Theor. Biol.*, **81**:389, 1979.

[170] B. Chopard, M. Droz, S. Cornell, and L. Frachebourg. Cellular automata approach to reaction-diffusion systems: theory and applications. In J.M. Perdang and A. Lejeune, editors, *Cellular Automata: Prospects in Astrophysical Applications*, pp. 157–186. World Scientific, 1993.

[171] Shang-Keng Ma. *Modern Theory of Critical Phenomena*. W.A. Benjamen, 1976.

[172] M. Doi. Stochastic theory of diffusion-controlled reaction. *J. Phys. A*, **9**:1479, 1976.

[173] P. Grassberger and M. Scheunert. Fock-space methods for identical classical objects. *Fortschrift. Phys.*, **28**:547, 1980.

[174] L. Peliti. Path integral approach to birth-death processes on a lattice. *J. de Physique*, **46**:1469, 1985.

[175] C. Itzykson and J.-B. Zuber. *Quantum Field Theory*. McGraw-Hill, 1985.

[176] L. Sasvàri and M. Droz. A renormalization-group approach to simple reaction-diffusion phenomena. *Phys. Rev. E*, **48**:R2343, 1993.

[177] B. P. Lee. Renormalisation group calculation for the reaction $kA \rightarrow 0$. *J. Phys. A*, **27**:2633, 1994.

[178] M. Droz and A. McKane. Equivalence between Poisson representation and Fock space formalism for birth–death processes. *J. Phys. A*, **27**:L467, 1994.

[179] C.W. Gardiner. *Handbook of Stochastic Methods*. Sprinder-Verlag, 1983.

[180] C.W. Gardiner and S. Chaturvedi. The Poisson representation. A new technique for chemical master equation. *J. Stat. Phys.*, **17**:469, 1978.

[181] D. Elderfield. Exact macroscopic dynamics in non-equilibrium chemical systems. *J. Phys. A*, **18**:2049, 1985.

[182] L. Frachebourg. *Sur le rôle des fluctuations en mécanique statistique hors de l'équilibre*. PhD thesis, Université de Genève, 1994.

[183] H.E. Stanley. *Introduction to Phase Transitions and Critical Phenomena*. Oxford University Press, 1971.

[184] B. Schmittmann and R.K.P. Zia. Statistical Mechanics of Driven Diffuse Systems. Volume 15 of *Phase Transitions and Critical Phenemena*. Academic Press, 1996.

[185] M. Droz, Z. Rácz, and J. Schmidt. One-dimensional kinetic Ising model with competing dynamics : Steady-state correlations and relaxation times. *Phys. Rev. A*, **39**:2141, 1989.

[186] K. G. Wilson and J. Kogut. Renormalization group and critical phenomena. *Phys. Rep. C*, **12**:75, 1974.

[187] T. M. Liggett. *Interacting Particle Systems*. Springer, 1985.

[188] W. Feller. *An Introduction to Probability Theory and its Applications*. J. Wiley, 1962.

[189] R. Dickman and M. Bruschka. Nonequilibrium critical poisoning in a single species model. *Phys. Lett. A*, **127**:132, 1987.

[190] R. Dickman. Nonequilibrium lattice models: series analysis of steady states. *J. Stat. Phys.*, **55**:997, 1989.

[191] Ph. Beney, M. Droz, and L. Frachebourg. On the critical behaviour of cellular automata models of nonequilibrium phase transitions. *J. Phys. A*, **23**:3353, 1990.

[192] R.M. Ziff, E. Gulari, and Y. Barshad. Kinetic Phase Transitions in an Irreversible Surface-Reaction Model. *Phys. Rev. Lett.*, **56**:2553, 1986.

[193] B. Chopard and M. Droz. Cellular automata approach to non equilibrium phase transitions in a surface reaction model : static and dynamic properties. *J. Phys. A*, **21**:205, 1987.

[194] R.M. Ziff, K. Fichthorn, and E. Gulari. Cellular automaton version of the AB_2 reaction model obeying proper stoichiometry. *J. Phys. A*, **24**:3727, 1991.

[195] F. Bagnoli, B. Chopard, M. Droz, and L. Frachebourg. Critical behavior of a diffusive model with one adsorbing state. *J. Phys. A*, **25**:1085, 1992.

[196] H.-P. Kaukonen and R.M. Nieminen. Computer simulations studies of a catalytic oxidation of carbon monoxide on platinum metals. *J. Chem. Phys.*, **91**:4380, 1989.

[197] M. Ehsasi, F.S. Rys, and W. Hirschwald. Steady and nonsteady rates of reaction in a heterogeneously catalyzed reaction: oxidation of CO on platinium, experiments and simulations. *J. Chem. Phys.*, **91**:4949, 1989.

[198] B. Sente. Contribution à l'étude des réactions de surface. PhD thesis, Université de Mons-Hainaut, Belgium, 1992.

[199] J. Mai and W. von Niessen. Cellular automata approach to a surface reaction. *Phys. Rev. A*, **44**:R6165, 1991.

[200] P. Grassberger. Directed percolation in $2 + 1$ dimensions. *J. Phys. A*, **22**:3673, 1989.

[201] M. Droz and L. Frachebourg. Damage spreading in cellular automata models of nonequilibrium phase transitions. *Phys. Lett. A*, **148**:447, 1990.

[202] W. Kinzel. Directed percolation. In G. Deutscher and J. Adler, editors, *Percolation Structures and Processes*, volume 5. Annals of the Israel Physical Society, 1983.

[203] P. Grassberger, F. Krause, and T. von der Twer. A new type of kinetic critical phenomenon. *J. Phys. A*, **17**:L105, 1984.

[204] J.L. Cardy and U.C. Täuber. Theory of branching and annihilating random walks. *Phys. Rev. Lett.*, **77**:4780, 1996.

[205] I. Jensen. Critical behavior of nonequilibrium models with infinitely many absorbing states. *Int. J. Mod. Phys. B*, **8**:3299, 1994.

[206] M.A. Munõs, G. Grinstein, R. Dickman, and R. Livi. Infinite numbers of adsorbing states: critical behavior. *Physica D*, **103**:485–490, 1997.

[207] F. Bagnoli, M. Droz, and L. Frachebourg. Ordering in a one-dimensional driven diffuse system with parallel dynamics. *Physica A*, **179**:269–276, 1991.

[208] M. Marder and J. Fineberg. How things break. *Physics Today*, pp. 24–29, September 1996.

[209] B. Chopard and P.O. Luthi. A lattice Boltzman model and its application. In S. Bandini and G. Mauri, editors, *ACRI'96 Proceedings of the Second Conference on Cellular Automata for Research and Industry*, pp. 13–24. Springer-Verlag, 1997.

[210] P. Ossadnick. Cellular Automaton for the Fracture of Elastic Media. In K. Schilling T. Lippert and P. Ueberholz, editors, *Science on the Connection Machine*, pp. 127–136. World Scientific, 1993.

[211] W. J. R. Hoeffer. The Transmission-Line Matrix method. Theory and applications. *IEEE Trans. on Microwave Theory and Techniques*, **MTT-33**(10):882–893, October 1985.

[212] H. J. Hrgovcić. Discrete representation of the n-dimensional wave equation. *J. Phys. A*, **25**:1329–1350, 1991.

[213] C. Vanneste, P. Sebbah, and D. Sornette. A wave automaton for time-dependent wave propagation in random media. *Europhys. Lett.*, **17**:715, 1992.

[214] D. Sornette, O. Legrand, F. Mortessagne, P. Sebbah, and C. Vanneste. The wave automaton for the time-dependent Schroedinger, classical wave and Klein-Gordon equations. *Phys. Lett. A*, **178**:292–300, May 1993.

[215] P.O. Luthi, B. Chopard, and J.-F. Wagen. Wave Propagation in Urban Microcells: a massively parallel approach using the TLM method. In J. Dongarra, K. Madsen, and J. Wasniewski, editors, *Applied Parallel Computing: Computations in Physics, Chemistry and Engineering Science*, pp. 429–435. Springer, 1996. Lecture Notes in Computer Science; Vol. 1041.

[216] D.H. Rothman and J.M. Keller. Immiscible Cellular Automaton Fluids. *J. Stat. Phys*, **52**:275–282, 1988.

[217] R. Holme and D.H. Rothman. Lattice Gas and Lattice Boltzmann Models of Miscible Fluids. *J. Stat. Phys.*, **68**:409–429, 1992.

[218] P. Calvin, P. Lallemand, Y. Pomeau, and G. Searby. Simulation of free boundaries in flow systems by lattice-gas models. *J. Fluid Mechanics*, **188**:437, 1988.

[219] V. Zehnlé and G. Searby. Lattice gas experiments on a non-exothermic diffusion flame in a vortex field. *J. de Physique*, **50**:1083–1097, 1989.

[220] W.B. Hardy. The Spreading of Fluid on Glass. *Phil. Mag.*, **38**:49–55, 1919.

[221] P. Ball. Spreading it about. *Nature*, **338**:624, April 1989.

[222] F. Heselot, N. Fraysse, and A.M. Cazabat. Molecular layering in the spreading of wetting liquid drops. *Nature*, **338**:640–642, April 1989.

[223] P. Thompson and M. Robbins. To slip or not to slip. *Physics World*, pp. 35–38, November 1990.

[224] P.G. de Gennes and A.M. Cazabat. Etalement d'une goutte stratifiée incompressible. *C.R. Acad. Sci.Paris, II*, **310**:1601–1606, 1990.

[225] D.B. Abraham, P. Collet, J. De Connick, and F. Dunlop. Langevin dynamics of spreading and wetting. *Phys. Rev. Lett.*, **65**:195–198, 1990.

[226] D.B. Abraham, J. Heiniö, and K. Kaski. Computer simulation studies of fluid spreading. *J. Phys. A*, **24**:L309, 1991.

[227] B. Chopard. Numerical simulations of a Langevin dynamics of wetting. *J. Phys. A*, **24**:L345–350, 1991.

[228] C. Appert and S. Zaleski. Lattice gas with a liquid-gas transition. *Phys. Rev. Lett.*, **64**:1–4, 1990.

[229] R.E. Rosensweig. Magnetic Fluids. *Scientific American*, pp. 124–132, October 1982.

[230] V. Sofonea. Lattice Boltzmann Approach to Collective-Particle Interactions in Magnetic Fluids. *Europhys. Lett.*, **25**:385–390, 1994.

[231] K. Binder and D.W. Heermann. *Monte Carlo Simulation in Statistical Physics*. Springer-Verlag, 1992.

[232] B.M. Boghosian, J. Yepez, F.J. Alexander, and N.H. Margolus. Integer lattice gases. *Phys. Rev. E*, **55**:4137–4147, 1997.

[233] Y.H. Qian, S. Succi, and S.A. Orszag. Recent Advances in Lattice Boltzmann Computing. In D. Stauffer, editor, *Annual Reviews of Computational Physics III*, pp. 195–242. World Scientific, 1996.

[234] M.E. Colvin, A.J.C. Ladd, and B.J. Alder. Maximally Discretized Molecular Dynamics. *Phys. Rev. Lett.*, **61**:381–384, 1988.

[235] E.H. Hauge. Time Correlations Functions from the Boltzmann Equation. *Phys. Rev. Lett.*, **72**:1501–1503, 1972.

[236] T. Uematsu, T. Nakata, K. Takeuchi, Y. Arisawa, and Y. Kaenada. Three-dimensional numerical simulation of snow drift. *Cold Regions Sci. Technol.*, **20**(1):25–39, 1991.

[237] B. Bang, A. Nielsen, P.-A. Sundsbø, and T. Wiik. Computer simulation of wind speed, wind pressure and snow accumulation around buildings (SNOW-SIM). *Energy and Buildings*, **21**:235–243, 1994.

[238] A. Clappier. Influence des particules en saltation sur la couche limite turbulente. Technical report, Ecole polytechnique Fédérale de Lausanne (Switzerland), LASEN, 1991.

Glossary

A model   A simple model discribing poisoning transitions similar to those observed on a catalytic surface. A-particles can be adsorbed on vacant sites with some rate p. An adsorbed particle having at least one nearest neighbor empty is desorbed with a probability $1 - p$. As a function of the value of p the stationary state is poisoned or not.

advection   Transport of some quantity ρ due to an underlying flow with speed $\vec{u}$, or due to an external drift. An advection term is usually written as $\vec{u} \cdot \nabla \rho$. Convection is often used as a synonym of advection.

annealed disorder   One possible type of disorder in statistical systems. The annealed degrees of freedom responsible for the disorder are in thermal equilibrium with the other degrees of freedom of the system. In some CA models, an annealed disorder can be produced by an independant random process at each time step.

BGK models   Lattice Boltzmann models where the collision term Ω is expressed as a deviation from a local equilibrium distribution $f^{(0)}$, namely $\Omega = (f^{(0)} - f)/\xi$, where f is the unknown particle distribution and ξ a relaxation time (which is a parameter of the model). BGK stands for Bhatnager, Gross and Krook who first considered such a collision term, but not specifically in the context of lattice systems.

Boltzmann equation   A balance equation which expresses how the average number of particles with a given velocity changes between $(t + dt, \vec{r} + d\vec{r})$ and $(t, \vec{r})$, due to inter-particle interactions and ballistic motion; t and $\vec{r}$ are the time and space coordinates, respectively.

CA   Abbreviation for cellular automata or cellular automaton.

cellular automaton   System composed of adjacent cells or sites (usually organized as a regular lattice) which evolves in discrete time

327

steps. Each cell is characterized by an internal state whose value belongs to a finite set. The updating of these states is made in parallel according to a local rule involving only a neighborhood of each cell.

chaos   Way to describe unpredictable and apparently random structures.

Chapman–Enskog method   Expansion around the local equilibrium distribution function $f^{(0)}$, often used to solve a Boltzmann equation. One expresses the solution as $f = f^{(0)} + \epsilon f^{(i)} + \epsilon^2 f^{(2)} + ...$ where ϵ is the expansion parameter. By identifying the same order in ϵ, the Boltzmann equation yields $f^{(i)}$ in terms of $f^{(i-1)}$. However, the relation is not invertible unless some assumptions (that are the essence of the Chapman–Enskog scheme) are made.

collision term   The right-hand side of a Boltzmann equation or of the microdynamics associated to an LGA model. Describes the balance of particles entering and leaving a lattice site due to collisions.

conservation law   A property of a physical system in which some quantity (such as mass, momentum or energy) is locally conserved during the time evolution. These conservation laws should be included in the microdynamics of a CA model because they are essential ingredients governing the macroscopic behavior of any physical system.

continuity equation   An equation of the form $\partial_t \rho + \mathrm{div} \rho \vec{u} = 0$ expressing the mass (or particle number) conservation law. The quantity ρ is the local density of particles and $\vec{u}$ the local velocity field.

coupled map lattice   A dynamical system defined on a spatial lattice. Each lattice site is characterized by an internal state taking a continuous range of possible values. The dynamics of each site is defined by a function (map) giving the evolution in terms of the local state and those of the nearest neighbors. For descrete time, it can be viewed as a CA with continuous states.

critical phenomena   The phenomena which occur in the vicinity of a continuous phase transition, and are characterized by very long correlation length.

diffusion   A physical process described by the equation $\partial_t \rho = D \nabla^2 \rho$, where ρ is the density of a diffusing substance. Microscopically, diffusion can be viewed as a random motion of particles.

directed percolation   Percolation phenomenon in which the connectivity between the lattice sites (bonds or sites) has a preferred direction.

DLA Abbreviation of Diffusion Limited Aggregation. Model of a physical growth process in which diffusing particles stick on an existing cluster when they hit it. Initially, the cluster is reduced to a single seed particle and grows as more and more particles arrive. A DLA cluster is a fractal object whose dimension is typically 1.72 if the experiment is conducted in a two-dimensional space.

driven diffusion Diffusion with a drift, possibly caused by an external field, in some preferred direction.

dynamical system A sytem of equations (differential equations or discretized equations) modeling the dynamical behavior of a physical system.

equilibrium states States characterizing a closed system or a system in thermal equilibrium with a heat bath.

equipartition Postulate according to which all the microstates of an isolated system formed by a large number of particles are realized with the same probability.

ergodicity Property of a system or process for which the time-averages of the observables converge, in a probabilistic sense, to their ensemble averages.

Euler equation Equation of fluid mechanics describing the evolution of the velocity field $\vec{u}$ of a perfect fluid (without viscosity). It reads

$$\partial_t \vec{u} + (\vec{u} \cdot \nabla)\vec{u} = -\frac{1}{\rho}\nabla P$$

where ρ is the density and P the pressure.

exclusion principle A restriction which is imposed on LGA or CA models to limit the number of particles per site and/or lattice directions. This ensures that the dynamics can be described with a cellular automata rule with a given maximum number of bits. The consequence of this exclusion principle is that the equilibrium distribution of the particle numbers follows a Fermi–Dirac-like distribution in LGA dynamics.

FCHC model Abbreviation for Face-Centered-Hyper-Cubic model. This is a four-dimensional lattice gas model (generalizing the FHP rule) whose projection onto the 3D space provides a model of hydrodynamic flows.

FHP model Abbreviation for the Frisch, Hasslacher and Pomeau lattice gas model which was the first serious candidate to simulate two-dimensional hydrodynamics on a hexagonal lattice.

FHP-III   An extension of the FHP lattice gas model, involving rest particles and more complicated collision rules in order to achieve lower viscosity and higher Reynolds number flows.

field theory   The theory that describes the dynamics of fields defined on space-time and obeying partial differential equations.

Fock space   A Hilbert space useful to study processes in which the number of particles is not conserved. The creation and annihilation of particles are described in terms of creation or annihilation operators acting on the vectors of this Hilbert space.

fractal   Mathematical object usually having a geometrical representation and whose spatial dimension is not an integer. The relation between the size of the object and its "mass" does not obey that of usual geometrical objects. A DLA cluster is an example of a fractal.

front   The region where some physical process occurs. Usually the front includes the locations in space that are first affected by the phenomena. For instance, in a reaction process between two spatially separated reactants, the front describes the region where the reaction takes place.

Galilean invariance   Fundamental physical properties characterizing a system that is invariant under a constant velocity shift (i.e. the system obeys the same equation whether or not it moves with respect to the observer). Galilean invariance is violated in several LGA and this is reflected by a factor $g \neq 1$ in front of the convective term $\vec{u}\nabla\vec{u}$ in the hydrodynamical description.

geometrical phase transition   Qualitative change of behavior (similar to a thermodynamical phase transition) in a system without thermal effects. Percolation and directed percolation are examples of geometrical phases transitions.

Glauber dynamics   Dynamical model aiming at describing the relaxation of an Ising spin system, from an initial state to the equilibrium state, with no conservation law.

Hamming distance   A way to define a distance between two microstates by counting the numbers of sites on which the two microstates differ.

HPP model   Abbreviation for the Hardy, de Pazzis and Pomeau model. The first two-dimensional LGA aimed at modeling the behavior of particles colliding on a square lattice with mass and momentum conservation. The HPP model has several physical drawbacks that have been overcome with the FHP model.

invariant   A quantity which is conserved during the evolution of a dynamical system. Some invariants are imposed by the physical laws (mass, momentum, energy) and others result from the model used to describe physical situations (spurious, staggered invariants). Collisional invariants are constant vectors in the space where the Chapman–Enskog expansion is performed, associated to each quantity conserved by the collision term.

Ising model   Hamiltonian model describing the ferromagnetic paramagnetic transition. Each local classical spin variables $s_i = \pm 1$ interacts with its neighbors.

isotropy   the property of continuous systems to be invariant under any rotations of the spatial coordinate system. Physical quantities defined on a lattice and obtained by an averaging procedure may or may not be isotropic, in the continuous limit. It depends on the type of lattice and the nature of the quantity. Second-order tensors are isotropic on a 2D square lattice but fourth-order tensors need a hexagonal lattice.

LB   Abbreviation for lattice Boltzmann.

lattice Boltzmann model   A physical model defined on a lattice where the variables associated to each site represent an average number of particles or the probability of the presence of a particle with a given velocity. Lattice Boltzmann models can be derived from cellular automata dynamics by an averaging and factorization procedure, or be defined *per se*, independently of a specific realization.

lattice gas   A system defined on a lattice where particles are present and follow a given dynamics. Lattice gas automata (LGA) are a particular class of such system where the dynamics is performed in parallel over all the sites and can be decomposed in two stages: (i) propagation: the particles jump to a nearest-neighbor site, according to their direction of motion and (ii) collision: the particles entering the same site at the same iteration interact so as to produce a new particle distribution. HPP and FHP are well-known LGA.

lattice spacing   The separation between two adjacent sites of a regular lattice. Throughout this book, it is denoted by the symbol λ.

LBGK model   Abbreviation for Lattice BGK models. See BGK for a definition.

LGA   Abbreviation for Lattice Gas Automaton. See lattice gas model for a definition.

local equilibrium   Situation in which a large system can be decomposed into subsystems, very small on a macroscopic scale but large on a microscopic scale such that each sub-system can be assumed to be in thermal equilibrium. The local equilibrium distribution is the function which makes the collision term of a Boltzmann equation vanish.

lookup table   A table in which all possible outcomes of a cellular automata rule are pre-computed. The use of a lookup table yields a fast implementation of a cellular automata dynamics since however complicated a rule is, the evolution of any configuration of a site and its neighbors is directly obtained through a memory access. The size of a lookup table grows exponentially with the number of bits involved in the rule.

Margolus neighborhood   A neighborhood made of two-by-two blocks of cells, typically in a two-dimensional square lattice. Each cell is updated according to the values of the other cells in the same block. A different rule may possibly be assigned dependent on whether the cell is at the upper left, upper right, lower left or lower right location. After each iteration, the lattice partition defining the Margolus blocs is shifted one cell right and one cell down so that at every other step, information can be exchanged across the lattice. Can be generalized to higher dimensions.

master equation   An equation describing the evolution of the probability of a state at a given time as the balance between transitions leading to this state and transitions removing the system from this state.

mean-field approximation   An approximation, often used in statistical mechanics, which approximates an interacting many-body problem by a sum of effective one-body problems.

microdynamics   The Boolean equation governing the time evolution of a LGA model or a cellular automata system.

Monte-Carlo method   Method for computing integrals in a large dimensional space. In statistical physics, the method is used to compute ensemble averages. The ensemble average is replaced by a time average. The dynamics defining the time evolution is given by a master equation such that only the most important states are visited during the time evolution.

Moore neighborhood   A neighborhood composed of the central cell and all eight nearest and next-nearest neighbors in a two-dimensional square lattice. Can be generalized to higher dimensions.

multiparticle models A discrete dynamics modeling a physical system in which an arbitrary number of particles is allowed at each site. This is an extension of an LGA where no exclusion principle is imposed.

multiphase flows A flow composed of several different fluid species or different phases of the same fluid. The different phases may be miscible or immiscible.

multiscale expansion A method for solving a differential equation in which phenomena at different scales are present. For instance, the time evolution can be governed by two scales (e.g. diffusive and convective processes) so that the time derivative is split as $\partial_t = \epsilon \partial_{t_1} + \epsilon^2 \partial_{t_2}$ where t_1 and t_2 are the two time scales. The solution is then obtained by identifying the same orders in ϵ.

multispeed models Designate a lattice gas model in which at least two velocity amplitudes (fast and slow particles) are allowed. For instance, on a 2D square lattice, a two-speed model may include particles of speed 1 which jump to the nearest neighbors in one time step, while speed-$\sqrt{2}$ particles jump to the next nearest neighbors (along the diagonal). Interactions conserving mass, momentum and energy are expected between the different particle populations. Multispeed models are useful in having a distinct mass and kinetic energy conservation laws.

multispin coding A way to code several lattice sites in the same computer word, using each bit to represent the state of a site. An N-states automaton requires N computer words to be represented. Only one dimension of the lattice can be multispin coded in a given implementation, for efficiency reasons. For 32-bit computers, this technique is expected to produce a speedup of 32 since 32 sites can be computed at once. This technique also saves on computer storage but is more involved from a programming point of view and less flexible when tuning a rule.

Navier–Stokes equation The equation describing the velocity field $\vec{u}$ in a fluid flow. For an incompressible fluid ($\partial_t \rho = 0$), it reads

$$\partial_t \vec{u} + (\vec{u} \cdot \nabla)\vec{u} = -\frac{1}{\rho}\nabla P + \nu\nabla^2\vec{u}$$

where ρ is the density and P the pressure. The Navier–Stokes equation expresses the local momentum conservation in the fluid and, as opposed to the Euler equation, includes the dissipative effects with a viscosity term $\nu\nabla^2\vec{u}$. Together with the continuity equation, this is the fundamental equation of fluid dynamics.

neighborhood   The set of all cells necessary to compute a cellular automaton rule. A neighborhood is usually composed of several adjacent cells organized in a simple geometrical structure. Moore, von Neumann and Margolus neighborhoods are typical examples.

occupation numbers   Boolean quantities indicating the presence or absence of a particle in a given physical state.

open system   A system communicating with the environment by exchange of energy or matter.

parallel   Refers to an action which is performed simultaneously at several places. A parallel updating rule corresponds to the updating of all cells at the same time as if resulting from the computations of several independent processors.

partitioning   A technique consisting in dividing space in adjacent domains (through a partition) so that the evolution of each block is uniquely determined by the states of the elements within the block.

percolation   A "geometrical phase transition" of connectivity in a system. Imagine an infinite lattice whose bonds (sites) are randomly occupied with probability p. When p is larger that a percolation threshold p_c, there is an infinite path of connected bonds (or contiguous sites) from one side of the system to the other. If $p < p_c$, clusters of connected bonds (sites) remain finite.

phase transition   Change of state obtained when varying a control parameter such as the one occurring in the boiling or freezing of a liquid, or in the change between ferromagnetic and paramagnetic states of a magnetic solid.

quenched disorder   One possible type of disorder in statistical systems. The quenched degrees of freedom responsible for the disorder are not in thermal equilibrium with the other degrees of freedom of the system. They are frozen or evolve at a much longer time scale that the other degrees of freedom.

random walk   A series of uncorrelated steps of length unity describing a random path with zero average displacement but characteristic size proportional to the square root of the number of steps.

rate equation   The evolution equation for the local particles densities in a reaction-diffusion systems in which the reaction terms are treated in mean-field approximation.

reaction-diffusion systems   Systems made of one or several species of particles which diffuse and react among themselves to produce some new species.

renormalization group   Method which systematically implements some form of coarse-graining to extract the properties of the large-scale phenomena, in physical systems where many scales are important.

Reynolds number   In fluid dynamics, a dimensionless quantity defined as the ratio $(u_\infty d)/v$, where u_∞ is the generic speed of the flow (e.g speed far away from obstacles), d the typical size characterizing the system (characteristic size of an obstacle) and v the kinematic viscosity of the fluid. For the same geometry, flows with identical Reynolds numbers are similar in the sense that they differ by scaling factors.

scaling hypothesis   A hypothesis concerning the analytical properties of the thermodynamic potentials and the correlation functions in a problem invariant under a change of scale.

scaling law   Relations among the critical exponents describing the power law behaviors of physical quantities in systems invariant under a change of scale.

self-organized criticality   Concept aimed at describing a class of dynamical systems which naturally drive themselves to a state where interesting physics occurs at all scales.

spatially extended systems   Physical systems involving many spatial degrees of freedom and which, usually, have a rich dynamics and show up complex behaviors. Coupled map lattices and cellular automata provides a way to model spatially extended systems.

spin   Internal degree of freedom associated to particles in order to describe their magnetic state. A widely used case is the one of classical Ising spins. To each particle, one associates an "arrow" which is allowed to take only two different orientations, up or down.

telegraphist equation   The partial differential equation $\partial_t \rho + (D/c^2)\partial_t^2 \rho = D\nabla^2\rho$, where ρ is the quantity which is considered, D a diffusion constant and c a propagation speed. The telegraphist equation describes a combination of diffusive phenomena and wave propagation. It prevents information from traveling infinitely fast.

thermal models   In the context of lattice models, it designates a model (such as a multispeed LGA) able to include thermal properties such as, for instance, temperature and heat conductivity.

thermohydrodynamics A branch of physics describing fluid flows and including local energy exchanges between the different fluid regions. An equation describing energy balance is added to the continuity and the Navier–Stokes equations. Using thermodynamical laws, this new equation can be written in terms of a local temperature function.

time step Interval of time separating two consecutive iterations in the evolution of a discrete time process, like a CA or a LB model. Throughout this book the time step is denoted by the symbol τ.

Turing patterns Inhomogeneous stationary state observed in some open reaction-diffusion systems. The diffusion is a destabilizing factor for a homogeneous stationary state.

universality The phenomenon whereby many microscopically different systems exhibit a critical behavior with quantitatively identical properties such as the critical exponents.

updating operation consisting of assigning a new value to a set of variables, for instance those describing the states of a cellular automata system. The updating can be done in parallel and synchronously as is the case in a CA dynamics or sequentially, one variable after another, as is usually the case for a Monte-Carlo dynamics. Parallel, asynchronous updating is less common but can be envisaged too. Sequential and parallel updating schemes may yield different results since the interdependencies between variables are treated differently.

viscosity A property of a fluid indicating how much momentum "diffuses" through the fluid in a inhomogeneous flow pattern. Equivalently, it describes the stress occurring between two fluid layers moving with different velocities. A high viscosity means that the resulting drag force is important and low viscosity means that this force is weak. Kinematic viscosity is usually denoted by v and dynamic viscosity is denoted by $\eta = v\rho$ where ρ is the fluid density.

von Neumann neighborhood On a two-dimensional square lattice, the neighborhood including a central cell and its nearest neighbors north, south, east and west.

Ziff model A simple model describing adsorption–dissociation–desorption on a catalytic surface. This model is based upon some of the known steps of the reaction $A - B_2$ on a catalyst surface (for example CO–O_2).

Index

Printed in the United States
By Bookmasters